# Isospectral Vibrating Systems

# Isospectral Vibrating Systems

## Ranjan Ganguli

**CRC Press**
Taylor & Francis Group
Boca Raton London New York

CRC Press is an imprint of the
Taylor & Francis Group, an **informa** business

First edition published 2021
by CRC Press
6000 Broken Sound Parkway NW, Suite 300, Boca Raton, FL 33487-2742

and by CRC Press
2 Park Square, Milton Park, Abingdon, Oxon, OX14 4RN

CRC Press is an imprint of Taylor & Francis Group, LLC

**Library of Congress Cataloging-in-Publication Data**

Names: Ganguli, Ranjan, author.
Title: Isospectral vibrating systems / Ranjan Ganguli.
Description: First edition. | Boca Raton : CRC Press, [2021] | Includes
bibliographical references and index.
Identifiers: LCCN 2021006026 (print) | LCCN 2021006027 (ebook) | ISBN
9780367725709 (hardback) | ISBN 9781003155379 (ebook)
Subjects: LCSH: Machinery--Vibration. | Vibration--Testing. | Spectral
theory (Mathematics)
Classification: LCC TJ177 .G36 2021 (print) | LCC TJ177 (ebook) | DDC
621.8/11--dc23
LC record available at https://lccn.loc.gov/2021006026
LC ebook record available at https://lccn.loc.gov/2021006027

ISBN: 978-0-367-72570-9 (hbk)
ISBN: 978-0-367-72572-3 (pbk)
ISBN: 978-1-003-15537-9 (ebk)

Typeset in Times
by KnowledgeWorks Global Ltd.

# Contents

# Preface

Mechanical systems are subject to vibration. Natural frequencies play an important role in vibration problems. The natural frequency can be considered as an extracted critical feature of the mechanical system which encapsulates its dynamics. The set of natural frequencies of a vibratory system is also known as its spectrum. Isospectral systems share the same natural frequencies and therefore have identical spectrum. The study of isospectral systems is a subset of the field of inverse problems in vibrations.

In this book, we introduce isospectral systems using multi-degree of freedom spring-mass systems. Most dynamical systems in engineering and science can be expressed as spring-mass systems. Isospectral spring-mass systems are found using linear algebraic methods. Another formulation of this problem in terms of an optimization problem is also presented. The firefly algorithm is used to find these isospectral spring-mass systems. We also use the optimization approach based on electromagnetism based methods to find isospectral beams.

In the later part of the book, several closed form solutions are presented using isospectral systems. For example, a uniform non-rotating Euler-Bernoulli beam has a well known closed form solution for its natural frequency. We show that there exist non-uniform rotating Euler-Bernoulli beams which are isospectral to such simple uniform non-rotating beams. Thus, these rotating beams have the same closed form solutions for the natural frequencies, a fact which is further demonstrated using the finite element method. Note that closed formed solutions of rotating beams are not available. Using this innovative approach, closed form solutions are also discovered for non-uniform Euler-Bernoulli beams and non-uniform Rayleigh beams. The method outlined in this book can be used for more such discoveries in structural dynamics.

The book also uses isospectral systems for making testing easier. Finding the natural frequencies of rotating beams is difficult as it requires a rotor test rig. Many universities and small companies are not able to afford such experimental facilities. However, if we can find isospectral non-rotating beams for a given rotating beam, then such tests can be performed at a fraction of the required cost and without the need for the rotor rig. This approach is discussed in the book and examples for helicopter rotor blades and wind turbine rotor blades are presented. Such analogs between a more complex structure and a less complex structure are also found for gravity loaded beams and stiff strings. These derivations presented in the book use differential equations and numerical analysis and provide the application oriented engineer with ideas for solving many difficult testing problems by searching for isospectral analogs in simpler vibratory systems.

Many systems can be modeled using the linear and cubic tapered beam models and spring-mass systems used in this book. However, to generalize the isospectral approach and disseminate it in the industry, lumped mass models of a general dynamic system is used and an isospectral system is found for this problem. It is shown that this method can be used to find the isospectral analog of gravity loaded beams, axially loaded beams and rotating beams with general cross sections.

The book will be useful for senior undergraduates, graduate students, research scientists and industry professionals working on problems involving vibration of mechanical systems. The book is also useful for applied mathematicians and mechanics researchers. Instead of always attacking problems directly, the readers will learn the art of oblique attack on the vibration problem through the inverse approach. This inverse problem inspired isospectral approach will spawn new insights and methods for the analysis, design and testing of vibratory systems.

Finally, I would like to acknowledge and thank my research collaborator V. Mani and my former students Rajdeep Dutta, Nirmal Ramachandran, Sandilya Kambampati, Rishi Agarwal, Korak Sarkar and Srivatsa Bhat for being part of the research in this area during the last decade.

Ranjan Ganguli
Bangalore, 2021

# 1 Introduction: Spring-Mass Systems

This chapter introduces the concept of isospectral systems, using the simple and tractable example of a multi-degree of freedom spring-mass system. Isospectral systems have the same frequency spectrum, and so they are dynamically similar in nature. Families of isospectral systems arise as multiple solutions to an inverse problem i.e. the spectral information is insufficient to determine a unique solution. Consider a typical forward problem where model parameters $p$ are used to calculate data $d$. This modeling process is a forward problem defined by: $d = F(p)$ where $F$ is the forward map. The inverse problem is then given by $p = F^{-1}d$ and is typically not unique, a fact which we will witness in this chapter and throughout this book.

Isospectral systems are often considered to be a type of inverse problem. Isospectral systems play a significant role in mechanics as they allow designers to choose systems with desirable mass and stiffness distributions, which have similar dynamic properties. In this chapter, we examine undamped chain-like spring-mass systems fixed at the one end and free at the other end, as shown in Fig. 1.1. This figure represents an undamped spring-mass system of $n$-degrees of freedom, with $n$ mass and $n$ massless spring elements. Such a system is representative of many dynamical systems and complex dynamical systems can be reduced to this form by using orthogonal mode shapes. Spring-mass systems are ubiquitous in the theory of vibration [54, 71, 127, 132, 141, 198]. Finite element models can also be reduced to spring-mass systems in many cases [20, 95, 191, 193], for example by modal transformation. The Jacobi matrix associated with this system is a symmetric tridiagonal positive-definite matrix of order $n$. The approach in this chapter frames the search of isospectral systems as a numerical analysis problem. The first approach used involves the solution of equations. The first approach is adapted from [1]. The second approach involves minimization of an objective function, a typical optimization problem. The second approach is adapted from Ref. [58].

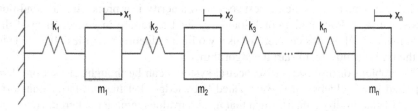

**Figure 1.1**
Schematic plot of $n$-degree of freedom spring-mass system

Seminal research work on the isospectral vibrating systems was conducted by Gladwell [74, 77], Gladwell and Morassi [79, 80], Gottlieb [86, 87], Dutta et al. [59], Ram [163], Jiménez, et al. [110, 111], Morassi [144] and Ram and Elhay [164]. Gladwell [74] found various methods for finding isospectral spring-mass system from a given system, and also discovered a pair of isospectral spring-mass systems analytically by interchanging $m_i \to k_{n-i+1}^{-1}$, $k_i \to m_{n-i+1}^{-1}$. Earlier research carried out by Boley and Golub [31], Chu [43], Gladwell and Zhu [82] and Nylen and Uhlig [151] presents a detailed study on the construction of isospectral systems as well as inverse problems. Inverse problems [75, 76] often reconstruct a system from response data. Inverse problems can be solved using an optimization approach [101, 107, 212] or by driving an error function to zero [46].

The determination of isospectral systems for discrete systems involves two important aspects: construction of the Jacobi matrix and the extraction of mass and stiffness elements from the Jacobi. The Jacobi matrix can be constructed using numerical methods or standard tridiagonalization methods. Boor and Golub [49] used an algorithm developed by Forsythe [66] to recreate a scalar Jacobi matrix. Gladwell and Willms [81] and Soto [183] applied the *Lanczos algorithm* and *Householder transformation* for tridiagonalization, respectively. Egana and Soto [62] applied a modified version of the *fast orthogonal reduction algorithm* to lower computational cost. Since the reconstructed spring-mass systems may not be unique, a scaling factor (for example, scaling factor can be $\sum_{i=1}^{n} m_i = 1$ [74]) should be introduced to generate the unique solution.

In earlier works, researchers have demonstrated the existence of isospectral systems using analytical as well as numerical approach. However, there remains considerable scope of exploring many more systems which are isospectral to a given one. Dutta et al. [59] explored a number of isospectral spring mass systems numerically using an evolutionary algorithm, FA (firefly algorithm). We will discuss this approach in later in this chapter.

The present chapter illustrates the isospectral vibrating systems as a solution to a nonlinear error function. The secant method is applied to find the roots of the error function with respect to the decision variables, which are selected to be the tridiagonal Jacobi matrix elements. The secant method can efficiently solve a nonlinear equation as well as an unconstrained optimization problem [18, 93, 136]. The secant method is a fast locally convergent method possessing super-linear convergence rate [3, 9]. It can converge to the nearest root with alacrity if a proper initial condition is provided [33]. Poor initial conditions may lead to failure in convergence of the secant method [9]. In this chapter, we use two linear algebraic theories [105, 151] to find the initial condition to start the algorithm.

The problem of discovering isospectral systems can be thought of as a problem in applied linear algebra. It is established knowledge that the similarity transform [105, 150] can modify a matrix such that its eigenvalues remain unaltered. However, the similarity transform does not guarantee that the structure of the modified matrix is preserved. A special type of similarity transform can only yield a nontrivial matrix with both the eigenvalues and the structure preserved. The present chapter focuses on determining distinct tridiagonal matrices possessing the same eigenvalues as that of a known tridiagonal matrix. The numerical example used in this chapter can be used as a pedagogical tool in courses on numerical analysis and mechanics. There is a need to illustrate the methods taught in courses on somewhat difficult problems [175].

Organization of this chapter is described in the following. In section 1.1, we present the mathematical model of spring-mass system. We will often use this model in the book to illustrate vibrating spring-mass systems. Section 1.2 presents the isospectral vibration problem formulation using the error function and explains the secant method used to obtain the zero of this error function. In section 1.3, we show numerical results of several isospectral systems. Section 1.4 presents an optimization approach which can be used to generate isospectral systems. Section 1.5 presents the firefly algorithm and Section 1.6 presents numerical results for 4-dof and 10-dof spring-mass isospectral systems. Finally, the chapter ends with section 1.7 as the summary.

## 1.1 MATHEMATICAL MODELING OF DISCRETE SYSTEM

The discrete system is one of the simplest models in the theory of vibration. Many actual systems such as automobiles, aircraft and bridges can be dynamically modeled as a discrete system. Consider a chain-like undamped spring-mass system consisting of $n$ masses $m_i > 0$ connected by $n$ massless springs of stiffness $k_i > 0$ for $i = 1, 2, ...., n$, with fixed-free condition i.e. $k_{n+1} = 0$, as shown in Fig. 1.1. The governing equation for free vibration of the discrete system, with $n$-degrees of freedom $(dof)$, is expressed as:

$$(K_n - \lambda M_n)x_n = 0, \quad x_n \neq 0 \tag{1.1}$$

where, $K_n$ is the stiffness matrix, $M_n$ is the mass matrix, $\lambda_i$ are the eigenvalues and $x_n^i = [x_{n,1}^i, x_{n,2}^i, ...., x_{n,n}^i]^T$ are the eigenvectors $\forall\ i = 1, 2, ...., n$. Eigenvalues are associated with the natural frequencies by a square law, and eigenvectors represent the vibrating modes of the mechanical system. The eigenvalues $\lambda_i$ in equation (1.1) are connected with the natural frequencies $\omega_i$ by the relation $\lambda_i = \omega_i^2$.

For a chain-like spring-mass system, $M_n$ is a diagonal matrix, $K_n$ is a tridiagonal matrix, and both of these matrices are symmetric positive-definite. The mass $(M_n)$ and stiffness $(K_n)$ matrices are expressed as,

$$M_n = \begin{bmatrix} m_1 & 0 & 0 & ... & ... & 0 \\ 0 & m_2 & 0 & 0 & ... & 0 \\ 0 & 0 & m_3 & 0 & ... & 0 \\ . & . & . & . & . & . \\ . & . & . & . & . & . \\ 0 & ... & 0 & 0 & m_{n-1} & 0 \\ 0 & ... & ... & 0 & 0 & m_n \end{bmatrix} \tag{1.2}$$

$$
K_n = \begin{bmatrix}
(k_1+k_2) & -k_2 & 0 & \cdots & & \cdots & 0 \\
-k_2 & (k_2+k_3) & -k_3 & 0 & & \cdots & 0 \\
0 & \cdot & \cdot & \cdot & \cdot & & \cdot \\
\cdot & \cdot & \cdot & \cdot & \cdot & \cdot & \cdot \\
\cdot & \cdot & \cdot & \cdot & \cdot & \cdot & 0 \\
0 & \cdots & 0 & -k_{n-1} & (k_{n-1}+k_n) & -k_n \\
0 & \cdots & & \cdots & 0 & -k_n & (k_n)
\end{bmatrix} \tag{1.3}
$$

The positive definiteness of the mass matrix $M_n$ enables it to be factorized as $M_n = M_n^{1/2} \times M_n^{1/2}$. After applying this factorization, the generalized eigenvalue equation (1.1) is transformed into:

$$
(A_n - \lambda I_n)y_n = 0, \quad y_n \neq 0 \tag{1.4}
$$

where, $A_n = M_n^{-1/2} K_n M_n^{-1/2}$ and $y_n = M_n^{1/2} x_n$. The transformed matrix $A_n$ in equation (1.4) is called the Jacobi matrix, and has the same eigenvalues as that of the system $(K_n, M_n)$ in equation (1.1). $A_n$ is a symmetric tridiagonal positive-definite matrix. The Jacobi matrix $(A_n)$ has the form:

$$
A_n = \begin{bmatrix}
a_1 & -b_1 & 0 & \cdots & & \cdots & 0 \\
-b_1 & a_2 & -b_2 & 0 & & \cdots & 0 \\
0 & -b_2 & a_3 & -b_3 & & 0 & \cdot \\
\cdot & \cdot & \cdot & \cdot & & \cdot & \cdot \\
\cdot & \cdot & \cdot & \cdot & & \cdot & 0 \\
0 & \cdots & 0 & -b_{n-2} & a_{n-1} & -b_{n-1} \\
0 & \cdots & & \cdots & 0 & -b_{n-1} & a_n
\end{bmatrix} \tag{1.5}
$$

where

$$
\begin{cases}
a_i = \dfrac{k_i+k_{i+1}}{m_i} > 0 \ \forall \ i = 1,2,...,n \\[2mm]
b_i = \dfrac{k_{i+1}}{\sqrt{m_i m_{i+1}}} > 0 \ \forall \ i = 1,2,...,n-1
\end{cases} \tag{1.6}
$$

Now consider a different (not scaled) Jacobi matrix $B_n$ of order $n$ with a similar tridiagonal structure as that of $A_n$. The new Jacobi matrix $B_n$ is

$$
B_n = \begin{bmatrix}
c_1 & -d_1 & 0 & \cdots & & \cdots & 0 \\
-d_1 & c_2 & -d_2 & 0 & & \cdots & 0 \\
0 & -d_2 & c_3 & -d_3 & & 0 & \cdot \\
\cdot & \cdot & \cdot & \cdot & & \cdot & \cdot \\
\cdot & \cdot & \cdot & \cdot & & \cdot & 0 \\
0 & \cdots & 0 & -d_{n-2} & c_{n-1} & -d_{n-1} \\
0 & \cdots & & \cdots & 0 & -d_{n-1} & c_n
\end{bmatrix} \tag{1.7}
$$

where, $c_i > 0 \ (i = 1,2,...,n)$ and $d_j > 0 \ (j = 1,2,...,n-1)$. The eigenvalues of $B_n$ are assumed to be $\mu_i \ \forall \ i = 1,2,...,n$. The new Jacobi matrix $B_n$ is labelled as isospectral to the given Jacobi matrix $A_n$, if the eigenvalues of $B_n$ are identical to that of $A_n$, i.e. $\mu_i = \lambda_i \ \forall \ i = 1, 2, ...., n$. In this chapter, we find the isospectral matrix (matrices) $B_n$ for a given $A_n$. The methodology to solve the isospectral problem numerically is described next.

## 1.2 ISOSPECTRAL VIBRATION PROBLEM FORMULATION

In this section, we present the error function, the application of secant method to locate its zero and the application of two linear algebraic theorems for estimating bounds on the matrix elements.

### 1.2.1 ERROR FUNCTION

The error function is given by

$$F_n = \sum_{i=1}^{n} (\lambda_i - \mu_i)^2 \tag{1.8}$$

where, $\{\lambda_i, \ \lambda_1 \leq \lambda_2 \leq ... \leq \lambda_n \}$ are the eigenvalues of Jacobi $A_n$, and $\{\mu_i, \ \mu_1 \leq \mu_2 \leq ... \leq \mu_n \}$ are the eigenvalues of Jacobi $B_n$. Let, $u_n$ and $v_n$ be two vectors of size $(n \times 1)$, which contain the eigenvalues of the matrices $A_n$ and $B_n$ in ascending order, respectively. Then, the error function in equation (1.8) can be expressed as $F_n = \|u_n - v_n\|^2$. Note that $F_n$ is always non-negative for symmetric positive-definite Jacobi matrices $A_n$ and $B_n$. The roots of $F_n$ provide the isospectral system(s) $B_n$ for a given $A_n$. We solve $F_n = 0$ using the secant method.

### 1.2.2 THE SECANT METHOD

In this section, we briefly describe the secant method and its application to obtain the isospectral system(s) $B_n$ for a given system $A_n$. We divide the objective function $F_n$ into $n$ functions $f_1, f_2, f_3, \ldots, f_{n-1}, f_n$. Here $f_i = (\lambda_i - \mu_i)^2$. Then, $F_n = \sum_{i=1}^{n} f_i$. Now, the secant method is applied to solve the set of $n$ equations ($f_1 = f_2 = .... = f_n = 0$) with respect to $n$ variables. The variables are chosen as $(n-1)$ super-diagonal (or sub-diagonal) elements and one (last) diagonal element of the Jacobi $B_n$. Let $P$ be a vector of size $(n \times 1)$, which contains the independent variables $d_1, d_2, ...., d_{n-1}$ (super-diagonal elements) and $c_n$ (last diagonal element). Specifically, $P$ is an $n$-dimensional solution point in the search space, i.e. $P = P_n = [d_1, ..., d_{n-1}, c_n]^T$.

The objective is to find the roots of the following equations using secant method.

$$f_i(d_1, d_2, \ldots, d_{n-1}, c_n) \quad = \quad 0, i = 1, n \tag{1.9}$$

The secant method updates a solution point in the search space according to the following rule.

$$P_n^{t+1} = P_n^t - (J_n^t)^{-1} \cdot f(P_n^t) \tag{1.10}$$

where, $P_n^t$ represents the position of solution point, $f(P_n^t)$ represents the function values at that point, and $J$ is the associated Jacobian matrix at the $t$'th iteration. These vectors and matrix have the form:

$$P_n^t = \begin{bmatrix} d_1^t \\ d_2^t \\ d_3^t \\ \vdots \\ d_{n-1}^t \\ c_n^t \end{bmatrix}, f(P_n^t) = \begin{bmatrix} f_1(d_1^t,\ldots,d_{n-1}^t,c_n^t) \\ f_2(d_1^t,\ldots,d_{n-1}^t,c_n^t) \\ f_3(d_1^t,\ldots,d_{n-1}^t,c_n^t) \\ \vdots \\ f_n(d_1^t,\ldots,d_{n-1}^t,c_n^t) \end{bmatrix} = \begin{bmatrix} f_1^t \\ f_2^t \\ f_3^t \\ \vdots \\ f_n^t \end{bmatrix} \quad (1.11)$$

$$J_n^t = \frac{\partial(f_1^t,f_2^t,\ldots,f_n^t)}{\partial(d_1^t,d_2^t,\ldots,c_n^t)} = \begin{bmatrix} \partial f_1/\partial d_1 & \partial f_1/\partial d_2 & \cdots & \partial f_1/\partial c_n \\ \partial f_2/\partial d_1 & \partial f_2/\partial d_2 & \cdots & \partial f_2/\partial c_n \\ \vdots & \vdots & \ddots & \vdots \\ \partial f_n/\partial d_1 & \partial f_n/\partial d_2 & \cdots & \partial f_n/\partial c_n \end{bmatrix} = \left[\frac{\partial f_i}{\partial P_j}\right]_{n\times n} \quad (1.12)$$

and

$$\frac{\partial f_i}{\partial P_j} \approx \frac{f_i(d_1,d_2,\cdots,d_j+h,\cdots,c_n) - f_i(d_1,d_2,\cdots,d_j,\cdots,c_n)}{h} \quad (1.13)$$

Here, $h$ is the step length of secant method. We solve the error function such that the error value goes below a certain tolerance level, i.e. $F_n$(error value) $\leq \varepsilon$(tolerance). The optimal solution point $(P_n^*)$ obtained by satisfying the tolerance criteria gives us the desired isospectral matrix $(B_n^*)$.

A good initial condition is given to the secant method by estimating bounds on Jacobi matrix elements with the help of linear algebraic theories.

### 1.2.3   ESTIMATING BOUNDS ON TRIDIAGONAL MATRIX ELEMENTS

Two linear algebraic theorems are applied to estimate of the bounds on matrix elements.

1.   Horn's majorization theorem [136]
2.   Gershgorin's disk theorem [185]

#### 1.2.3.1   Horn's majorization theorem

Horn's majorization theorem or Schur-Horn theorem [105] is used to estimate the bounds on the diagonal elements of the tridiagonal matrix $B_n$. The theorem is stated below.

**Theorem**: The diagonal vector (vector containing the diagonal values) of a Hermitian matrix majorizes the eigenvalue vector (vector containing the eigenvalues) of that matrix. Specifically, the sum of $k$ smallest eigenvalues should be smaller than or equal to the sum of $k$ smallest elements of the diagonal vector of a Hermitian matrix for $k = 1, 2, ...., n - 1$, where $n$ is the order of the matrix.

The theorem postulates that for a known set of eigenvalues $\lambda_i$, there can exist multiple sets of diagonal elements $c_i$ satisfying the majorization condition. We use the majorization theorem to generate diagonal elements of $B_n$ within some range of values such that they obey the above condition with respect to the eigenvalues of given $A_n$. Let the eigenvalues of $A_n$ be ordered as $\lambda_1 \leq \lambda_2 \leq \ ... \ \leq \lambda_{n-1} \leq \lambda_n$, and the diagonal elements of $B_n$ be ordered as $c_1 < c_2 < ... < c_{n-1} < c_n$. Then, Horn's majorization theorem gives the following inequalities.

$$\begin{cases} c_1 \geq \lambda_1 \\ c_1 + c_2 \geq \lambda_1 + \lambda_2 \\ c_1 + c_2 + c_3 \geq \lambda_1 + \lambda_2 + \lambda_3 \\ \quad \vdots \\ c_1 + c_2 + c_3 + ... + c_{n-1} \geq \lambda_1 + \lambda_2 + \lambda_3 + ... + \lambda_{n-1} \end{cases} \tag{1.14}$$

Note that the last diagonal element of $B_n$ can be determined using it's trace. Recall from matrix theory that for $B_n$ to have the same set of eigenvalues as $A_n$, both of them should have equal trace. Hence,

$$\begin{cases} c_n = trace(B_n) - (c_1 + c_2 + ... + c_{n-1}) \\ \Rightarrow c_n = (\lambda_1 + \lambda_2 + ... + \lambda_n) - (c_1 + c_2 + ... + c_{n-1}) \end{cases} \tag{1.15}$$

The inequalities (equation (1.53)) along with an equality (equation (1.15)) facilitate the search for bounds on diagonal elements of $B_n$. These conditions ensure significant information about the lowest and highest diagonal elements, which are lowest diagonal value($c_1$) $\geq$ lowest eigenvalue($\lambda_1$), and highest diagonal value($c_n$) $\leq$ highest eigenvalue($\lambda_n$). An algorithm is developed to generate diagonal elements $c_1$, $c_2$, ... $c_{n-1}$, $c_n$ within estimated bounds in accordance with the above conditions. The algorithm will be described later in this chapter.

### 1.2.3.2 Gershgorin's disk theorem

The Gershgorin's disk theorem enables us to predict the lower bounds on the superdiagonal (or sub-diagonal) elements of matrix $B_n$. We now present the Gershgorin's disk theorem described in literature [185], and its application to find the bounds.

**Theorem**: Every eigenvalue of $B_n$ lies inside at least one of the circles $C_1, C_2, ..., C_n$, where $C_i$ has its center at the diagonal element $c_i$ with a radius $r_i = \sum_{j \neq i} |B_{ij}|$ equal to the absolute sum of the off-diagonal elements along the $i$'th row for $i = 1, 2, ..., n$.

On analyzing this theorem, we obtain

$$|B_{ii} - \lambda| \leq \sum_{j \neq i} |B_{ij}| \ \forall \ i = \{1, 2, ..., n\} \tag{1.16}$$

The sum of absolute values of the off-diagonal elements in a row of a matrix is greater than or equal to the absolute difference between diagonal element in that row and its nearest eigenvalue. In other words, the radius of the Gershgorin circle corresponding to a row must be greater than or equal to the absolute difference between diagonal element in that row and its nearest eigenvalue. Application of this theorem to the tridiagonal matrix $B_n$ gives us lower bounds on the sum of off-diagonal elements in a particular row. As $B_n$ is a tridiagonal matrix, the sum $\sum_{j \neq i} |B_{ij}|$ will only include super-diagonal and sub-diagonal elements, thus providing us with bounds on them or their sum. The diagonal elements $\{\lambda_i\}$ and the eigenvalues $\{c_i\}$ are assumed to be in ascending order. We obtain the following inequalities on applying Gershgorin's disk theorem to the tridiagonal matrix $B_n$.

$$|c_1 - \lambda_1| \leq d_1$$
$$|c_2 - \lambda_2| \leq d_1 + d_2$$
$$|c_3 - \lambda_3| \leq d_2 + d_3$$
$$\vdots \qquad\qquad (1.17)$$
$$|c_{n-1} - \lambda_{n-1}| \leq d_{n-2} + d_{n-1}$$
$$|c_n - \lambda_n| \leq d_{n-1}$$

Thus, Gershgorin's disk theorem produces a set of inequalities (equation (1.17)) on solving which we get the bounds on super-diagonal elements. We develop an algorithm to address the above conditions (equation (1.17)). The algorithm is described later in this chapter.

The eigenvalue vector should be ordered in the same fashion as the diagonal vector. While performing numerical calculations, the difference should be considered between corresponding pairs of eigenvalues and diagonal elements. Specifically, if diagonal values are 58.5960, 7.5593, 17.9090 and 30.2690 and eigenvalues are 0.9050, 15.4773, 38.3013 and 59.6497, then, the pairs of diagonal values and eigenvalues that will be subtracted are –( 58.5960 – 59.6497 ), ( 7.5593 – 0.9050 ), ( 17.9090 – 15.4773 ) and ( 30.2690 – 38.3013 ).

We generate the initial point $P_n^0 = [d_1^0, d_2^0, ..., d_{n-1}^0, c_n^0]^T$ as the starting point for the secant method. The bounds on the variables, i.e. matrix elements, are developed using linear algebraic theories. A pseudocode of the method is given next. This method uses the secant method to solve a nonlinear error function, and utilizes the theorems presented in section earlier to generate a good starting point for the secant method.

Pseudocode
*main()*
Known: $m_i$ and $k_i$
Form $A_n$
Calculate $\lambda_i$
Sort $\lambda_i$
$c_i^0 \leftarrow$ majorization($\lambda_i$,n)
$d_j^0 \leftarrow$ gershbound($c_i,\lambda_i$,n)
Initial point: $P_n^0 = [d_1^0, d_2^0, ..., d_{n-1}^0, c_n^0]^T$

Start count: $t = 1$
**while** $(t \leq \text{MaxIter})$
$P_n^t = [d_1^t, d_2^t, ..., d_{n-1}^t, c_n^t]^T$
Form $B_n^t$ using $c_i^t$ and $d_j^t$
Calculate $\mu_i^t$
Sort $\mu_i^t$
Calculate $f(P_n^t)$
Calculate $J$
Move point: $P_n^{t+1} \leftarrow P_n^t - (J_n^t)^{-1} \cdot f(P_n^t)$
Error function: $F_n^{t+1} = \text{sum}(f(P_n^{t+1}))$
**if** $(F_n^{t+1} \leq \varepsilon)$
$P_n^* \leftarrow P_n^{t+1}$
Break
**end** if
Increase count: $t = t + 1$
**end** while
Solution: $P_n^*$
Isospectral matrix: $B_n^*$
(*Note: $i = 1, ..., n$ and $j = 1, ..., n-1$*)

'majorization' function:
$c_i = majorization\ (\lambda_i, n)$

(Order of the eigenvalues $\lambda_i$: ascending)
$c_1 \leftarrow \lambda_1 + \text{rand} \times (\lambda_2 - \lambda_1)$
**for** $k = 2 : n - 1$
$c_k \leftarrow (\lambda_1 + \lambda_2 + ... + \lambda_k) - (c_1 + c_2 + ... + c_{k-1}) + \text{rand} \times (\lambda_{k+1} - \lambda_k)/2$
**end** for
$c_n \leftarrow \text{sum}(\lambda(1 : n)) - \text{sum}(c(1:n-1))$
(Order of the diagonal elements $c_i$: ascending)

'gershbound' function:
$d_j = gershbound(c_i, \lambda_i, n)$

(Rearrange $\lambda_i$ according to the order of $c_i$)
$d_1 \leftarrow |c_1 - \lambda_1| + \text{rand} \times (20)$
$d_{n-1} \leftarrow |c_n - \lambda_n| + \text{rand} \times (20)$
**for** $k = 2 : n - 2$
$p_k \leftarrow |c_k - \lambda_k| - |c_{k-1} - \lambda_{k-1}|$
$q_{n-k} \leftarrow |c_{n-k+1} - \lambda_{n-k+1}| - |c_{n-k+2} - \lambda_{n-k+2}|$
**end** for
**for** $k = 2 : n - 2$
$d_k \leftarrow \max(p_k, q_k, 0) + \text{rand} \times (20)$

**Table 1.1**

**Specification of the 10-dof Original Spring-mass System $A_{10}$**

| Elements (sequence) | Mass (kg) | Stiffness (kN/m) | Eigenvalues | Natural Frequency (rad/sec) |
|:---:|:---:|:---:|:---:|:---:|
| 1 | 1.00 | 10.00 | 0.0776 | 0.2785 |
| 2 | 2.00 | 20.00 | 1.9796 | 1.4070 |
| 3 | 3.00 | 30.00 | 5.8363 | 2.4158 |
| 4 | 4.00 | 40.00 | 11.2697 | 3.3570 |
| 5 | 5.00 | 50.00 | 17.7637 | 4.2147 |
| 6 | 6.00 | 60.00 | 24.7004 | 4.9699 |
| 7 | 7.00 | 70.00 | 31.4077 | 5.6043 |
| 8 | 8.00 | 80.00 | 37.2010 | 6.0993 |
| 9 | 9.00 | 90.00 | 41.4177 | 6.4357 |
| 10 | 10.00 | 100.00 | 46.6360 | 6.8291 |

**end** for

## 1.3   ISOSPECTRAL SPRING-MASS SYSTEMS

The objective now is to investigate the isospectral spring-mass systems $(B_n)$ for a given system $(A_n)$. In this section, we show the numerical results for isospectral spring-mass systems. Two systems are defined as isospectral if the error function value becomes zero. A root of the error function corresponds to a spring-mass system isospectral to the given one. Consider a chain-like undamped spring-mass system of 10 dof as shown in Table 1.1.

The Jacobi matrix $A_n$ associated with the given spring-mass system is as follows:

$$
A_{10} = \begin{bmatrix}
a_1 & -b_1 & 0 & 0 & 0 & 0 & 0 & 0 & 0 & 0 \\
-b_1 & a_2 & -b_2 & 0 & 0 & 0 & 0 & 0 & 0 & 0 \\
0 & -b_2 & a_3 & -b_3 & 0 & 0 & 0 & 0 & 0 & 0 \\
0 & 0 & -b_3 & a_4 & -b_4 & 0 & 0 & 0 & 0 & 0 \\
0 & 0 & 0 & -b_4 & a_5 & -b_5 & 0 & 0 & 0 & 0 \\
0 & 0 & 0 & 0 & -b_5 & a_6 & -b_5 & 0 & 0 & 0 \\
0 & 0 & 0 & 0 & 0 & -b_6 & a_7 & -b_7 & 0 & 0 \\
0 & 0 & 0 & 0 & 0 & 0 & -b_7 & a_8 & -b_8 & 0 \\
0 & 0 & 0 & 0 & 0 & 0 & 0 & -b_8 & a_9 & -b_9 \\
0 & 0 & 0 & 0 & 0 & 0 & 0 & 0 & -b_9 & a_{10}
\end{bmatrix}
\tag{1.18}
$$

The tridiagonal matrix elements $a_i$ and $b_i$ are found by using equation (1.6). Thus, the original Jacobi matrix $A_{10}$ is obtained. The eigenvalue and the natural frequencies for system $A_{10}$ are given in Table 1.1. The Jacobi matrix related to a different spring-mass system is as follows:

$$B_{10} = \begin{bmatrix} c_1 & -d_1 & 0 & 0 & 0 & 0 & 0 & 0 & 0 & 0 \\ -d_1 & c_2 & -d_2 & 0 & 0 & 0 & 0 & 0 & 0 & 0 \\ 0 & -d_2 & c_3 & -d_3 & 0 & 0 & 0 & 0 & 0 & 0 \\ 0 & 0 & -d_3 & c_4 & -d_4 & 0 & 0 & 0 & 0 & 0 \\ 0 & 0 & 0 & -d_4 & c_5 & -d_5 & 0 & 0 & 0 & 0 \\ 0 & 0 & 0 & 0 & -d_5 & c_6 & -d_6 & 0 & 0 & 0 \\ 0 & 0 & 0 & 0 & 0 & -d_6 & c_7 & -d_7 & 0 & 0 \\ 0 & 0 & 0 & 0 & 0 & 0 & -d_7 & c_8 & -d_8 & 0 \\ 0 & 0 & 0 & 0 & 0 & 0 & 0 & -d_8 & c_9 & -d_9 \\ 0 & 0 & 0 & 0 & 0 & 0 & 0 & 0 & -d_9 & c_{10} \end{bmatrix} \quad (1.19)$$

Now, we find the $B_{10}$ matrix elements such that eigenvalues are the same as those of $A_{10}$. The error function for 10-$dof$ is given by

$$F_{10} = (\lambda_1 - \mu_1)^2 + (\lambda_2 - \mu_2)^2 + \cdots + (\lambda_{10} - \mu_{10})^2 \quad (1.20)$$

The secant method drives this error function to zero with respect to the $n$ ($n = 10$) matrix element variables, which are $d_1, d_2, d_3, d_4, \cdots, d_9$, and $c_{10}$. We generate the diagonal elements of $B_{10}$ randomly within appropriate bounds using $(n-1) = 9$ inequalities and one equality, as described earlier in section 1.2.3.1. The first $(n-1)$ diagonal values are generated using the inequalities, whereas the last diagonal value is generated using an equality. Applying the Horn's majorization theorem, the following inequalities for 10-dof system $B_{10}$ are obtained.

$$\begin{cases} c_1 \geq 0.07756 \\ c_1 + c_2 \geq 2.05716 \\ c_1 + c_2 + c_3 \geq 7.89346 \\ c_1 + c_2 + c_3 + c_4 \geq 19.16316 \\ c_1 + c_2 + c_3 + c_4 + c_5 \geq 36.92686 \\ c_1 + c_2 + c_3 + c_4 + c_5 + c_6 \geq 61.62726 \\ c_1 + c_2 + c_3 + c_4 + c_5 + c_6 + c_7 \geq 93.03496 \\ c_1 + c_2 + c_3 + c_4 + c_5 + c_6 + c_7 + c_8 \geq 130.23596 \\ c_1 + c_2 + c_3 + c_4 + c_5 + c_6 + c_7 + c_8 + c_9 \geq 171.65366 \end{cases} \quad (1.21)$$

The equality emanates from the trace condition, which is given as:

$$c_{10} = \sum_{i=1}^{10} a_i - \sum_{i=1}^{9} c_i \quad (1.22)$$

The *majorization* algorithm helps to generate the diagonal elements. Now, we choose a randomly generated set of $(n-1)$ diagonal elements satisfying the inequalities (1.21): $\{c_1, ...., c_9\} = \{15.1659, 20.7370, 20.8689, 21.0343, 21.1194, 21.1694,$

$21.7410, 22.4272, 24.3582\}$. The last diagonal element is calculated using equation (1.22), which is $c_{10} = 29.6686$. The diagonal elements can be arranged in an ascending, descending or arbitrary manner.

Once the diagonal values and eigenvalues of the system $B_{10}$ are known, the Gershgorin's disk theorem is applied to impose bounds on the super-diagonal elements of $B_{10}$. Application of Gershgorin's disk theorem, as described earlier in section 1.2.3.2, produces the following inequalities.

$$\begin{cases} |10.0000 - 11.2697| \leq d_1 \ or, \ d_1 \geq 1.2697 \\ |21.1111 - 17.7637| \leq d_1 + d_2 \ or, \ d_1 + d_2 \geq 3.3474 \\ |21.2500 - 24.7004| \leq d_2 + d_3 \ or, \ d_2 + d_3 \geq 3.4504 \\ |21.4286 - 24.7004| \leq d_3 + d_4 \ or, \ d_3 + d_4 \geq 3.2718 \\ |21.6667 - 24.7004| \leq d_4 + d_5 \ or, \ d_4 + d_5 \geq 3.0337 \\ |22.0000 - 24.7004| \leq d_5 + d_6 \ or, \ d_5 + d_6 \geq 2.7004 \\ |22.5000 - 24.7004| \leq d_6 + d_7 \ or, \ d_6 + d_7 \geq 2.2004 \\ |23.3333 - 24.7004| \leq d_7 + d_8 \ or, \ d_7 + d_8 \geq 1.3671 \\ |25.0000 - 24.7004| \leq d_8 + d_9 \ or, \ d_8 + d_9 \geq 0.2996 \\ |30.0000 - 31.4077| \leq d_9 \ or, \ d_9 \geq 1.4077 \end{cases} \qquad (1.23)$$

We follow the *gershbound* algorithm explained earlier as pseudocode to generate the super-diagonal elements of $B_{10}$. We select a randomly generated set of $(n-1)$ super-diagonal elements satisfying the above inequalities (1.23), which is: $\{d_1, \ldots, d_9\} = \{11.2539, 13.0307, 2.3237, 10.8621, 7.2830, 17.6685, 7.0375, 12.4904, 15.6340\}$.

Note that the variables associated with the secant method are the $(n-1)$ super-diagonal elements and the last diagonal element. The previous discussion shows the approach for assigning initial values to the variables in the secant method. Thus, a better initial condition is provided for the secant method. Starting with a good set of initial conditions, the secant method converged to the desired values of the six ($n = 10$) variables through an iterative process, and the error function equation (1.20) gets driven to zero.

The proposed methodology finds different isospectral systems for different sets of diagonal values. For a particular set of diagonal values, we get the following isospectral system $B_{10}^*$ using the secant method.

$$A_{10} = \begin{bmatrix} 30.0000 & -14.1421 & 0 & 0 & 0 & 0 & 0 & 0 & 0 & 0 \\ -14.1421 & 25.0000 & -12.2474 & 0 & 0 & 0 & 0 & 0 & 0 & 0 \\ 0 & -12.2474 & 23.3333 & -11.5470 & 0 & 0 & 0 & 0 & 0 & 0 \\ 0 & 0 & -11.5470 & 22.5000 & -11.1803 & 0 & 0 & 0 & 0 & 0 \\ 0 & 0 & 0 & -11.1803 & 22.0000 & -10.9545 & 0 & 0 & 0 & 0 \\ 0 & 0 & 0 & 0 & -10.9545 & 21.6667 & -10.8012 & 0 & 0 & 0 \\ 0 & 0 & 0 & 0 & 0 & -10.8012 & 21.4286 & -10.6904 & 0 & 0 \\ 0 & 0 & 0 & 0 & 0 & 0 & -10.6904 & 21.2500 & -10.6066 & 0 \\ 0 & 0 & 0 & 0 & 0 & 0 & 0 & -10.6066 & 21.1111 & -10.5409 \\ 0 & 0 & 0 & 0 & 0 & 0 & 0 & 0 & -10.5409 & 10.0000 \end{bmatrix} \qquad (1.24)$$

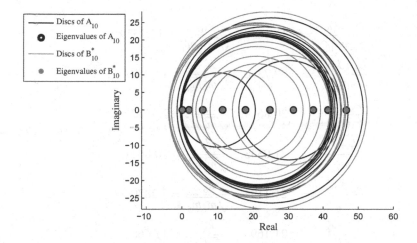

**Figure 1.2**

Eigenvalue locations of the known and the isospectral matrices

$$B_{10}^* = \begin{bmatrix} 15.1659 & -11.2539 & 0 & 0 & 0 & 0 & 0 & 0 & 0 & 0 \\ -11.2539 & 20.7370 & -13.0307 & 0 & 0 & 0 & 0 & 0 & 0 & 0 \\ 0 & -13.0307 & 20.8689 & -2.3237 & 0 & 0 & 0 & 0 & 0 & 0 \\ 0 & 0 & -2.3237 & 21.0343 & -10.8621 & 0 & 0 & 0 & 0 & 0 \\ 0 & 0 & 0 & -10.8621 & 21.1194 & -7.2830 & 0 & 0 & 0 & 0 \\ 0 & 0 & 0 & 0 & -7.2830 & 21.1694 & -17.6685 & 0 & 0 & 0 \\ 0 & 0 & 0 & 0 & 0 & -17.6685 & 21.7410 & -7.0375 & 0 & 0 \\ 0 & 0 & 0 & 0 & 0 & 0 & -7.0375 & 22.4272 & -12.4904 & 0 \\ 0 & 0 & 0 & 0 & 0 & 0 & 0 & -12.4904 & 24.3582 & -15.6340 \\ 0 & 0 & 0 & 0 & 0 & 0 & 0 & 0 & -15.6340 & 29.6686 \end{bmatrix} \quad (1.25)$$

The eigenvalues of $B_{10}^*$ are 0.0776, 1.9795, 5.8363, 11.2697, 17.7637, 24.7004, 31.4077, 37.2010, 41.4177 and 46.6361. The eigenvalue locations for both known and obtained isospectral systems ($A_{10}$ and $B_{10}^*$) are described in Fig. 1.2. The error function obtained for the above isospectral system is $8.44 \times 10^{-9}$ which can be interpreted to be close to zero. The mass and stiffness values for the isospectral system $B_{10}^*$ are extracted from the Jacobi matrix using the following equations.

We now choose $m_1^*$ and use equation (1.26) to calculate rest of the mass and stiffness values. The calculated mass and spring values $m_1^*$, $m_2^*$, $\cdots$, $m_{10}^*$ ; and $k_1^*$, $k_2^*$, $\cdots$, $k_{10}^*$ are shown in Table 1.2 as result (1).

$$\frac{k'_{10}}{m'_{10}} = 15.1659; \quad \frac{k'_{10}}{\sqrt{m'_{10}m'_9}} = 11.2539$$

$$\frac{k'_9 + k'_{10}}{m'_9} = 20.7370; \quad \frac{k'_9}{\sqrt{m'_9 m'_8}} = 13.0307$$

$$\frac{k'_8 + k'_9}{m'_8} = 20.8689; \quad \frac{k'_8}{\sqrt{m'_8 m'_7}} = 2.3237$$

$$\frac{k'_7 + k'_8}{m'_7} = 21.0343; \quad \frac{k'_7}{\sqrt{m'_7 m'_6}} = 10.8621$$

$$\frac{k'_6 + k'_7}{m'_6} = 21.1194; \quad \frac{k'_6}{\sqrt{m'_6 m'_5}} = 7.2830$$

$$\frac{k'_5 + k'_6}{m'_5} = 21.1694; \quad \frac{k'_5}{\sqrt{m'_5 m'_4}} = 17.6685$$

$$\frac{k'_4 + k'_5}{m'_4} = 21.7410; \quad \frac{k'_4}{\sqrt{m'_4 m'_3}} = 7.0375$$

$$\frac{k'_3 + k'_4}{m'_3} = 22.4272; \quad \frac{k'_3}{\sqrt{m'_3 m'_2}} = 12.4904 \qquad (1.26)$$

$$\frac{k'_2 + k'_3}{m'_2} = 24.3582; \quad \frac{k'_2}{\sqrt{m'_2 m'_1}} = 15.6340$$

$$\frac{k'_1 + k'_2}{m'_1} = 29.6686$$

Multiple isospectral systems are obtained for different known systems, $A_{10}$ and $A'_{10}$. Tables 1.1 and 1.3 show the properties of the known systems, whereas Tables 1.2 and 1.4 show the isospectral systems $B^*_{10}$ and $B'^*_{10}$ for $A_{10}$ and $A'_{10}$, respectively.

We obtain a number of isospectral systems with a wide range of mass and stiffness values. The best isospectral system with lowest error value, determined by the secant method, is shown in Table 1.2 as result (2). The difference in traces of the original and the isospectral systems is lowest in result (4) in Table 1.2, although the corresponding system does not possess lowest error function value. The trace of a matrix is related with just the first coefficient of the characteristic polynomial of that matrix [31].

Furthermore, we modify the parameters associated with the secant method; the step size is reduced to $h = 10^{-13}$ and the tolerance value is set as $\varepsilon = 10^{-27}$. In this case, the lowest error function ($F_{10}$) value is of the order of $10^{-28}$. Table 1.5 presents the overall best results of isospectral systems found numerically. Figure 1.3 displays the variations in error function values with iterations. It is clear that a high level of closeness is obtained for the discrete isospectral system by the proposed method. The pseudocode provided in this chapter permits easy implementation of the method.

## Table 1.2
## 4 Isospectral Spring-mass Systems (10-dof) for a Given $A_{10}$

| Isospectral Systems | Intial Conditions | Eigenvalues $(\mu)$ | Error function values $(\times 10^{-9})$ | Difference in trace $(\times 10^{-4})$ | Mass (kg) | Stiffness $(kNm^{-1})$ |
|---|---|---|---|---|---|---|
| 1. | 16.487060 | 0.077578 | 8.444516 | 1.977710 | 11.181995 | 51.097872 |
| | 6.468557 | 1.979521 | | | 9.913242 | 118.486084 |
| | 9.982375 | 5.836252 | | | 4.505328 | 87.082942 |
| | 1.118780 | 11.269716 | | | 1.978422 | 6.937575 |
| | 2.623553 | 17.763749 | | | 5.151448 | 34.676137 |
| | 2.287198 | 24.700364 | | | 20.105210 | 74.118646 |
| | 6.885814 | 31.407727 | | | 19.684122 | 351.495112 |
| | 6.770493 | 37.201043 | | | 5.998215 | 76.470721 |
| | 19.805601 | 41.417716 | | | 3.601348 | 58.051767 |
| | 29.668533 | 46.636057 | | | 1 | 29.668601 |
| 2. | 13.594606 | 0.077530 | 5.617327 | −0.198874 | 27.041418 | 16.524651 |
| | 10.42278 | 1.979533 | | | 31.067791 | 319.497268 |
| | 6.648723 | 5.836215 | | | 37.950450 | 334.791892 |
| | 5.034487 | 11.269793 | | | 38.955447 | 468.875037 |
| | 4.661935 | 17.763745 | | | 23.975515 | 362.628314 |
| | 3.056971 | 24.700392 | | | 12.744125 | 150.874610 |
| | 8.786058 | 31.407781 | | | 11.481964 | 126.177777 |
| | 4.778904 | 37.201048 | | | 8.306799 | 126.578721 |
| | 18.630956 | 41.417746 | | | 3.846943 | 65.543215 |
| | 29.321885 | 46.636067 | | | 1 | 29.321886 |
| 3. | 17.611664 | 0.077530 | 8.848102 | −2.170468 | 7.832367 | 9.640351 |
| | 5.532742 | 1.979533 | | | 11.067891 | 98.791968 |
| | 10.346210 | 5.836215 | | | 9.985454 | 126.822792 |
| | 3.391365 | 11.269793 | | | 11.119247 | 80.802637 |
| | 0.371651 | 17.763745 | | | 14.895715 | 151.702814 |
| | 2.106551 | 24.700392 | | | 18.152825 | 163.163610 |
| | 8.925986 | 31.407781 | | | 14.692964 | 232.796521 |
| | 6.810413 | 37.201048 | | | 7.336999 | 97.416815 |
| | 20.159619 | 41.417746 | | | 4.051043 | 70.695815 |
| | 29.327835 | 46.636067 | | | 1 | 29.327786 |
| 4. | 9.963645 | 0.077541 | 6.204503 | −0.106240 | 154.547418 | 66.859051 |
| | 14.268778 | 1.979542 | | | 176.916191 | 1303.397268 |
| | 7.040023 | 5.836271 | | | 149.338150 | 2323.591892 |
| | 4.667087 | 11.269775 | | | 82.796647 | 812.582737 |
| | 2.942435 | 17.763756 | | | 103.260215 | 935.345614 |
| | 1.658571 | 24.700366 | | | 81.910125 | 1284.774610 |
| | 6.224158 | 31.407797 | | | 32.201064 | 517.272977 |
| | 7.084504 | 37.201063 | | | 16.935999 | 231.400121 |
| | 19.376056 | 41.417721 | | | 8.370643 | 183.529415 |
| | 30.048185 | 46.636077 | | | 1 | 30.048186 |

We select the following 10 degree of freedom baseline system from [59]. The 10 mass values are 1, 2, 4, 6, 7, 10, 5, 3, 8 and 9 kg and the 10 given stiffness values are 20, 10, 40, 60, 30, 70, 50, 80, 100, 90 kN/m. This system yields a Jacobi matrix as shown below:

$$
A^{(base)} = \begin{bmatrix}
30.0000 & -7.0711 & 0 & 0 & 0 & 0 & 0 & 0 & 0 & 0 \\
-7.0711 & 25.0000 & -14.1421 & 0 & 0 & 0 & 0 & 0 & 0 & 0 \\
0 & -14.1421 & 25.0000 & -12.2474 & 0 & 0 & 0 & 0 & 0 & 0 \\
0 & 0 & -12.2474 & 15.0000 & -4.6291 & 0 & 0 & 0 & 0 & 0 \\
0 & 0 & 0 & -4.6291 & 14.2857 & -8.3666 & 0 & 0 & 0 & 0 \\
0 & 0 & 0 & 0 & -8.3666 & 12.0000 & -7.0711 & 0 & 0 & 0 \\
0 & 0 & 0 & 0 & 0 & -7.0711 & 26.0000 & -20.6559 & 0 & 0 \\
0 & 0 & 0 & 0 & 0 & 0 & -20.6559 & 60.0000 & -20.4124 & 0 \\
0 & 0 & 0 & 0 & 0 & 0 & 0 & -20.4124 & 23.7500 & -10.6066 \\
0 & 0 & 0 & 0 & 0 & 0 & 0 & 0 & -10.6066 & 10.0000
\end{bmatrix} \quad (1.27)
$$

We use the firefly algorithm to find an isospectral matrix. This method is a swarm optimization method and will be discussed in the next section. The firefly algorithm approach uses a global search augmented by a local search to find the minima of a non-linear error function. Decision variables associated with this optimization problem are the off-diagonal elements and the perturbation elements of the system Jacobi matrix. One of the isospectral systems determined by the evolutionary optimization approach is:

$$
A^{(firefly)} = \begin{bmatrix}
30.0000 & -5.8663 & 0 & 0 & 0 & 0 & 0 & 0 & 0 & 0 \\
-5.8663 & 25.0000 & -14.2761 & 0 & 0 & 0 & 0 & 0 & 0 & 0 \\
0 & -14.2761 & 25.0000 & -12.9828 & 0 & 0 & 0 & 0 & 0 & 0 \\
0 & 0 & -12.9828 & 15.0000 & -4.9882 & 0 & 0 & 0 & 0 & 0 \\
0 & 0 & 0 & -4.9882 & 14.2857 & -9.9463 & 0 & 0 & 0 & 0 \\
0 & 0 & 0 & 0 & -9.9463 & 13.6749 & -6.3886 & 0 & 0 & 0 \\
0 & 0 & 0 & 0 & 0 & -6.3886 & 26.0000 & -12.1309 & 0 & 0 \\
0 & 0 & 0 & 0 & 0 & 0 & -12.1309 & 60.0000 & -26.7554 & 0 \\
0 & 0 & 0 & 0 & 0 & 0 & 0 & -26.7554 & 23.7500 & -7.5277 \\
0 & 0 & 0 & 0 & 0 & 0 & 0 & 0 & -7.5277 & 8.3251
\end{bmatrix} \quad (1.28)
$$

The error function for this case is $5.7 \times 10^{-7}$ and the computational time requirement is 30 seconds in a PC (Core i5 CPU @ 2.50 GHz, 8 GB RAM).

The second approach involves using the secant method-based iterative technique for solving a non-linear error function by setting it to zero. The decision variables here are the off-diagonal terms of the Jacobian and the last diagonal element of the system Jacobian matrix. One of the isospectral systems generated by this approach is:

$$
A^{(secant)} = \begin{bmatrix}
0.2962 & -0.5796 & 0 & 0 & 0 & 0 & 0 & 0 & 0 & 0 \\
-0.5796 & 2.1731 & -1.5678 & 0 & 0 & 0 & 0 & 0 & 0 & 0 \\
0 & -1.5678 & 6.8917 & -4.7456 & 0 & 0 & 0 & 0 & 0 & 0 \\
0 & 0 & -4.7456 & 11.4456 & -2.1760 & 0 & 0 & 0 & 0 & 0 \\
0 & 0 & 0 & -2.1760 & 17.9443 & -2.5529 & 0 & 0 & 0 & 0 \\
0 & 0 & 0 & 0 & -2.5529 & 22.8302 & -3.1365 & 0 & 0 & 0 \\
0 & 0 & 0 & 0 & 0 & -3.1365 & 28.1512 & -0.4145 & 0 & 0 \\
0 & 0 & 0 & 0 & 0 & 0 & -0.4145 & 34.0413 & -6.6808 & 0 \\
0 & 0 & 0 & 0 & 0 & 0 & 0 & -6.6808 & 54.6537 & -17.1839 \\
0 & 0 & 0 & 0 & 0 & 0 & 0 & 0 & -17.1839 & 62.6085
\end{bmatrix} \quad (1.29)
$$

The error function value for this case is $5.1 \times 10^{-9}$ and the computational time is 18 seconds, which is less than the 30 seconds needed by the firefly method.

## 1.4　OPTIMIZATION APPROACH FOR ISOSPECTRAL SYSTEMS

The earlier sections presented an approach to find isospectral systems based on finding roots of nonlinear equations. In this section, we construct the problem of finding isospectral mass-spring inline systems as an optimization problem where the

**Table 1.3**
**Specification of the 10-dof Original Spring-mass System $A'_{10}$**

| Elements (sequence) | Mass (kg) | Stiffness (kN/m) | Eigenvalues | Natural Frequency (rad/sec) |
|---|---|---|---|---|
| 1 | 20.00 | 50.00 | 0.158281 | 0.397846 |
| 2 | 11.00 | 70.00 | 1.315909 | 1.147131 |
| 3 | 8.00 | 110.00 | 3.028554 | 1.740274 |
| 4 | 5.00 | 150.00 | 6.193149 | 2.488604 |
| 5 | 9.00 | 90.00 | 8.765148 | 2.960599 |
| 6 | 21.00 | 180.00 | 16.613713 | 4.075992 |
| 7 | 10.00 | 130.00 | 24.110128 | 4.910221 |
| 8 | 5.00 | 100.00 | 37.446824 | 6.119381 |
| 9 | 17.00 | 60.00 | 44.562668 | 6.675528 |
| 10 | 13.00 | 40.00 | 69.390304 | 8.330084 |

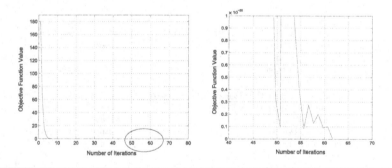

**Figure 1.3**
Function error as iterations progress

## Table 1.4
## 4 Isospectral Spring-mass Systems (10-dof) for a Given $A'_{10}$

| Isospectral Systems | Intial Conditions | Eigenvalues ($\mu$) | Error Function Values ($\times 10^{-9}$) | Difference in Trace ($\times 10^{-4}$) | Mass (kg) | Stiffness (kNm$^{-1}$) |
|---|---|---|---|---|---|---|
| 1. | 7.285358 | 0.158264 | 9.143250 | 1.907354 | 94.802743 | 179.627380 |
| | 3.186265 | 1.315888 | | | 26.553548 | 117.085291 |
| | 8.403732 | 3.028567 | | | 1.976296 | 11.520619 |
| | 7.862527 | 6.193167 | | | 1.200285 | 8.190117 |
| | 9.127747 | 8.765112 | | | 0.782000 | 6.284117 |
| | 2.732385 | 16.613731 | | | 0.745550 | 10.809971 |
| | 1.972599 | 24.110258 | | | 0.923436 | 6.345479 |
| | 2.801878 | 37.446862 | | | 2.881720 | 19.625669 |
| | 27.825608 | 44.562666 | | | 3.663818 | 70.052777 |
| | 45.834757 | 69.390373 | | | 1 | 45.834829 |
| 2. | 7.374151 | 0.158241 | 8.378386 | −0.240737 | 20.457465 | 64.246540 |
| | 3.346810 | 1.315879 | | | 14.561503 | 49.367762 |
| | 3.872903 | 3.028531 | | | 33.681522 | 36.395119 |
| | 3.952944 | 6.193123 | | | 22.235049 | 207.994703 |
| | 2.899519 | 8.765173 | | | 11.404871 | 99.408359 |
| | 4.627165 | 16.613747 | | | 6.426992 | 85.064807 |
| | 3.360787 | 24.110303 | | | 2.853640 | 58.042631 |
| | 3.242694 | 37.446804 | | | 4.425941 | 25.67708 |
| | 26.376335 | 44.562688 | | | 4.944995 | 111.936851 |
| | 47.864918 | 69.390331 | | | 1 | 47.864958 |
| 3. | 3.868876 | 0.158261 | 9.116261 | −2.149996 | 7552.375595 | 2388.551147 |
| | 5.213302 | 1.315899 | | | 4278.182338 | 7423.642926 |
| | 13.018553 | 3.028571 | | | 778.206805 | 6173.705160 |
| | 6.697510 | 6.193111 | | | 453.301680 | 5004.434694 |
| | 7.923846 | 8.765161 | | | 409.624521 | 2356.404836 |
| | 2.245795 | 16.613743 | | | 306.815709 | 6451.663303 |
| | 3.939026 | 24.110281 | | | 39.278467 | 803.776962 |
| | 7.087564 | 37.446826 | | | 8.817649 | 224.689504 |
| | 27.649735 | 44.562674 | | | 2.367865 | 27.059556 |
| | 45.835169 | 69.390330 | | | 1 | 45.835109 |
| 4. | 6.264336 | 0.158234 | 9.318423 | −1.393437 | 520.883012 | 164.616164 |
| | 7.311091 | 1.315891 | | | 215.422331 | 839.454009 |
| | 6.793200 | 3.028529 | | | 89.393897 | 358.118302 |
| | 8.206397 | 6.193184 | | | 70.944182 | 750.491263 |
| | 10.377049 | 8.765176 | | | 33.453211 | 351.548791 |
| | 1.312665 | 16.613681 | | | 18.917702 | 390.423277 |
| | 4.390122 | 24.110290 | | | 5.169046 | 79.695860 |
| | 10.955004 | 37.446819 | | | 2.826685 | 51.028374 |
| | 25.480857 | 44.562692 | | | 2.670112 | 28.414058 |
| | 47.359281 | 69.390356 | | | 1 | 47.359305 |

## Table 1.5

## Numerically Found Best Isospectral Spring-mass Systems (10-dof), $B_{10}^*$ and $B_{10}^{\prime *}$

| Isospectral Systems | Intial Conditions | Eigenvalues ($\mu$) | Error Function Values ($\times 10^{-28}$) | Difference in Trace ($\times 10^{-14}$) | Mass (kg) | Stiffness (kNm$^{-1}$) |
|---|---|---|---|---|---|---|
| $B_{10}^*$ | 7.285358 | 0.077561 | 4.047349 | 2.842171 | 24.190499 | 22.847054 |
| | 3.186265 | 1.979556 | | | 32.800811 | 334.671122 |
| | 8.403732 | 5.836304 | | | 60.808681 | 342.742516 |
| | 7.862527 | 11.269680 | | | 77.335830 | 927.150791 |
| | 9.127747 | 17.763739 | | | 39.331374 | 700.554985 |
| | 2.732385 | 24.700387 | | | 9.923807 | 134.460701 |
| | 1.972599 | 31.407692 | | | 8.703728 | 78.858474 |
| | 2.801878 | 37.201021 | | | 7.210738 | 113.796460 |
| | 27.825608 | 41.417697 | | | 3.230841 | 49.310462 |
| | 45.834757 | 46.636045 | | | 1 | 29.159029 |
| $B_{10}^{\prime *}$ | 8.596094 | 0.077561 | 5.711353 | −1.136868 | 13.489848 | 46.706041 |
| | 3.672250 | 1.979556 | | | 6.694101 | 26.583294 |
| | 6.399894 | 5.836304 | | | 4.236208 | 10.953621 |
| | 4.043261 | 11.269680 | | | 2.659300 | 21.635587 |
| | 4.159258 | 17.763739 | | | 2.663678 | 17.183128 |
| | 2.043748 | 24.700387 | | | 1.979570 | 23.787711 |
| | 0.654171 | 31.407692 | | | 1.319267 | 20.414507 |
| | 1.204713 | 37.201021 | | | 4.304511 | 18.167876 |
| | 24.103663 | 41.417697 | | | 5.198721 | 118.139713 |
| | 47.726787 | 46.636045 | | | 1 | 47.726788 |

objective function is a nonnegative error function. The problem becomes analogous to finding multiple solutions of a least square problem. Previously, researchers ([42],[38]) had proposed several numerical techniques for the least squares solution of inverse eigenvalue problems. An iterative method, called *lift and projection* (LP), was presented by Chu [42] to construct a Hermitian matrix from its diagonal elements and eigenvalues. Chu also proposed the projected gradient method, a continuous approach to solve the inverse eigenvalue problem. Later, Chen and Chu [38] presented an efficient hybrid method called LP-Newton, by first applying the LP method to attain a low order of accuracy and then switching to a comparatively faster locally convergent method (Newton method) to reach a high order of accuracy.

Recently, stochastic optimization algorithms (Fouskakis and Draper [67]; Ji and Klinowski [109]) have become ubiquitous because of their attractive search mechanisms. These algorithms search from a population of solution points instead of from a single solution point. They need fitness function information instead of gradient based knowledge and use probabilistic transition rules in their search for the optimal solution. In this section, we use the firefly algorithm (FA) augmented by local search (LS) to minimize the objective function. FA (Yang [213, 214, 215]; Lukasik and Zak [135]) is a nature inspired computational algorithm motivated by the sociobiology of fireflies. It is demonstrated by Yang that FA can capture multiple optimum points of a multimodal function. A multimodal function is a function with more than one minimum point.

## THE OPTIMIZATION PROBLEM

Let $p_n$ and $q_n$ be two vectors of size $(n \times 1)$, containing the eigenvalues of the matrices $A_n$ and $B_n$, respectively. The eigenvalues of $A_n$ and $B_n$ are arranged in ascending order $\lambda_1 \leq \lambda_2 \leq ... \leq \lambda_n$ and $\mu_1 \leq \mu_2 \leq ... \leq \mu_n$, in the vectors $p_n$ and $q_n$, respectively. We need an objective function to cast the optimization problem. We construct a vector $e_n$, which is the difference between $p_n$ and $q_n$ vectors. Let $F_n = \|e_n\|_2^2 = \|p_n - q_n\|_2^2$ ($\|.\|_2$ denotes 2-norm). We will now call $F_n$ as the *error function* which is always nonnegative for symmetric positive definite Jacobi matrices $A_n$ and $B_n$. The objective function in the optimization problem is therefore the error function ($F_n$) which can be expressed as

$$F_n = (\lambda_1 - \mu_1)^2 + (\lambda_2 - \mu_2)^2 + .... + (\lambda_n - \mu_n)^2 \tag{1.30}$$

The system $B_n$ is called isospectral to the given system $A_n$ if $\mu_i = \lambda_i$ for all $i$. Specifically, $B_n$ is isospectral to $A_n$ if the error function $F_n = 0$.

Therefore, to find the isospectral matrix (matrices) $B_n$ for a given $A_n$, we minimize the error function (objective function). The optimization problem is given as

$$\begin{aligned} \min_{c_i, d_i} \quad & F_n = (\lambda_1 - \mu_1)^2 + (\lambda_2 - \mu_2)^2 + .... + (\lambda_n - \mu_n)^2 \\ subject\ to\ & 0 < c_i \leq C\ (i = 1,...,n),\ \ 0 < d_i \leq D\ (i = 1,...,n-1) \end{aligned} \tag{1.31}$$

where $C$ and $D$ are the upper bounds on the diagonal and super-diagonal (or subdiagonal) elements of the isospectral matrix $B_n$. Once the values of $c_i$ and $d_i$ are known, we can form the $B_n$ matrix and calculate its eigenvalues. The values of $C$ and $D$ are related to the availability of physically real springs and masses. The number of decision variables in the aforementioned optimization problem is $n + (n-1) = (2n-1)$ i.e. $n$ diagonal elements and $(n-1)$ super-diagonal elements.

A theoretical result from linear algebra is applied to reduce the computational effort in the optimization problem. It is known that $Trace(B_n) = Trace(A_n)$, when matrix $B_n$ is isospectral to matrix $A_n$. We use this result as a constraint in the optimization problem

$$a_1 + a_2 + ... + a_n = c_1 + c_2 + ... + c_n \tag{1.32}$$

A simple numerical example is invoked to illustrate this optimization problem which yields isospectral systems.

## A 2-DOF MOTIVATING EXAMPLE

Consider a 2-dof system with masses $m_1 = 6.25$ kg, $m_2 = 4$ kg and stiffness $k_1 = 92.5$ kN/m, $k_2 = 20$ kN/m. The mass matrix $M_2$ of the system is given by

$$M_2 = \begin{bmatrix} m_1 & 0 \\ 0 & m_2 \end{bmatrix} = \begin{bmatrix} 6.25 & 0 \\ 0 & 4 \end{bmatrix} \tag{1.33}$$

and the stiffness matrix $K_2$ is given by

$$K_2 = \begin{bmatrix} (k_1 + k_2) & -k_2 \\ -k_2 & (k_2) \end{bmatrix} = \begin{bmatrix} 112.5 & -20 \\ -20 & 20 \end{bmatrix} \tag{1.34}$$

Using the aforementioned mass and stiffness matrices, the Jacobi matrix for this system is obtained as:

$$A_2 = \begin{bmatrix} \frac{k_1+k_2}{m_1} & -\frac{k_2}{\sqrt{m_1 m_2}} \\ -\frac{k_2}{\sqrt{m_1 m_2}} & \frac{k_2}{m_2} \end{bmatrix} = \begin{bmatrix} a_1 & -b_1 \\ -b_1 & a_2 \end{bmatrix} = \begin{bmatrix} 18 & -4 \\ -4 & 5 \end{bmatrix}$$

The eigenvalues of this dynamic system $(A_2)$ are $\lambda_1 = 3.8678$ and $\lambda_2 = 19.1322$.

Consider another Jacobi matrix $B_2$. The structure of the $B_2$ matrix is

$$B_2 = \begin{bmatrix} c_1 & -d_1 \\ -d_1 & c_2 \end{bmatrix}$$

We assume some random values of the elements of $B_2$ such that

$$B_2 = \begin{bmatrix} 17 & -3 \\ -3 & 7 \end{bmatrix}$$

This Jacobi matrix has eigenvalues $\mu_1 = 6.1690$ and $\mu_2 = 17.8310$. The error function value becomes $F_2 = (\lambda_1 - \mu_1)^2 + (\lambda_2 - \mu_2)^2 = 6.9886 \neq 0$. Therefore, the above Jacobi is not isospectral to the given one.

For the matrix $B_2$ to be isospectral to the given matrix $A_2$, the eigenvalues of $B_2$ must be identical to the eigenvalues of $A_2$. The constraint on the $B_2$ matrix is:

- Sum of the diagonal elements (trace) of $B_2$ should be equal to sum of the diagonal elements (trace) of $A_2$; i.e., $c_1 + c_2 = 23$.

We seek to find the isospectral $B_2$ matrix. Consider $c_1 = a_1 - \varepsilon$ and $c_2 = a_2 + \varepsilon$ ; i.e., $18 - \varepsilon$ and $5 + \varepsilon$. This satisfies the "trace" constraint mentioned above. The value of $d_1$ needs to be found. Since isospectral matrices have the same determinant:

$$\begin{aligned} c_1 c_2 - d_1^2 &= a_1 a_2 - b_1^2 = 74 \\ (a_1 - \varepsilon)(a_2 + \varepsilon) - d_1^2 &= 74 \\ (18 - \varepsilon)(5 + \varepsilon) - d_1^2 &= 74 \end{aligned} \qquad (1.35)$$

From the above equation, we get

$$\begin{aligned} d_1^2 &= (18 - \varepsilon)(5 + \varepsilon) - 74 \\ d_1 &= \sqrt{(18 - \varepsilon)(5 + \varepsilon) - 74} \end{aligned} \qquad (1.36)$$

Let us assume $\varepsilon = 2$ and obtain $d_1 = 6.1644$ where $c_1 = 16$ and $c_2 = 7$. This corresponds to an isospectral Jacobi matrix $B_2^*$ given by

$$B_2^* = \begin{bmatrix} \frac{k_1'+k_2'}{m_1'} & -\frac{k_2'}{\sqrt{m_1' m_2'}} \\ -\frac{k_2'}{\sqrt{m_1' m_2'}} & \frac{k_2'}{m_2'} \end{bmatrix} = \begin{bmatrix} 16 & -6.1644 \\ -6.1644 & 7 \end{bmatrix}$$

where, $m'_1, m'_2$ are the masses and $k'_1, k'_2$ are the stiffness of this isospectral system $B^*_2$. We can see that the eigenvalues of $B^*_2$ are $\mu^*_1 = \lambda_1$ and $\mu^*_2 = \lambda_2$. Therefore, the error (objective) function value becomes $F^*_2 = 0$. The values of $m'_1, m'_2, k'_1$ and $k'_2$ are obtained as follows: We know that

$$\frac{k'_2}{m'_2} = 7 \tag{1.37}$$

$$\frac{k'_2}{\sqrt{m'_1 m'_2}} = 6.1644 \tag{1.38}$$

$$\frac{k'_1 + k'_2}{m'_1} = 16 \tag{1.39}$$

If we choose $m'_2 = 1$, then from equation (1.37), we get $k'_2 = 7$. Using this $m'_2$ and $k'_2$ in equation (1.38), we obtain $m'_1 = 1.2895$. Now, we know $m'_1$ and $k'_2$. From equation (1.39), we obtain $k'_1 = 13.6316$. The mass matrix $M'_2$ and stiffness matrix $K'_2$ for this isospectral system $B^*_2$ are:

$$M'_2 = \begin{bmatrix} m'_1 & 0 \\ 0 & m'_2 \end{bmatrix} = \begin{bmatrix} 1.2895 & 0 \\ 0 & 1 \end{bmatrix}$$

$$K'_2 = \begin{bmatrix} (k'_1 + k'_2) & -k'_2 \\ -k'_2 & (k'_2) \end{bmatrix} = \begin{bmatrix} 20.6316 & -7 \\ -7 & 7 \end{bmatrix} \tag{1.40}$$

This system is obtained using $\varepsilon = 2$ and $m'_2 = 1$kg. In Table 1.6, we present ten 2-dof spring-mass systems isospectral to a given system. The given 2-dof system has the masses $m_1 = 6.25$ kg, $m_2 = 4$ kg and stiffness $k_1 = 92.5$ kN/m, $k_2 = 20$ kN/m. These are obtained for different values of $\varepsilon$ for $m'_2 = 1$kg. We can also use a different value of $m'_2$ and obtain additional isospectral systems.

For this 2-dof system, the isospectral system can be described in a pictorial manner as shown in Fig. 1.4. In this figure, the X-axis represents the value of $c_1$. The value of $c_2$ is obtained as $(23 - c_1)$. The Y-axis represents the value of $d_1^2$. This figure shows the variation of sub-diagonal element $d_1$ with the diagonal elements $(c_1, c_2)$ of the Jacobi matrix $B_2$. Each point on the curve $PQR$ in Fig. 1.4 refers to a 2-dof spring-mass system which is isospectral to the given system $A_2$. The isospectral system corresponding with $\varepsilon = 2$ and $m'_2 = 1$ is shown by the point Q in Fig. 1.4. This figure shows the value of $d_1^2$ for all values of $\varepsilon$.

We also see from Fig. 1.4 that $d_1$ is real if the values of $c_1$, $c_2$ are within the boundary points P(3.8678) and R(19.1322). The boundary points $P$ and $R$ are the eigenvalues of the system $A_2$. Therefore, the value of $\varepsilon$ should be selected such that the diagonal elements are within the eigenvalue range (minimum and maximum eigenvalues). The number of decision variables in this 2-dof optimization problem is 3; i.e. 2 diagonal elements $c_1$ and $c_2$, and 1 super-diagonal element ($d_1$). By selecting the value of $\varepsilon$, the number of decision variables diminishes to one ($d_1$). Therefore, we are able to obtain all the isospectral systems using equation (1.36).

For a general n-dof optimization problem, the number of decision variables is $n + (n - 1) = (2n - 1)$ i.e. $n$ diagonal elements and $n - 1$ super-diagonal elements. Even

## Table 1.6
**Ten 2-dof Spring-mass Systems Isospectral to a Given System**

| Isospectral Systems | Chosen $\varepsilon$ | Mass (kg) $m_1'$ | $m_2'$ | Stiffness (kN/m) $k_1'$ | $k_2'$ |
|---|---|---|---|---|---|
| 1 | 1.00 | 1.2857 | 1.0000 | 15.8571 | 6.0000 |
| 2 | 2.00 | 1.2895 | 1.0000 | 13.6316 | 7.0000 |
| 3 | 3.00 | 1.3913 | 1.0000 | 12.8696 | 8.0000 |
| 4 | 4.00 | 1.5577 | 1.0000 | 12.8077 | 9.0000 |
| 5 | 5.00 | 1.7857 | 1.0000 | 13.2143 | 10.0000 |
| 6 | 6.00 | 2.0862 | 1.0000 | 14.0345 | 11.0000 |
| 7 | 7.00 | 2.4828 | 1.0000 | 15.3103 | 12.0000 |
| 8 | 8.00 | 3.0179 | 1.0000 | 17.1786 | 13.0000 |
| 9 | 9.00 | 3.7692 | 1.0000 | 19.9231 | 14.0000 |
| 10 | 10.00 | 4.8913 | 1.0000 | 24.1304 | 15.0000 |

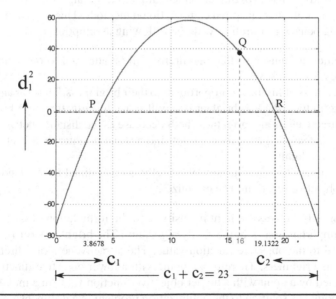

## Figure 1.4
Relation between sub-diagonal and diagonal elements

in the general n-dof optimization problem, the selection of $\varepsilon$ will reduce the number of decision variables to $n-1$ (sub-diagonal elements). Therefore, we should use an optimization method to obtain the isospectral system $B_n$ for a given system $A_n$. The objective function in the optimization problem is the error function given in equation (6.10). The error function corresponds to the square of errors in the eigenvalues of a given Jacobi matrix $A_n$ and another Jacobi matrix $B_n$. It is known that the elements of a Jacobi matrix are related to the mass and stiffness values of a spring-mass system.

For a general n-dof system, the objective function is related to the decision variables through a complicated and unknown function. Gradient information is also onerous to extract from the objective function as analytical functional representation is not possible. Therefore, a gradient-free stochastic optimization technique is selected (firefly algorithm) to solve this optimization problem. The constraint on trace defined in equation (1.32) helps to lessen the computational time and effort.

## 1.5   FIREFLY ALGORITHM

We will now use the firefly algorithm to find isospectral spring-mass systems using the optimization approach. In this section, we briefly discuss the firefly algorithm (FA) developed by Yang [214] and elaborate on its application to our problem of obtaining the isospectral system $B_n$ for a given system $A_n$. Firefly algorithm is a population-based stochastic optimization technique. Firefly algorithm is a recent inclusion to the cast of nature inspired computational methods. This algorithm is based on the flashing behavior of fireflies with the following assumptions:

1.   All fireflies are unisex so that one firefly will be attracted to other fireflies regardless of their sex.
2.   Attractiveness of fireflies is proportional to their brightness. Thus for any two flashing fireflies, the less brighter one will move towards the more brighter one. Attractiveness and brightness both decrease as the distance between the fireflies increases. If a firefly does not find anyone brighter than itself, then it moves randomly.
3.   The brightness or light intensity of a firefly is determined by the landscape of the objective function to be optimized.

The variation of brightness (or light intensity) and the formulation of attractiveness are the two important issues in the firefly algorithm. The brightness (or light intensity) is related to the objective function value. The attractiveness of a firefly is determined by its brightness. Therefore, a point with a lower objective function value will be attracted by a point with a higher objective function value in a maximization problem. A point is a solution to the optimization problem in a $d$-dimensional search space, where $d$ represents the number of decision variables.

Assume that there are $N$ fireflies in the search space, each associated with a co-ordinate $x_i(t)$ which represents the point (location) of firefly $i$ at iteration $t$. The location is a possible solution to the optimization problem. Each location (solution) $x_i$, $i = 1, 2, ...., N$ is a $d$ dimensional vector. The brightness (light intensity) $I$ of a firefly at location $x_i$ can be chosen as

$$I(x_i) = f(x_i) \qquad (1.41)$$

where $f$ is the objective function value at that point $x_i$. The attractiveness $\beta$ varies with the distance $r_{ij}$ between fireflies $i$ and $j$ and is given by

$$\beta(t) = \beta_0 e^{-\gamma r_{ij}^2(t)} \qquad (1.42)$$

where $\beta_0$ is the attractiveness when the distance $r_{ij} = 0$. The light intensity decreases with the distance between two fireflies and the light is also absorbed in the media. The absorption coefficient is $\gamma$ in the above equation. The distance between any two fireflies $i$ and $j$ located at points $x_i$ and $x_j$ (at iteration $t$) is the Cartesian distance

$$r_{ij}(t) = \|x_i(t) - x_j(t)\|_2 = \sqrt{\sum_{k=1}^{d} (x_{i,k} - x_{j,k})^2} \qquad (1.43)$$

A firefly $i$ at location $x_i$ will be attracted to another firefly $j$ because the firefly $j$ is brighter than the firefly $i$. The movement of firefly $i$ towards firefly $j$ is given by

$$x_i(t+1) = x_i(t) + \beta_0 e^{-\gamma r_{ij}^2(t)}(x_j(t) - x_i(t)) + \alpha(rand - 0.5) \qquad (1.44)$$

Equation (1.44) can be expressed as

$$x_i(t+1) = x_i(t) + \beta(t)(x_j(t) - x_i(t)) + \alpha(rand - 0.5) \qquad (1.45)$$

In the above equation, $t$ is the iteration number and the second term $(\beta(t)(x_j(t) - x_i(t)))$ emanates from attraction. The third term $(\alpha(rand - 0.5))$ is randomization with control parameter $\alpha$. Also, 'rand' is a random number, between 0 to 1, obtained from the uniform distribution. The third term is useful in the exploration of the search space in an efficient manner. The reason for incorporating $-0.5$ is to allow the movement within the range of $-0.5\alpha$ to $+0.5\alpha$. This value is recommended in earlier firefly algorithm implementations.

The control parameters of the firefly algorithm (FA) are $\alpha, \beta_0$ and $\gamma$. The influence of other solutions is managed by the value of attractiveness which depends on two parameters: initial attractiveness $\beta_0$ and absorption coefficient $\gamma$. In general $\beta_0 \in [0, 1]$ has been suggested (Yang [213]). When $\beta_0 = 0$, the value of $\beta(t) = 0$ and the movement of firefly $i$ in equation (1.45) is random and is called non-cooperative search. When $\beta_0 = 1$, the value of $\beta(t)$ in equation (1.45) is a finite value depending on the distance between fireflies $i$ and $j$. The movement of firefly $i$ depends on the position of firefly $j$ and is called cooperative search. The value of $\gamma$ defines the nature of variation of the attractiveness with respect to the distance. Using $\gamma = 0$ corresponds to the constant attractiveness whereas $\gamma \to \infty$ corresponds to the complete random search. Moreover, $\gamma$ is related to the scale of the design variables. The value of $\gamma$ can be derived from

$$\gamma = \frac{\gamma_0}{\max\|x_j - x_i\|_2}$$

where, $\gamma_0 \in [0, 1]$. The randomness control parameter can be chosen from $\alpha \in [0, 1]$.

## 1.5.1   LOCAL SEARCH

Local search is a technique applied in stochastic optimization algorithms to accelerate convergence to the optimal solution. We introduce the local search procedure described in the literature (Birbil and Fang [26]) into the firefly algorithm. In the local search procedure, information about the points in the neighborhood of the point $x_i$ is collected. If any of the neighborhood points is better than the original point $x_i$, then this better point is stored instead of the original point $x_i$. A better neighborhood point means that the objective function value of this point is better than the objective function value at point $x_i$. A better objective function value implies more (less) in maximization (minimization) problem. The local search procedure described by Birbil and Fang [26] is followed in this chapter. Here, LISTER and $\delta$ are the two parameters that are used in the local search. The maximum number of iterations that are used in the local search is LISTER, and the multiplier for the neighborhood search is denoted by $\delta$.

We are aware that the solution to the optimization problem is represented as point $x_i$ in $d$-dimensional space, where $d$ is the number of decision variables. In the local search algorithm, the maximum feasible step length $\delta$ is computed as $Max_d(u_d - l_d)$, where $u$ and $l$ are the upper and lower bounds on decision variables. For a given point $x_i$, the local search is performed by considering one dimension after another. For a given dimension $k$, $x_i^k$ represents the value of the point $x_i$ in dimension $k$. This $x_i^k$ is moved to a new position in dimension $k$ and is denoted as $\overline{x}_i^k$. The value of $\overline{x}_i^k$ is calculated by using two random numbers $\xi_1 = rand(0,1)$, $\xi_2 = rand(0,1)$ and $\delta$. If the objective function value of point $\overline{x}_i$ is better than the objective function value of point $x_i$, the point $x_i$ is replaced by the point $\overline{x}_i$. More details about this local search algorithm is given in the literature (Birbil and Fang [26]). The pseudocode of the *firefly algorithm* and the *local search* procedure are given below:

*Localsearch($\delta$, LSITER)*

- **for** $k = 1 : d$ all $d$ dimensions
      generate 1st random number $\xi_1 = rand(0,1)$
      **while** (*counter* < *LSITER*)
            generate 2nd random number $\xi_2 = rand(0,1)$
            **if** ($\xi_1 > 0.5$)
                  $\overline{x}_i^k \leftarrow x_i^k + \xi_2(\delta)$
            **else**
                  $\overline{x}_i^k \leftarrow x_i^k - \xi_2(\delta)$
            **end if**
            **if** ($f(\overline{x}_i) < f(x_i)$)
                  $x_i \leftarrow \overline{x}_i$
                  *counter* $\leftarrow (LSITER - 1)$
            **end if**
                  increase counter
      **end while**
- **end for**

*The firefly algorithm*

- Initialize the population of $N$ fireflies (solutions) $x_i$, $i = 1, 2, ..., N$

- Evaluate the objective function value $f(x_i)$ at each solution $x_i$ in the population
- Assign light intensity $I(x_i)$ using equation (1.41)
- Define control parameters $\alpha, \beta_0, \gamma$
- **While** (*counter < MaxIteration*)
    **for** $i = 1 : N$ all $N$ fireflies
        **for** $j = 1 : N$ all $N$ fireflies
            **if** $(I_j < I_i)$
                Calculate the distance $r_{ij}$ using equation (1.43)
                Vary the attractiveness $\beta(t)$ using equation (1.42)
                Move firefly $i$ towards $j$ using equation (1.45)
            **end if**
            Evaluate new solutions and update the light intensity
            Find the best solution
            Call *Localsearch*($\delta$, *LSITER*) to update the best solution
        **end for** $j$
    **end for** $i$
    increase counter
- **End While**
- Postprocess results and visualization

## 1.6   ISOSPECTRAL SPRING MASS SYSTEMS - OPTIMIZATION APPROACH

In this section, the computational results obtained using the firefly algorithm with local search procedure are presented. The objective is to find the isospectral system $B_n$ for a given system $A_n$, where $n$ represents the number of degrees of freedom of the system. The objective function is the error function given in equation (6.10). Results obtained for the 4-dof and 10-dof systems using the firefly algorithm are presented.

In the firefly algorithm with local search, initially all the fireflies ($N$) are placed randomly in the search space in a well-dispersed manner. The parameters used in the firefly algorithm are: Population size $N = 20$; initial attractiveness $\beta_0 = 1.0$; light absorption coefficient $\gamma = 0.04$; and randomness control parameter $\alpha = 0.2$. In the local search procedure, $\delta = 0.02$ and $LSITER = 20$ are used.

The system $B_n$ is isospectral to system $A_n$ when the objective function value given by equation (6.10) is zero. Therefore, the algorithm is terminated when the objective function value is less than a given tolerance. The tolerance values used in this computation are $10^{-8}$ for 4-dof system and $10^{-6}$ for 10-dof system.

### 1.6.1   COMPUTATIONAL RESULTS FOR 4-DOF SYSTEM

Consider a 4-dof mass-spring in-line system i.e. $n = 4$. The original Jacobi matrix($A_4$) takes the form

$$A_4 = \begin{bmatrix} a_1 & -b_1 & 0 & 0 \\ -b_1 & a_2 & -b_2 & 0 \\ 0 & -b_2 & a_3 & -b_3 \\ 0 & 0 & -b_3 & a_4 \end{bmatrix} \tag{1.46}$$

Consider another Jacobi matrix $(B_4)$ of the form

$$
B_4 = \begin{bmatrix}
c_1 & -d_1 & 0 & 0 \\
-d_1 & c_2 & -d_2 & 0 \\
0 & -d_2 & c_3 & -d_3 \\
0 & 0 & -d_3 & c_4
\end{bmatrix}
\tag{1.47}
$$

We want to find the elements of $B_4$ such that it is isospectral to $A_4$. The constraint on trace defined in equation (1.32) is applied to reduce the computational time and effort. The trace constraint for this 4-dof system is

$$
a_1 + a_2 + a_3 + a_4 = c_1 + c_2 + c_3 + c_4
\tag{1.48}
$$

The methodology used for the 2-dof system is applied to modify the diagonal elements of $B_4$. The diagonal elements of $B_4$ are adjusted in the following manner by considering two variables $\varepsilon_1$ and $\varepsilon_2$:

$$
diag(B_4) = \{c_1, c_2, c_3, c_4\} = \{a_1 + \varepsilon_1 \,,\, a_2 - \varepsilon_1 \,,\, a_3 + \varepsilon_2 \,,\, a_4 - \varepsilon_2\}
\tag{1.49}
$$

$$
diag(B_4) = \{c_1, c_2, c_3, c_4\} = \{a_1 + \varepsilon_1 \,,\, a_2 + \varepsilon_2 \,,\, a_3 - \varepsilon_1 \,,\, a_4 - \varepsilon_2\}
\tag{1.50}
$$

$$
diag(B_4) = \{c_1, c_2, c_3, c_4\} = \{a_1 + \varepsilon_1 \,,\, a_2 + \varepsilon_2 \,,\, a_3 - \varepsilon_2 \,,\, a_4 - \varepsilon_1\}
\tag{1.51}
$$

There exist many other ways of modifying the diagonal elements; for example, $\{c_1, c_2, c_3, c_4\} = \{a_1 + \varepsilon_1 \,,\, a_2 - \varepsilon_1 \,,\, a_3 - \varepsilon_2 \,,\, a_4 + \varepsilon_2\}$. We have used the above three modifications in our computational algorithm. The key idea is to keep the trace of matrix $B_4$ equal to the trace of matrix $A_4$ given in (1.48). Thus, for the 4-dof system, the optimization problem is expressed as

$$
\begin{array}{c}
\min \\
\varepsilon_1, \varepsilon_2, d_1, d_2, d_3
\end{array}
F_4 = (\lambda_1 - \mu_1)^2 + (\lambda_2 - \mu_2)^2 + (\lambda_3 - \mu_3)^2 + (\lambda_4 - \mu_4)^2
$$
$$
subject\ to \quad 0 < \varepsilon_1 \leq E_1 \,,\, 0 < \varepsilon_2 \leq E_2 \,,\, 0 < d_1, d_2, d_3 \leq D
\tag{1.52}
$$

Note that $\varepsilon_1, \varepsilon_2 \neq 0$ guarantees a nontrivial solution. The upper bounds $(E_1, E_2)$ on the variables $\varepsilon_1, \varepsilon_2$ can be established using Horn's majorization theorem. Applicability of the Horn's majorization theorem to determine the bounds is explained next.

### 1.6.1.1  Applicability of the Horn's majorization theorem

We illustrate the applicability of Horn's majorization theorem or Schur-Horn theorem (Horn & Johnson 1985) to estimate the bounds on the modification variables.

According to the theorem, the sum of $k$ smallest elements of the diagonal vector (vector containing diagonal elements) of a Hermitian matrix should be greater than or equal to the sum of $k$ smallest elements of eigenvalue vector (vector containing all the eigenvalues) of the matrix for $k = 1, 2, \ldots, n-1$, where $n$ is the order of the matrix.

We explain the applicability of the theorem with the help of an example of 4-dof spring-mass system. $A_4$ is the given Jacobi matrix as shown by equation (1.46). The eigenvalues of $A_4$ are ordered as $\lambda_1 \leq \lambda_2 \leq \lambda_3 \leq \lambda_4$. $B_4$ is the modified Jacobi matrix as shown by equation (1.47). Without loss of generality, we assume $c_4 < c_3 < c_2 < c_1$ for our ease of understanding. Then Horn's majorization theorem can be expressed mathematically as

$$\begin{cases} c_4 \geq \lambda_1 \\ c_3 + c_4 \geq \lambda_1 + \lambda_2 \\ c_2 + c_3 + c_4 \geq \lambda_1 + \lambda_2 + \lambda_3 \end{cases} \tag{1.53}$$

We consider the modified Jacobi matrix $B_4$ in a form so that $diag\ (B_4) = \{c_1, c_2, c_3, c_4\} = \{a_1 + \varepsilon_1\ ,\ a_2 - \varepsilon_1\ ,\ a_3 + \varepsilon_2\ ,\ a_4 - \varepsilon_2\}$. Using these inequalities (1.53), we obtain the the the upper bounds $(E_1, E_2)$ on the variables $\varepsilon_1, \varepsilon_2$

$$E_2 = (a_4 - \lambda_1) \tag{1.54}$$
$$E_1 = (a_2 + a_3 + a_4) - (\lambda_1 + \lambda_2 + \lambda_3) \tag{1.55}$$

We also obtain the condition $(a_3 + a_4 \geq \lambda_1 + \lambda_2)$ to be satisfied.

Now consider a 4-dof spring-mass system with mass and stiffness values given as: $m_1 = 1$ kg, $m_2 = 2$ kg, $m_3 = 3$ kg, $m_4 = 4$ kg, and $k_1 = 50$ kN/m, $k_2 = 60$ kN/m, $k_3 = 70$ kN/m, $k_4 = 80$ kN/m. The Jacobi matrix $A_4$ with these values is given by

$$A_4 = \begin{bmatrix} a_1 & -b_1 & 0 & 0 \\ -b_1 & a_2 & -b_2 & 0 \\ 0 & -b_2 & a_3 & -b_3 \\ 0 & 0 & -b_3 & a_4 \end{bmatrix} = \begin{bmatrix} 110.000 & -42.4264 & 0 & 0 \\ -42.4264 & 65.000 & -28.5774 & 0 \\ 0 & -28.5774 & 50.000 & -23.094 \\ 0 & 0 & -23.094 & 20.000 \end{bmatrix}$$

The eigenvalues of the given system $A_4$ are 2.2394, 30.6069, 73.8387 and 138.3150. The corresponding natural frequencies are 1.4964, 5.5323, 8.5929 and 11.7607 rad/sec.

For 4-dof spring-mass system, the optimization problem outlined in equation (1.52) has five decision variables: subdiagonal elements $d_1, d_2, d_3$ and the values of $\varepsilon_1, \varepsilon_2$. The bounds on the subdiagonal elements are presumed to be: $b_i - 10 \leq d_i \leq b_i + 10$ for $i = 1, 2, 3$. Applying Horn's majorization theorem, bounds on the values of $\varepsilon_1, \varepsilon_2$ are: $0 < \varepsilon_1 \leq 28.3$ and $0 < \varepsilon_2 \leq 17.76$, as we use the modifications provided by equation (1.49).

The firefly algorithm is invoked on the optimization problem given in equation (1.52) and gives an optimal $B_4$ for certain values of $\varepsilon_1$ and $\varepsilon_2$. For different values of $\varepsilon_1$ and $\varepsilon_2$, we obtain many $B_4$ matrices which are isospectral to $A_4$. A sample of 10 isospectral spring-mass systems constructed using the firefly algorithm for the given system $A_4$ is presented in Table 1.7. The mass and stiffness values of the isospectral systems are labelled as: $m'_1, m'_2, m'_3, m'_4$, and $k'_1, k'_2, k'_3, k'_4$, in Table 1.7.

By applying the firefly algorithm, the elements of $B_4$ can be obtained such that this matrix is isospectral to the given system $A_4$. The mass and stiffness values of $B_4$ are extracted in a manner analogous to that explained for the 2-dof system. We will now explain this for the 4-dof system. Consider one of the $B_4$ matrices obtained by

## Table 1.7
## 10 Isospectral Spring-mass Systems (4-dof) for a Given $A_4$

| Isospectral | Mass | (kg) | | | Stiffness | (kN/m) | | | Error Function |
|---|---|---|---|---|---|---|---|---|---|
| Systems | $m'_1$ | $m'_2$ | $m'_3$ | $m'_4$ | $k'_1$ | $k'_2$ | $k'_3$ | $k'_4$ | Values ($\times 10^{-8}$) |
| 1. | 0.235271 | 0.381719 | 0.489879 | 1.000000 | 15.963134 | 11.720827 | 10.163733 | 16.194461 | 0.960402 |
| 2. | 0.231464 | 0.402794 | 0.537667 | 1.000000 | 15.124556 | 12.006839 | 11.267998 | 17.148507 | 0.524389 |
| 3. | 0.241068 | 0.383153 | 0.701362 | 1.000000 | 16.637594 | 11.212928 | 15.246264 | 15.943470 | 0.686016 |
| 4. | 0.364332 | 0.247392 | 0.340877 | 1.000000 | 30.033649 | 11.052560 | 7.962913 | 8.136238 | 0.471836 |
| 5. | 0.263995 | 0.282657 | 0.470098 | 1.000000 | 19.494049 | 10.303199 | 10.399849 | 11.755608 | 0.590045 |
| 6. | 0.277360 | 0.302321 | 0.606813 | 1.000000 | 23.080943 | 9.821031 | 11.876505 | 13.230098 | 0.294504 |
| 7. | 0.244027 | 0.425525 | 0.785641 | 1.000000 | 15.146293 | 12.361252 | 17.654474 | 17.276599 | 0.257451 |
| 8. | 0.272795 | 0.426948 | 1.016386 | 1.000000 | 18.068137 | 12.508579 | 20.467849 | 17.913302 | 0.182957 |
| 9. | 0.264768 | 0.302594 | 0.561481 | 1.000000 | 21.005799 | 9.982614 | 11.586811 | 12.960207 | 0.983247 |
| 10. | 0.288428 | 0.326439 | 0.843608 | 1.000000 | 22.315792 | 10.715922 | 15.022602 | 15.476857 | 0.914785 |

## Table 1.8
## Specification of the 10-dof Original Spring-mass System

| Elements (sequence) | 1 | 2 | 3 | 4 | 5 | 6 | 7 | 8 | 9 | 10 |
|---|---|---|---|---|---|---|---|---|---|---|
| Mass (kg) | 1.00 | 2.00 | 4.00 | 6.00 | 7.00 | 10.00 | 5.00 | 3.00 | 8.00 | 9.00 |
| Stiffness (kN/m) | 20.00 | 10.00 | 40.00 | 60.00 | 30.00 | 70.00 | 50.00 | 80.00 | 100.00 | 90.00 |
| Eigenvalues | 0.0830 | 1.5954 | 4.5320 | 13.5961 | 17.6032 | 22.6093 | 29.6266 | 31.0239 | 43.6842 | 76.6821 |
| Natural frequency (rad/sec) | 0.2881 | 1.2631 | 2.1288 | 3.6873 | 4.1956 | 4.7549 | 5.4430 | 5.5699 | 6.6094 | 8.7568 |

the firefly algorithm given below:

$$
B_4 = \begin{bmatrix} c_1 & -d_1 & 0 & 0 \\ -d_1 & c_2 & -d_2 & 0 \\ 0 & -d_2 & c_3 & -d_3 \\ 0 & 0 & -d_3 & c_4 \end{bmatrix} = \begin{bmatrix} 114.459 & -41.5207 & 0 & 0 \\ -41.5207 & 60.541 & -20.1496 & 0 \\ 0 & -20.1496 & 62.602 & -16.2052 \\ 0 & 0 & -16.2052 & 7.398 \end{bmatrix}
$$

The eigenvalues of $B_4$ are 2.2394, 30.6069, 73.8386 and 138.3150. Hence, the error (objective) function value becomes $F_4 = 10^{-8} \simeq 0$. The values of $m'_1, m'_2, m'_3, m'_4$ and $k'_1, k'_2, k'_3$ and $k'_4$ are generated in a manner similar to that described for the 2-dof system. The steps are given below:

$$
\frac{k'_4}{m'_4} = 7.3980; \quad \frac{k'_4}{\sqrt{m'_3 m'_4}} = 16.2052
$$

$$
\frac{k'_3 + k'_4}{m'_3} = 62.6020; \quad \frac{k'_2}{\sqrt{m'_2 m'_3}} = 20.1496
$$

$$
\frac{k'_2 + k'_3}{m'_2} = 60.5410; \quad \frac{k'_1}{\sqrt{m'_1 m'_2}} = 41.5207
$$

$$
\frac{k'_1 + k'_2}{m'_1} = 114.4590 \tag{1.56}
$$

## Table 1.9

**Mass Elements of 10 Isospectral Spring-mass Systems (10-dof) Found by Stochastic Optimization**

| Isospectral Systems | $m_1'$ | $m_2'$ | $m_3'$ | $m_4'$ | $m_5'$ | $m_6'$ | $m_7'$ | $m_8'$ | $m_9'$ | $m_{10}'$ | Error Function Values ($\times 10^{-6}$) |
|---|---|---|---|---|---|---|---|---|---|---|---|
| 01. | 0.137245 | 0.130885 | 0.208212 | 0.257843 | 0.244474 | 0.366447 | 0.281827 | 0.222784 | 0.750188 | 1.000000 | 0.850105 |
| 02. | 0.157598 | 0.126783 | 0.187211 | 0.217976 | 0.203955 | 0.251621 | 0.183095 | 0.204934 | 0.705162 | 1.000000 | 0.724819 |
| 03. | 0.120963 | 0.133802 | 0.224301 | 0.292502 | 0.304913 | 0.461073 | 0.324372 | 0.241886 | 0.772715 | 1.000000 | 0.985663 |
| 04. | 0.579199 | 3.478019 | 8.720931 | 13.219689 | 10.545401 | 8.069486 | 1.050142 | 0.490487 | 1.223074 | 1.000000 | 0.565763 |
| 05. | 0.289229 | 1.451808 | 3.523125 | 5.415582 | 4.897102 | 4.330386 | 0.909850 | 0.468213 | 1.121614 | 1.000000 | 0.832081 |
| 06. | 0.132882 | 0.123653 | 0.193845 | 0.238996 | 0.248725 | 0.326142 | 0.230271 | 0.219558 | 0.725241 | 1.000000 | 0.956835 |
| 07. | 0.108772 | 0.144302 | 0.256205 | 0.353089 | 0.413945 | 0.589086 | 0.363700 | 0.265698 | 0.794136 | 1.000000 | 0.735440 |
| 08. | 0.216167 | 0.937214 | 2.214455 | 3.427681 | 3.349517 | 3.255572 | 0.855465 | 0.450021 | 1.068888 | 1.000000 | 0.998943 |
| 09. | 0.170765 | 0.136593 | 0.203059 | 0.234239 | 0.187501 | 0.270345 | 0.233993 | 0.200943 | 0.725921 | 1.000000 | 0.851394 |
| 10. | 0.123624 | 0.125910 | 0.204211 | 0.259867 | 0.282924 | 0.381797 | 0.261938 | 0.230126 | 0.740779 | 1.000000 | 0.994481 |

## Table 1.10

**Stiffness Elements of 10 Isospectral Spring-mass Systems (10-dof) Found by Stochastic Optimization**

| Isospectral Systems | $k_1'$ | $k_2'$ | $k_3'$ | $k_4'$ | $k_5'$ | $k_6'$ | $k_7'$ | $k_8'$ | $k_9'$ | $k_{10}'$ | Error Function Values ($\times 10^{-6}$) |
|---|---|---|---|---|---|---|---|---|---|---|---|
| 01. | 3.164262 | 0.953100 | 2.319015 | 2.886282 | 0.981365 | 2.511118 | 1.949193 | 5.378303 | 7.988740 | 9.828236 | 0.850105 |
| 02. | 3.732706 | 0.995240 | 2.174326 | 2.505947 | 0.763688 | 2.149954 | 1.426772 | 3.333692 | 8.962330 | 7.785275 | 0.724819 |
| 03. | 2.720841 | 0.908057 | 2.436991 | 3.170525 | 1.217007 | 3.138894 | 2.497924 | 5.935755 | 8.577423 | 9.774572 | 0.985663 |
| 04. | 9.049860 | 8.326118 | 78.624355 | 139.398921 | 58.896414 | 91.752173 | 18.597432 | 8.706270 | 20.722940 | 8.325077 | 0.565763 |
| 05. | 4.710359 | 3.966518 | 32.328678 | 55.749441 | 25.484287 | 44.474307 | 13.616271 | 10.039823 | 18.052965 | 8.585359 | 0.832081 |
| 06. | 3.080368 | 0.906084 | 2.185249 | 2.660885 | 0.924055 | 2.629155 | 1.887854 | 4.099186 | 9.074302 | 8.150169 | 0.956835 |
| 07. | 2.370961 | 0.892188 | 2.715362 | 3.689766 | 1.606570 | 4.306929 | 3.316989 | 6.139217 | 9.802675 | 9.058054 | 0.735440 |
| 08. | 3.646452 | 2.838556 | 20.591806 | 34.769571 | 16.645648 | 31.204594 | 11.852654 | 10.389444 | 16.611798 | 8.774293 | 0.998943 |
| 09. | 4.043872 | 1.079067 | 2.335759 | 2.740715 | 0.772867 | 1.905716 | 1.364573 | 4.719249 | 7.337351 | 9.903267 | 0.851394 |
| 10. | 2.824514 | 0.884221 | 2.263528 | 2.841756 | 1.056257 | 2.985517 | 2.213462 | 4.596932 | 9.210635 | 8.382860 | 0.994481 |

If we select $m_4' = 1$, then from equations (1.56), all the values of $m_1', m_2', m_3'$ and $k_1', k_2', k_3'$ and $k_4'$ can be calculated. This spring-mass system is shown in Table 1.7 as isospectral system (01).

## COMPUTATIONAL RESULTS FOR 10-DOF SYSTEM

Now, we scale up the same numerical method for a ten-dimensional problem. Consider a 10-dof chain like mass-spring system as shown in Table 1.8. Now, we change only two diagonal elements of the original Jacobi matrix ($A_{10}$), in one particular manner. We use one $\varepsilon$ and apply to the 6th and 10th diagonal terms; i.e., the element $a_6$ is changed to ($a_6 + \varepsilon$), and the element $a_{10}$ is changed to ($a_{10} - \varepsilon$). The number of decision variables for this 10-dof case is 10; subdiagonal elements $d_1, ..., d_9$ and $\varepsilon$. Bounds on the decision variables are imposed as $b_i - 10 \leq d_i \leq b_i + 10$ for $i = 1, 2, ...., 9$ and $0 < \varepsilon \leq 10$. The results generated for this 10-dof dynamic system using the firefly algorithm are shown in Tables 1.9 and 1.10.

## 1.7  SUMMARY

In this chapter, we present the problem of finding isospectral discrete systems as a root finding problem for a nonlinear error function, which is solved numerically using the secant method. The nonlinear error function $F_n$ calculates the difference in eigenvalues of the tridiagonal Jacobi matrices $A_n$ and $B_n$. The variables used in the secant method are the Jacobi matrix elements, first $(n-1)$ super-diagonal elements and last diagonal element. The Horn's majorization and Gershgorin's disk theorems are applied to create good starting points for the secant method. The numerical approach finds multiple solutions with respect to a stopping criteria based on tolerance value, and results in a multitude of distinct isospectral spring-mass systems. The results show that systems with different combinations of mass-spring values (low and high orders) can be isospectral in nature. The present chapter shows that a simple algorithm, with proper initial conditions, has the potential to reveal several isospectral systems. Isospectral systems exhibit similar dynamic behavior as they have identical natural frequencies. The proposed methodology has applications in dynamics of continuous as well as discrete systems, of which most can be represented with the help of spring-mass systems.

The problem of finding isospectral spring-mass systems is also cast as an optimization problem. The objective is to minimize an error function defined using the Jacobi matrices of $A_n$ and $B_n$. Again, we use the fact that the trace of two isospectral Jacobi matrices $A_n$ and $B_n$ should be identical. We use the *firefly algorithm* augmented with *local search* to minimize the error function. This stochastic optimization technique is applied to the four- and ten-degree of freedom systems. The hybrid algorithm captures multiple optimum points. The isospectral spring-mass systems (scaled) are constructed using the optimum points and all generated systems are presented. Several isospectral systems are obtained using this methodology. Numerical results are presented for four and ten degree of freedom systems. The results obtained confirm that the optimization methodology is also useful to obtain isospectral systems and is an alternative to the nonlinear equation solution approach presented early in this chapter. The firefly algorithm starts with a population of solutions and converges to an optimal solution. The decision variables in this approach correspond to the stiffness values ($k_i$) and the mass values ($m_i$) for $i = 1, 2, ..., n$. Once the values of $k_i$ and $m_i$ are known, the eigenvalues can be computed and used to obtain the objective function value. In this firefly method approach, gradient information is not needed to obtain the optimal solution.

# 2 Discrete Models of Beams

In this chapter, we consider discrete system models of continuous systems such as beams and find their isospectral counterparts. This approach allows us to use matrix theory to determine the isospectral systems and to extend the ideas seeded in Chapter 1. The analysis of vibration of slender, beam-type components under the influence of axially-distributed forces is of practical significance in aerospace, mechanical, civil and marine engineering [47]. For example, some spacecraft structural components, such as booms for supporting solar arrays, exist in a state of preload. To design of such beam-type components requires the determination of normal modes and frequencies [179]. A flexible missile or launch-vehicle under the effect of engine thrust is typically idealized as a beam under an end-axial force. A key risk for the vehicle's flight control system is the likelihood of interaction between flight control and structural bending modes. However, accurate determination of natural vibration frequencies of such structures is required in the design of flight control systems [19, 56, 205, 208]. The application of carbon nanotubes as mechanical sensors is based on the shift in natural resonant frequency of a nanotube resonator when exposed to strain resulting from external axial loading. A model based on the Euler-Bernoulli beam theory is typically applied to analyze transverse vibration of such nanotubes [190, 222].

Modern high-rise buildings, high-mast towers etc. represent examples of structures modeled as vertical beams under axial-loads due to self-weight. Characterization of the dynamic properties of such structures is needed to reduce the effects of earthquake and seismic excitations [103, 197]. A flexible beam attached radially to a rotating rigid hub is subject to axial centrifugal forces. Such rotating beams can be used to model several structures such as helicopter and wind turbine rotor blades, rotating space booms and aircraft propellers. Typically, vibration control of rotating beams is onerous compared to non-rotating beams and demands accurate assessment of natural frequencies. Several research papers on the modelling and control of rotating flexible beam-type components use the Euler-Bernoulli beam theory [53, 125, 211, 224]. Several other examples of slender structures of practical value, for which axial loads and dynamics should be considered together, are enumerated in [195]. Therefore, precise estimation of natural frequencies of structures modeled as Euler-Bernoulli beams under the influence of axially-distributed forces is very important in engineering. Some of the material in the current chapter and the next three chapters is adapted from Refs. [165] and [166].

## 2.1 DISCRETE MODELS OF AXIALLY LOADED BEAMS

Dynamic characteristics of a beam emanate from the nature of axial-force distribution along the length of the beam. In this chapter, we consider axially-loaded *cantilever* beams (ALCBs) and introduce a method to analyze the vibration of such beams. This approach is based on a discrete model generated by discretizing the given ALCB into idealized point-masses, mass-less rigid links and rotational springs. This method is adaptable to construct a family of *unloaded* cantilever beams (UCBs) *isospectral* to a given ALCB. Recall that two dynamic systems are called isospectral if they have the same vibration spectrum although their material and/or geometric properties may be different. Given an ALCB, we can determine its transverse vibration frequencies from the discrete model presented in this chapter. We can then find a family of unloaded cantilever beams with a natural frequency spectrum identical to that of the given ALCB. This frequency-matching is accomplished by modifying mass and stiffness distributions of the given ALCB such that the change in its frequency on account of axial-forces, is countered. Further details of this methodology are discussed in subsequent chapters although the basic steps involved in this methodology are listed below:

1. Create the discrete model of the given ALCB.
2. Cconvert this discrete model into a family of models corresponding to UCBs using known matrix-factorization techniques that preserve eigenvalues.
3. Transform the discrete model of UCBs into realistic, continuous beams.

In this chapter, cross-sections of beams are assumed to be doubly-symmetric although any specified variation along the length is permitted. The beams are assumed to be slender so that the Euler-Bernoulli beam theory is applicable. Though the method presented here is valid for any general distribution of axial-forces along the beam, three typical categories of force distributions, as listed below, are considered (refer Fig. 2.1). We study the vibration of both homogeneous and non-homogeneous cantilever beams under the influence of axial-forces in each category. Moreover, practically realizable isospectral UCBs are constructed in each case.

1. *Constant tip-force applied at the free-end.* The tip-force produces a constant axial-force distribution along the length of the beam. It is assumed that the externally applied axial-load, tension or compression, deflects with the beam but its line of action remains parallel to the axis of the undeflected beam. A uniform beam and three different types of stepped-beams are considered.
2. *Constant tip-force combined with gravity load due to self-weight of vertical beams.* A uniform beam under the influence of these loads has a linear axial-force distribution. We consider both hanging and standing vertical beams (columns) under axial tensile and compressive loads. A uniform beam and three different types of tapered beams are analyzed.
3. *Centrifugal forces due to rotation of the beam.* For a rotating uniform beam, axial-force distribution displays a quadratic variation. Typically, rotating beams undergo flapping (out-of-plane), lead-lag (in-plane) and torsional vibrations. For doubly-symmetric, untwisted cross-sections; torsional motion is absent and other motions are uncoupled. We analyze flapping vibrations (about $x$-axis) of a uniform beam (with and without hub offset), a tapered beam (typically used as a model for helicopter rotor blades) and a beam with general polynomial variation of mass and stiffness (used as a model for wind turbine rotor blades).

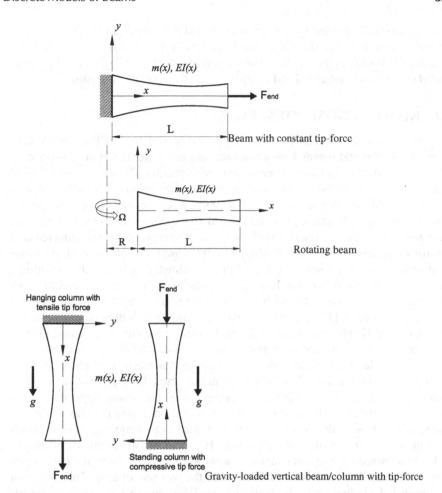

**Figure 2.1**
Schematic representation of axially-loaded beams

In each case we calculate the beam natural frequencies using the discrete model. These predicted frequency values are compared with published results. We then construct UCBs with a rectangular cross-section varying in width and depth, such that these are isospectral to the given ALCB. Characteristics of these UCBs are dependent on three-parameters. Two of these are scaling parameters. The third parameter can be tailored to yield the desired mass and stiffness *distributions* on the UCBs. The natural frequencies of the UCBs so constructed are calculated using a conventional FEM code and compared with those of the corresponding ALCB.

An additional advantage of the discrete model is that it can be used to determine critical combinations of loading parameters that precipitate onset of dynamic instability in the ALCB. For example, the critical buckling load of a standing column under the influence of gravity can be determined. An iterative algorithm is introduced for this purpose. The stability problem is evaluated in terms of a vibration problem, exploiting the fact that onset of dynamic instability is the limiting case at which the fundamental frequency of a system goes to zero. Critical buckling loads of gravity-loaded columns are calculated and compared with the theoretical results.

## 2.2  NEED FOR ISOSPECTRAL BEAMS

We now discuss the need to construct UCBs isospectral to ALCBs. For the case of rotating beams and beams with a constant tip-force, determination of isospectral UCBs is important to facilitate experimental investigations. There are some practical limitations related to modal experiments involving such beams. Several researchers have reported difficulties in matching experimental results with theoretical predictions. For example, Senatore [177] introduced an experimental method to estmiate the natural frequencies of rotating uniform beams. Experiments were conducted on a non-rotating uniform beam with calibrated weights used to apply lumped parameter axial loads. These axial loads were used to approximately simulate the centrifugal force field on the rotating beam. However, the first mode frequencies obtained from the experiment did not correlate with the theoretical results. Results of similar studies were also reported in [123], although it did not match well with theoretical predictions. Choi and Han [40] experimentally studied the active vibration control of flexible rotating beams using piezoactuators. Variation of natural frequencies with rotating speed of a glass/epoxy composite beam was experimentally determined although a comparison with theoretical estimates was not reported. Livingston et al. [134] introduced a non-linear least squares parameter estimation technique applied to experimental modal data to estimate axial load on a square rod loaded in uniaxial tension. The specimen was loaded in steps and at each load step, frequency measurements were made after impacting with a hammer. He obtained two estimates of the axial load-one based on frequency measurements while loading the beam and another while unloading–but observed some scatter in the two sets of data. This deviation was attributed to changes in the boundary condition during the course of experiment. Piana [158] performed an experimental study on the fundamental frequency evolution in slender elastic beams subjected to displacements imposed on one end (to each value of imposed displacement corresponds an internal compressive axial load). Unfortunately, the zero-frequency condition, as axial load approaches the critical buckling load, could not be demonstrated. Instead, two different trends were noted in the variation of frequencies with axial load. In the first phase, the frequency decreased with an increase in the compressive load. After reaching a minimum value, the frequency increased with increasing compressive load.

Some researchers were able to achieve satisfactory results when experiments were carried out under carefully controlled conditions. For example, Chandra and Chopra [34] performed experimental studies on rotating composite box beams (rotation speeds upto 1000 rpm) under vacuum environment with excitation induced by means of piezoelectric devices. Good correlation (within 10%) between theoretical results and experiment was observed. Nevertheless, it is obvious that performing modal experiments involving axially-loaded beams is difficult. However, if we can

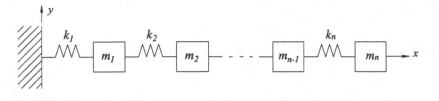

**Figure 2.2**
Discrete spring-mass system

find UCBs with the same frequency spectrum as a given ALCB, then these beams can be used as substitutes in modal experiments.

UCBs isospectral to gravity-loaded beams are useful from a different perspective. Gravitational loading significantly alters the dynamic characteristic of flexible structures for space applications [108, 196]. A key problem associated with the design and qualification of large structures for space applications is the limitation in testing these structures and their substructures on earth. It is possible that such structures may be required to have an operational natural frequency spectrum (i.e. the spectrum while operating in space) identical to the spectrum obtained from ground-based modal experiments on some other components. In ground, components are under the influence of gravity. Frequency spectrum obtained from a ground-based modal test can be affected by self-weight of components. However, in space, under conditions of zero gravity, self-weight becomes irrelevant. The technique presented here can be applied to effectively design space-based components with a desired natural frequency spectrum obtained from ground-based tests.

Isospectral UCBs also provide an insight into the stiffening or softening effects of the axial-loads. An understanding of these effects is important in the dynamic design of beam-type components. Therefore, determination of isospectral UCBs has considerable practical significance.

## 2.3 DISCRETE MODELS FOR VIBRATION ANALYSIS

Discrete models for analysis of longitudinal and transverse vibrations of cantilever bars and beams, buttressed with equations governing kinematics of these models, are presented in [78]. These models and equations form the basis of this chapter and are discussed briefly in this section.

### 2.3.1 DISCRETE MODEL FOR LONGITUDINAL VIBRATION

The spring-mass system shown in Fig. 2.2 can be used as a model to analyze free, longitudinal vibration of a bar with one end cantilevered and the other end free. Numerical values of the stiffness, $k_r$ and mass, $m_r$, of discrete springs and masses in the model are decided based on dimensions and material properties of the given bar.

Let $u_r$, $\{r = 1, 2, \ldots, n-1, n\}$ be the longitudinal displacement of the $r^{th}$ discrete mass in the model. Force, $f_r$ developed along positive $x-$axis in the $r^{th}$ spring can be written as

$$f_r = k_r (u_r - u_{r-1})$$  (2.1)

Equation (2.1) written for all discrete springs yields the following matrix equation,

$$\mathbf{f} = \hat{\mathbf{K}}\mathbf{E}^T\mathbf{u}$$  (2.2)

where $\mathbf{f}$ is the force vector $\{f_1, f_2, \ldots, f_n\}^T$, $\hat{\mathbf{K}} = diag(k_r)$ is the diagonal matrix containing stiffness of the discrete springs as diagonal elements, $\mathbf{u}$ is the displacement vector $\{u_1, u_2, \ldots, u_n\}^T$ and $\mathbf{E}$ is the square matrix defined as

$$\mathbf{E} = \begin{bmatrix} 1 & -1 & 0 & 0 & \ldots & 0 \\ 0 & 1 & -1 & 0 & \ldots & 0 \\ \vdots & \vdots & \vdots & \ddots & \vdots & \vdots \\ 0 & 0 & \ldots & 0 & 1 & -1 \\ 0 & 0 & \ldots & 0 & 0 & 1 \end{bmatrix} \text{ and } \mathbf{E}^{-1} = \begin{bmatrix} 1 & 1 & 1 & 1 & \ldots & 1 \\ 0 & 1 & 1 & 1 & \ldots & 1 \\ \vdots & \vdots & \vdots & \ddots & \vdots & \vdots \\ 0 & 0 & 0 & \ldots & 1 & 1 \\ 0 & 0 & \ldots & 0 & 0 & 1 \end{bmatrix}$$  (2.3)

Applying Newton's second law of motion to the $r^{th}$ discrete mass yields,

$$m_r \ddot{u}_r = f_{r+1} - f_r$$  (2.4)

Equation (2.4), written for all discrete masses yields

$$\mathbf{M}\ddot{\mathbf{u}} = -\mathbf{E}\mathbf{f}$$  (2.5)

where $\mathbf{M} = diag(m_r)$ is the diagonal matrix containing the discrete masses as diagonal elements. Combining equations (2.2) and (2.5), we obtain the matrix equation

$$\mathbf{M}\ddot{\mathbf{u}} + \mathbf{K}\mathbf{u} = 0$$  (2.6)

where

$$\mathbf{K} = \mathbf{E}\hat{\mathbf{K}}\mathbf{E}^T$$  (2.7)

is called the stiffness matrix of the system. Matrix $\mathbf{M}$ is diagonal and positive-definite. Matrix $\mathbf{K}$ is symmetric, positive definite and tridiagonal. Using $\mathbf{M}$ and $\mathbf{K}$ matrices, natural frequencies of longitudinal vibration of the system are obtained by formulating it as a generalized eigenvalue problem of the form $(\mathbf{K} - \lambda\mathbf{M})\mathbf{x} = 0$. Substituting $\mathbf{M} = \mathbf{D}^2$, this generalized eigenvalue problem is transformed into a simple eigenvalue problem of the form $(\mathbf{A} - \lambda\mathbf{I})\mathbf{z} = 0$ where

$$\mathbf{A} = \mathbf{D}^{-1}\mathbf{K}\mathbf{D}^{-1}$$  (2.8)

and $\mathbf{z} = \mathbf{D}\mathbf{x}$. Eigenvalues of $\mathbf{A}$ yield the square of natural frequencies of longitudinal vibration. Matrix $\mathbf{A}$, called the *mass reduced stiffness matrix* (MRSM), is symmetric, positive-definite and tridiagonal with strictly negative co-diagonals. Such a matrix falls under the category of *Jacobi matrices*. Combining equations (2.7) and (2.8), the expression for MRSM of a discrete spring-mass system is given by the relation

$$\mathbf{A} = \mathbf{D}^{-1}\mathbf{E}\hat{\mathbf{K}}\mathbf{E}^T\mathbf{D}^{-1}$$  (2.9)

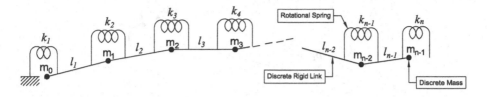

**Figure 2.3**
Discrete model of transverse vibration of a beam

## 2.3.2   DISCRETE MODEL FOR TRANSVERSE VIBRATION

A compact discrete model (Fig. 2.3) of *unloaded* cantilever beams and equations governing its kinematics were presented in [78]. These results are briefly discussed in this section. In this model, mass and stiffness of any general beam, which are continuously distributed along its length, are lumped into discrete units of $n$ masses linked by $(n-1)$ massless rigid links of specified lengths. The rigid links are connected by $n$ rotational springs of specified stiffness. For a cantilever beam, stiffness of the end-spring, $k_n$ is zero since that end is free. Moreover, mass $m_0$ does not oscillate due to the constraint against lateral deflection imposed by the fixed support. Therefore, we effectively have a system with $(n-1)$ discrete masses, $(n-1)$ discrete rigid links and $(n-1)$ discrete rotational springs. This model could be used to get $(n-1)$ natural frequencies of transverse vibration. Numerical values of the discrete elements in the model can be calculated based on dimensions and material-property of the beam. It should be noted that, for analysis of ALCBs, we use a similar, but different discrete model. The approach to be adopted for transforming a continuous beam into a discrete model was *not* presented in [78]. Studies by De Rosa & Lippiello [50] and Eddanguir & Benamar [61] address this aspect of the problem. Further details will be provided later in this chapter.

A set of four first-order difference equations governs the kinematics of this discrete model. In the subsequent analysis, $\Phi_r$ and $\tau_r$ denote the shear force (SF) and the bending moment (BM), respectively, at a section of the beam with $\Phi_n$ and $\tau_{n-1}$ representing SF and BM applied at the free end of the beam. Here $u_r$ denotes the transverse displacement of a point-mass $m_r$.

Considering equilibrium of the $r^{th}$ rigid link (Fig. 2.4) connecting masses $m_{r-1}$ and $m_r$,

$$\Phi_r = \frac{\tau_{r-1} - \tau_r}{l_r} \tag{2.10}$$

Equation (2.10) can be written for all rigid rods in the discrete model which leads to the following matrix equation.

$$\mathbf{\Phi} = \mathbf{L}^{-1}\mathbf{E}\boldsymbol{\tau} - \frac{\tau_{n-1}}{l_{n-1}}\mathbf{e}_{n-1} \tag{2.11}$$

where $\mathbf{L} = diag(l_r)$ is the diagonal matrix with lengths of the rigid links as diagonal elements, $\boldsymbol{\tau}$ is the vector $\{\tau_0, \tau_1, \tau_2, \ldots, \tau_{n-2}\}^T$ of the bending moments at the ends of each rigid link and $\mathbf{e}_{n-1}$ is the unit vector $\{0, 0, 0, \ldots, 1\}^T$ with $(n-1)$ elements.

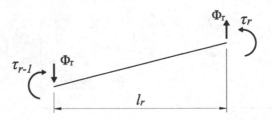

**Figure 2.4**
Free-body diagram of a discrete rigid link in discrete model of a UCB

The relation between the angular rotation $\theta_r$ of a rigid link and the lateral displacements, $u_r$ and $u_{r-1}$, of discrete masses attached at its ends is

$$\theta_r = \frac{u_r - u_{r-1}}{l_r} \tag{2.12}$$

which when expressed in matrix form, leads to the second difference equation

$$\boldsymbol{\theta} = \mathbf{L}^{-1}\mathbf{E}^T\mathbf{u} \tag{2.13}$$

where $\boldsymbol{\theta} = \{\theta_1, \theta_2, \theta_3, \ldots, \theta_{n-1}\}^T$ and $\mathbf{u}$ is the vector $\{u_1, u_2, u_3, \ldots, u_{n-1}\}^T$ of transverse displacements of discrete masses of the model.

The third difference equation is obtained by applying Newton's second law of motion to each discrete mass. When applied to the $r^{th}$ discrete mass, we obtain

$$m_r\ddot{u}_r = \Phi_r - \Phi_{r-1} \tag{2.14}$$

In matrix form, we obtain the third difference equation as

$$\mathbf{M}\ddot{\mathbf{u}} = -\mathbf{E}\boldsymbol{\Phi} + \Phi_n\mathbf{e}_{n-1} \tag{2.15}$$

The fourth and final difference equation is obtained by considering equilibrium of the sectional bending moment and the restoring torque developed in the corresponding rotational spring. For the $r^{th}$ rotational spring,

$$\tau_r = k_r\left(\theta_r - \theta_{r-1}\right) \tag{2.16}$$

In matrix form

$$\boldsymbol{\tau} = \hat{\mathbf{K}}\mathbf{E}^T\boldsymbol{\theta} \tag{2.17}$$

The aforementioned set of four difference equations governs the transverse vibration of the discrete model of a UCB. These difference equations can be combined and simplified to yield

$$\mathbf{M}\ddot{\mathbf{u}} + \mathbf{E}\mathbf{L}^{-1}\mathbf{E}\hat{\mathbf{K}}\mathbf{E}^T\mathbf{L}^{-1}\mathbf{E}^T\mathbf{u} = \frac{\tau_{n-1}}{l_{n-1}}\mathbf{E}\mathbf{e}_{n-1} + \Phi_n\mathbf{e}_{n-1} \tag{2.18}$$

When $\tau_{n-1}$ and $\Phi_n$ equal zero, this equation is of the form $\mathbf{M}\ddot{\mathbf{u}} + \mathbf{K}\mathbf{u} = 0$ which represents the general form of an equation representing free vibration of a system. Stiffness matrix, $\mathbf{K}$, is given by

$$\mathbf{K} = \mathbf{E}\mathbf{L}^{-1}\mathbf{E}\hat{\mathbf{K}}\mathbf{E}^T\mathbf{L}^{-1}\mathbf{E}^T \tag{2.19}$$

As described in section 2.3.1, using $\mathbf{K}$ and $\mathbf{M}$ matrices, we can find the spectrum of transverse vibration of the cantilever beam. The $\mathbf{M}$ matrix is diagonal and positive definite. The $\mathbf{K}$ matrix is symmetric, positive definite and pentadiagonal. In this case, the MRSM is formulated as

$$\mathbf{A} = \mathbf{D}^{-1}\mathbf{EL}^{-1}\mathbf{E}\hat{\mathbf{K}}\mathbf{E}^T\mathbf{L}^{-1}\mathbf{E}^T\mathbf{D}^{-1} \tag{2.20}$$

Eigenvalues of the MRSM $\mathbf{A}$ yield the square of natural frequencies of transverse vibration of the UCB. Matrix $\mathbf{A}$ is symmetric, positive definite and pentadiagonal. Elements in the first co-diagonal of this matrix are strictly negative while those in the second co-diagonal are strictly positive. Such matrices can be classified as *Sign Oscillatory* matrices.

## 2.4  CONSTRUCTING DISCRETE MODELS FROM AN MRSM (MASS REDUCED STIFFNESS MATRIX)

In the previous section, the method to derive the MRSM of discrete models was enunciated. Now we ponder upon the reverse problem i.e. to factorize a matrix $\mathbf{A}'$ that has properties characteristic of a realistic MRSM and obtain discrete models of spring-mass systems or UCBs that have $\mathbf{A}'$ as the MRSM. These problems are originally discussed in [77, 78] and are critical to the current study.

### 2.4.1  CONSTRUCTING SPRING-MASS SYSTEMS FROM MRSM

Equation (2.8), defines the relationship between the MRSM, mass matrix and stiffness matrix of a spring-mass system. In the present case, $\mathbf{A}'$, the MRSM of an unknown spring-mass system is known. The mass matrix, $\mathbf{M}'$ and the diagonal matrix, $\hat{\mathbf{K}}'$, that defines stiffness of springs in this model, need to be constructed such that when these matrices are combined using the expression on RHS of equation (2.9) with $\mathbf{D} = \sqrt{\mathbf{M}'}$, we get back matrix $\mathbf{A}'$. Gladwell [78] described a factorization procedure to accomplish this. The stiffness matrix, $\mathbf{K}'$ of any spring-mass system, as shown in Fig. 2.2 has the property

$$\mathbf{K}'\{1,1,1,\ldots,1\}^T = \{k_1',0,0,\ldots,0,0\}^T \tag{2.21}$$

where $k_1'$ is the unknown stiffness of the first spring in the model. In terms of physical interpretation, this expression essentially means that a force, of magnitude equal to stiffness of the first spring, applied on the first mass causes every mass in the model to undergo unit displacement. From equation (2.8) and equation (2.21),

$$\mathbf{D}'\mathbf{A}'\mathbf{D}'\{1,1,1,\ldots,1\}^T = \{k_1',0,0,\ldots,0,0\}^T \tag{2.22}$$

where $\mathbf{D}' = diag(d_r') = \sqrt{\mathbf{M}'}$. This relation can be simplified to obtain,

$$\mathbf{A}'\mathbf{d}' = \{p_1,0,0,\ldots,0,0\}^T \tag{2.23}$$

where $\mathbf{d}' = \{d_1',d_2',d_3'\ldots,d_n'\}^T$ and $p_1 = k_1'(d_1')^{-1}$. For convenience, it can be assumed that $p_1 = 1$. Equation (2.23) represents a set of linear equations where matrix

$\mathbf{A}'$ is known. Solution to this set of equations yields $\mathbf{d}'$, the vector containing square roots of the discrete masses in the model. Specifically, $\mathbf{D}' = diag(\mathbf{d}')$ and $\mathbf{M}' = (\mathbf{D}')^2$. All values in $\mathbf{d}'$ will be positive because $\mathbf{A}'$ is a Jacobi matrix (a characteristic property of realistic MRSMs of spring-mass systems) and the inverse of a Jacobi matrix is a strictly positive matrix i.e. a matrix with strictly positive values for all elements. Product of such a matrix with a positive vector will always produce another vector with all entries positive. Thus, elements of $\mathbf{d}'$ satisfy the required positivity condition.

The mass matrix, $\mathbf{M}'$ of the unknown spring-mass system is now created. Parameter $p_1$ represents a scaling factor. Its value can be fixed based on some auxiliary condition like a limiting value for the sum total of all masses in the model. The next step in the construction procedure is to determine matrix $\hat{\mathbf{K}}'$. For this, equation (2.9) can be modified to obtain

$$\hat{\mathbf{K}}' = \mathbf{E}^{-1}\mathbf{D}'\mathbf{A}'\mathbf{D}'\left(\mathbf{E}^T\right)^{-1} \tag{2.24}$$

Therefore, the discrete model of a spring-mass system that has the given matrix $\mathbf{A}'$ as the MRSM, is completely defined. Value of $p_1$ uniquely identifies this system. The procedure to be followed to construct discrete models of a UCB from the MRSM is now elucidated.

### 2.4.2  CONSTRUCTING DISCRETE MODEL OF A UCB (UNLOADED CANTILEVER BEAM) FROM MRSM

Gladwell [77] presented a matrix-factorization technique using which a given matrix $\mathbf{A}'$ that has properties characteristic of a realistic MRSM of a beam, could be decomposed into mass, length and spring-stiffness matrices corresponding to discrete models of a family of UCBs, each with $\mathbf{A}'$ as its MRSM. This technique, makes use of the relationship between static deflection, applied force and stiffness matrix of the discrete model of any given UCB to derive expressions for $\mathbf{M}'$, $\hat{\mathbf{K}}'$ and $\mathbf{L}'$ matrices. Equation (2.21) gives a similar relation for the case of spring-mass system. For a UCB, the relation is given by

$$\mathbf{K}'\{1,1,1,\ldots,1\}^T = \{f_1,-f_2,0,\ldots,0,0\}^T \tag{2.25}$$

Physically, it means that there can exist forces, which when applied in opposite directions on the first two discrete masses of the model, can cause all discrete masses to undergo unit deflection. From equation (2.8) and equation (2.25),

$$\mathbf{D}'\mathbf{A}'\mathbf{D}'\{1,1,1,\ldots,1\}^T = \{f_1,-f_2,0,\ldots,0,0\}^T \tag{2.26}$$

This relation can be simplified to obtain the vector $\mathbf{d}'$, whose elements are square roots of point-masses in the discrete model using the relation

$$\{d_1',d_2',d_3',\ldots,d_{n-1}'\}^T = (\mathbf{A}')^{-1}\{g_1,-g_2,0,0,\ldots,0\}^T \tag{2.27}$$

where $d_r' = \sqrt{m_r'}$ and $g_r = f_r/d_r'$. Arbitrary values may be chosen for $g_1$ and $g_2$ such that positive values are obtained for all $d_r'$. Since $\mathbf{A}'$ is a realistic MRSM, it should be a Sign Oscillatory (SO) matrix. Inverse of such a matrix should be an Oscillatory matrix, which implies that $\mathbf{H}' = (\mathbf{A}')^{-1}$ is a strictly positive matrix and all its minors are non-negative. (Note that Jacobi matrices, as seen in section 2.4.1, are a sub-class of SO matrices.) Mathematical proofs of these theorems are provided in [77]. Using these properties of $\mathbf{H}'$, it can be surmised that if we choose $g_1 = 1$ and $g_2$ equal to a value such that $d'_{n-1}$ is positive, then all $d'_i$ will be positive. To satisfy this condition for positivity of all elements of $\mathbf{d}' = \{d'_1, d'_2, d'_3, \ldots, d'_{n-1}\}^T$, $g_2$ should be selected such that $0 < g_2 < \dfrac{\mathbf{H}'(n-1,1)}{\mathbf{H}'(n-1,2)}$ where $\mathbf{H}'(i,j)$ denotes the element at $(i,j)$ location of $\mathbf{H}'$. In the current study, we choose

$$g_2 = (1 - v) \times \frac{\mathbf{H}'(n-1,1)}{\mathbf{H}'(n-1,2)} \tag{2.28}$$

where $0 < v < 1$. Thus, $g_2$ becomes a parameter for determining characteristics of the discrete model. We have

$$\mathbf{D}' = diag\{d'_1, d'_2, d'_3, \ldots, d'_{n-1}\}^T \tag{2.29}$$

and the mass matrix associated with the discrete model to be constructed is

$$\mathbf{M}' = \mathbf{D}'^2 \tag{2.30}$$

The next step is to construct $\hat{\mathbf{K}}'$ and $\mathbf{L}'$ matrices. For this, Gladwell [77] formed matrix $\mathbf{C}'$ defined as

$$\mathbf{C}' = \mathbf{E}^{-1} \mathbf{D}' \mathbf{A}' \mathbf{D}' (\mathbf{E}^T)^{-1} \tag{2.31}$$

and mathematically proved that this is a Jacobi matrix. Therefore, $\mathbf{C}'$ could be considered as the MRSM of a spring-mass system similar to the one depicted in Fig. 2.2 but with $(n-1)$ elements. In section 2.4.1 we saw that MRSM of a spring-mass system could be decomposed into positive diagonal matrices $\mathbf{D}'$ and $\hat{\mathbf{K}}'$. Thus $\mathbf{C}'$ could also be decomposed in an analogous manner.

The matrices obtained by decomposing $\mathbf{C}'$ in this manner can, in fact, be considered as $\mathbf{L}'$ and $\hat{\mathbf{K}}'$ matrices of the discrete model of a UCB. This can be explained as follows. In equation (2.9) giving the expression for MRSM of a spring-mass system, let $\mathbf{A}$ be replaced with $\mathbf{C}'$ and $\mathbf{D}$ be replaced with $\mathbf{L}'$. The following equation is obtained.

$$\mathbf{C}' = (\mathbf{L}')^{-1} \mathbf{E} \hat{\mathbf{K}}' \mathbf{E}^T (\mathbf{L}')^{-1} \tag{2.32}$$

Combining equations (2.31) and (2.32), we get

$$\mathbf{A}' = (\mathbf{D}')^{-1} \mathbf{E} (\mathbf{L}')^{-1} \mathbf{E} \hat{\mathbf{K}}' \mathbf{E}^T (\mathbf{L}')^{-1} \mathbf{E}^T (\mathbf{D}')^{-1} \tag{2.33}$$

Equation (2.20), provides the expression for MRSM of a UCB. Comparing equations (2.33) and (2.20), it is deduced that for a UCB with $\mathbf{A}'$ as MRSM, factors $\mathbf{M}'$, $\mathbf{L}'$ and $\hat{\mathbf{K}}'$ can be considered as the mass, length and spring-stiffness matrices of its discrete model. Expression for $\mathbf{M}'$ is obtained from equations (2.27), (2.29) and (2.30). Matrices $\hat{\mathbf{K}}'$ and $\mathbf{L}'$ are obtained by decomposing $\mathbf{C}'$ matrix using the methods presented in section 2.4.1 after replacing $\mathbf{A}'$ with $\mathbf{C}'$ and $\mathbf{D}'$ with $\mathbf{L}'$. Following expressions are obtained.

$$\{l_1', l_2', l_3', \ldots, l_{n-1}'\}^T = (\mathbf{C}')^{-1}\{1, 0, 0, \ldots, 0\}^T \tag{2.34}$$

$$\mathbf{L}' = diag\{l_1', l_2', l_3', \ldots, l_{n-1}'\}^T \tag{2.35}$$

$$\hat{\mathbf{K}}' = \mathbf{E}^{-1}\mathbf{L}'\mathbf{C}'\mathbf{L}'(\mathbf{E}^T)^{-1} \tag{2.36}$$

Having determined matrices $\mathbf{M}', \mathbf{L}'$ and $\hat{\mathbf{K}}'$ the discrete model of a UCB that has $\mathbf{A}'$ as its MRSM, is completely defined.

### 2.4.2.1  Role of parameters

Thus far, only one parameter - $g_2$, is applied for defining characteristics of the discrete model constructed from a given MRSM. This parameter has to be fixed ensuring positivity of $\mathbf{D}'$ matrix. Theoretically, parameter $g_2$ has a range of permissible values with $\nu$ between 0 and 1. In chapters 3, 4 and 5, where we peruse the results of numerical studies, it is seen that parameter $g_2$ plays a significant role in determining *distributions* of mass and stiffness on the isospectral models. Furthermore, the range of values of $g_2$ that ensure realistic discrete models, is limited.

Two additional parameters - $\alpha$ and $\beta$, which are scaling parameters, can be introduced. If we have a new set of mass, length and stiffness matrices given by the relations

$$\mathbf{M}'' = \alpha^2 \mathbf{M}' \qquad \hat{\mathbf{K}}'' = \beta^2 \hat{\mathbf{K}}' \qquad \mathbf{L}'' = \frac{\beta}{\alpha}\mathbf{L}'$$

then this system also has $\mathbf{A}'$ as its MRSM. $\alpha$ and $\beta$ can be fixed based on certain limiting values for the total length and total mass of the constructed discrete model.

Thus, we have three parameters that define characteristics of the discrete model of a UCB, constructed from a given MRSM $\mathbf{A}'$. For different combinations of permissible values of these parameters, we obtain different discrete models. Since natural frequencies of a model are determined by eigenvalues of the MRSM, all these models derived from the same MRSM *are isospectral*. In short, we obtain a three-parameter *family of isospectral discrete models*, each with $\mathbf{A}'$ as the MRSM.

This chapter enunciates the mathematical formulation required for the next three chapters. The procedure to be followed for discretizing different types of ALCBs and generating a discrete model is elucidated. Kinematic equations of the models are derived and MRSM is formulated. Moreover, the procedure to be followed for determining a family of UCBs isospectral to a given ALCB is explained. Also, the iterative algorithm to be adopted to estimate the critical buckling loads of ALCBs (with emphasis on gravity-loaded beams/columns) from the discrete model is discussed.

## 2.5  DISCRETE MODEL OF BEAMS WITH CONSTANT TIP-FORCE

The discrete model of a beam with a constant tip force is shown in Fig. 2.5. This model is similar to the one introduced by Gladwell [77] and described in

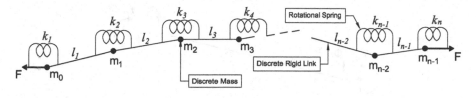

**Figure 2.5**
Discrete model of transverse vibration of a beam with tip-force

section 2.3.2. However, it includes an additional constant axial load, $F$, applied at one end of the beam and reacted at the other.

In the discrete model, mass and stiffness, which are continuously distributed along the length of any general beam, are lumped into discrete units of $n$ masses linked by $(n-1)$ massless rigid links of specified lengths. The rigid links are connected by $n$ rotational springs of specified stiffness. A continuous beam is thus converted into a discrete system with finite degrees of freedom. To model a given beam on which there is no rapid variation in beam-geometry over a local region, lengths of all rigid links can be considered equal. Let $L$ denote the total length of the beam, $l_r$ denote the length of each rigid link and $BS_r$ denote the beam-segment between mid-points of adjacent rigid links to which mass $m_r$ (or spring of stiffness $k_{r+1}$) is attached $\{r = 0, 1, 2 \ldots, n-1\}$. For any given (uniform or non-uniform) beam, we assume that over each segment $BS_r$, cross-section of the beam is uniform with dimensions evaluated at the location of the discrete mass, $m_r$. Mass of $m_r$ is equal to the mass of $BS_r$. Stiffness, $k_{r+1}$, of each discrete rotational spring, except those at either ends, is equal to the stiffness of $BS_r$. Stiffness, $k_1$ and $k_n$, of springs at either ends of the discrete model is obtained as the combined stiffness, $\left( \frac{1}{k_{support}} + \frac{1}{k_{segment}} \right)^{-1}$, of the end-support and that of $BS_0$ or $BS_{n-1}$. For a cantilever beam, $k_1$ would be equal to $k_{segment}$ since $k_{support} = \infty$ and $k_n$ would be equal to zero since $k_{support} = 0$. Also, mass $m_0$ does not oscillate due to the constraint against lateral deflection imposed by the clamped support. Thus, we effectively have a system with $(n-1)$ discrete masses, $(n-1)$ discrete rigid links and $(n-1)$ discrete rotational springs. This model could be used to obtain $(n-1)$ natural frequencies of transverse vibration.

Numerical values for mass, spring-stiffness and link-lengths should be selected based on dimensions and material property of the beam. The approach to fix these values for different types of beams is outlined in section 2.8. First, we look at the equations governing kinematics of the model in Fig. 2.5.

### 2.5.1 KINEMATIC EQUATIONS

In section 2.3.2, we derived a set of four difference equations governing kinematics of the discrete model of a UCB. Except for the equation derived by considering equilibrium of a rigid link in the model, the other equations, (2.13), (2.15) and (2.17), are valid in the present case of a beam with a tip-force. The equilibrium equation for the present case is derived as follows.

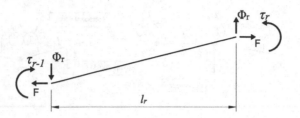

## Figure 2.6
Free-body diagram of a discrete rigid link

Considering equilibrium of the $r^{th}$ rigid link (Fig. 2.6) connecting masses $m_{r-1}$ and $m_r$,

$$\Phi_r = \frac{\tau_{r-1} - \tau_r}{l_r} + \frac{F}{l_r}(u_r - u_{r-1}) \tag{2.37}$$

Equation (2.37) can be written for all rigid rods in the discrete model which yields the following matrix equation.

$$\Phi = \mathbf{L}^{-1}\mathbf{E}\tau + F\mathbf{L}^{-1}\mathbf{E}^T\mathbf{u} - \frac{\tau_{n-1}}{l_{n-1}}\mathbf{e}_{n-1} \tag{2.38}$$

Here positive $F$ denotes a tensile axial load at the free end of the beam. Compared to equation (2.11), the additional term, $F\mathbf{L}^{-1}\mathbf{E}^T\mathbf{u}$, represents the effect of an axial load applied at the free end of the cantilever beam. Now, the four difference equations, (2.38), (2.13), (2.15) and (2.17) can be combined and simplified to obtain

$$\mathbf{M}\ddot{\mathbf{u}} + [\mathbf{E}\mathbf{L}^{-1}\mathbf{E}\hat{\mathbf{K}}\mathbf{E}^T\mathbf{L}^{-1}\mathbf{E}^T + F\mathbf{E}\mathbf{L}^{-1}\mathbf{E}^T]\mathbf{u} = \frac{\tau_{n-1}}{l_{n-1}}\mathbf{E}\mathbf{e}_{n-1} + \Phi_n\mathbf{e}_{n-1} \tag{2.39}$$

When $\tau_{n-1}$ and $\Phi_n$ equal zero, this equation is of the form $\mathbf{M}\ddot{\mathbf{u}} + \mathbf{K}\mathbf{u} = 0$ which is the general form of an equation representing free vibration of a system. Stiffness matrix, $\mathbf{K}$, is defined as

$$\mathbf{K} = \mathbf{E}\mathbf{L}^{-1}\mathbf{E}\hat{\mathbf{K}}\mathbf{E}^T\mathbf{L}^{-1}\mathbf{E}^T + F\mathbf{E}\mathbf{L}^{-1}\mathbf{E}^T \tag{2.40}$$

Using equation (2.8), MRSM of the discrete model is formulated as

$$\mathbf{A} = \mathbf{D}^{-1}\left[\mathbf{E}\mathbf{L}^{-1}\mathbf{E}\hat{\mathbf{K}}\mathbf{E}^T\mathbf{L}^{-1}\mathbf{E}^T + F\mathbf{E}\mathbf{L}^{-1}\mathbf{E}^T\right]\mathbf{D}^{-1} \tag{2.41}$$

The matrix $\mathbf{M}$ is diagonal and unconditionally positive definite. The matrix $\mathbf{K}$ is symmetric and pentadiagonal. Positive definiteness of $\mathbf{K}$ depends on the end-axial force applied on the beam. It can be proved that $\mathbf{K}$ ceases to be positive definite when a compressive axial force, $F$, in excess of the critical buckling load of the beam is applied. MRSM $\mathbf{A}$ is symmetric and pentadiagonal. It is positive definite as long as $\mathbf{K}$ is positive definite.

## 2.6   DISCRETE MODEL OF GRAVITY-LOADED BEAMS

In this section, we include the effect of gravity-load to the beam with a tip-force. The discrete model of such a beam includes axial forces, $W_r$, $\{r = 0, 1, 2 \ldots, n -$

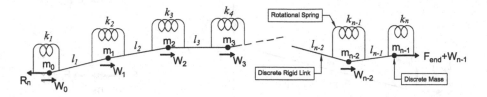

**Figure 2.7**
Discrete model of transverse vibration of a gravity-loaded beam with tip-force

1}, applied to each discrete mass (Fig. 2.7). These axial forces represent weight of the corresponding discrete mass i.e. $W_r = m_r g$, where $g$ denotes acceleration due to gravity. The force, $F_{end}$ represents the tip-force and $R_n$ denotes the reaction force at the clamped end.

### 2.6.1  KINEMATIC EQUATIONS

The kinematics of this model is governed by a set of five difference equations. Equations (2.13), (2.15) and (2.17), are valid here too. The other two equations are derived in this section.

Considering equilibrium of $r^{th}$ rigid link (Fig. 2.8) between masses $m_{r-1}$ and $m_r$, we get

$$\Phi_r = \frac{\tau_{r-1} - \tau_r}{l_r} + \frac{F_r}{l_r}(u_r - u_{r-1}) \tag{2.42}$$

Equation (2.42) can be written for all rigid links in the discrete model leading to the equation,

$$\mathbf{\Phi} = \mathbf{L}^{-1}\mathbf{E}\boldsymbol{\tau} + \mathbf{L}^{-1}\mathbf{F}\mathbf{E}^T\mathbf{u} - \frac{\tau_{n-1}}{l_{n-1}}\mathbf{e}_{n-1} \tag{2.43}$$

where $\mathbf{F}$ is the diagonal matrix with axial forces, $F_r$ on each of the $(n-1)$ rigid links as diagonal elements.

The equilibrium equation for mass, $m_r$ (Fig. 2.8) is

$$W_r = m_r g = F_r - F_{r+1} \tag{2.44}$$

Similar equations for all discrete masses in the model can be amalgamated to obtain the second governing difference equation.

$$\mathbf{f} = \mathbf{E}^{-1}[\mathbf{w} + F_{end}\mathbf{e}_{n-1}] = \mathbf{E}^{-1}[\mathbf{Mg} + F_{end}\mathbf{e}_{n-1}] \tag{2.45}$$

where $\mathbf{f}$ is the vector $\{F_1, F_2, F_3, \ldots, F_{n-1}\}^T$ of axial forces on each of the $(n-1)$ rigid links i.e. $\mathbf{F} = diag(\mathbf{f})$, $\mathbf{g}$ is the vector $g \times \{1, 1, 1, \ldots, 1\}^T$ with $(n-1)$ elements, $\mathbf{w}$ is the vector whose elements denote weights of discrete masses in the model.

The additional term in equation (2.43), $\mathbf{L}^{-1}\mathbf{F}\mathbf{E}^T\mathbf{u}$, represents the effect of axial forces (due to self-weight and applied tip force) on the transverse vibrations of the

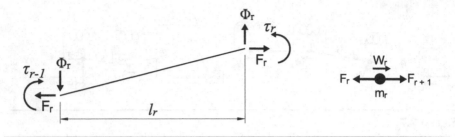

**Figure 2.8**
FBD of a typical a) rigid link b) mass, of a gravity-loaded beam with tip-force

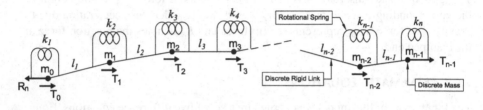

**Figure 2.9**
Discrete model of transverse vibration of a rotating beam

beam. Equation (2.45) yields an expression using which the matrix $\mathbf{F}$ can be evaluated as $\mathbf{F} = diag(\mathbf{f})$. These difference equations can be combined with equations (2.13), (2.15) and (2.17) and simplified to obtain

$$\mathbf{M}\ddot{\mathbf{u}} + [\mathbf{EL}^{-1}\mathbf{E}\hat{\mathbf{K}}\mathbf{E}^{T}\mathbf{L}^{-1}\mathbf{E}^{T} + \mathbf{EL}^{-1}\mathbf{FE}^{T}]\mathbf{u} = \frac{\tau_{n-1}}{l_{n-1}}\mathbf{E}\mathbf{e}_{n-1} + \Phi_{n}\mathbf{e}_{n-1} \qquad (2.46)$$

When $\tau_{n-1}$ and $\Phi_{n}$ equal zero, this equation is of the form $\mathbf{M}\ddot{\mathbf{u}} + \mathbf{K}\mathbf{u} = 0$. The stiffness matrix is given by

$$\mathbf{K} = \mathbf{EL}^{-1}\mathbf{E}\hat{\mathbf{K}}\mathbf{E}^{T}\mathbf{L}^{-1}\mathbf{E}^{T} + \mathbf{EL}^{-1}\mathbf{FE}^{T} \qquad (2.47)$$

Again, the matrix $\mathbf{K}$ is symmetric and pentadiagonal. Positive definiteness of $\mathbf{K}$ depends on the applied external tip force and the self-weight of the beam. It can be shown that the $\mathbf{K}$ matrix is no longer positive definite when the combined effect of the applied external force and self-weight is large enough to cause compressive buckling of the beam. Using equation (2.8), MRSM of the discrete model is formulated as

$$\mathbf{A} = \mathbf{D}^{-1}\left[\mathbf{EL}^{-1}\mathbf{E}\hat{\mathbf{K}}\mathbf{E}^{T}\mathbf{L}^{-1}\mathbf{E}^{T} + \mathbf{EL}^{-1}\mathbf{FE}^{T}\right]\mathbf{D}^{-1} \qquad (2.48)$$

## 2.7  DISCRETE MODEL OF ROTATING BEAMS

The discrete model of a rotating beam is shown in Fig. 2.9. Axial forces, $T_{r}$, $\{r = 0, 1, 2 \dots, n-1\}$, representing centrifugal forces on account of rotation of the beam, are applied to each discrete mass. Centrifugal force on $BS_{r}$ is lumped on $m_{r}$ and denoted as $T_r$ i.e. the beam segment, $BS_r$, which has a finite length, is idealized as a point-mass, $m_r$, located at a distance of $(R+x_r)$ from the axis of rotation ($R$ denotes hub-radius and $x_r$ denotes distance between the clamped end and the location of $m_r$).

### 2.7.1 KINEMATIC EQUATIONS

The kinematics of this model is also governed by a set of five difference equations. As in all previous cases, equations (2.13), (2.15) and (2.17), retain their validity here. Moreover, equation (2.43) can be directly applied, although the definition of $\mathbf{F}$ is different in this case. The difference equation yields the expression for $\mathbf{F}$ as derived below.

The equilibrium equation for mass, $m_r$ yields

$$T_r = m_r \Omega^2 (R+x_r) = F_r - F_{r+1} \tag{2.49}$$

Similar equations for all discrete masses in the model can be combined to obtain

$$\mathbf{f} = \Omega^2 \mathbf{E}^{-1} \mathbf{M} \mathbf{r} \tag{2.50}$$

where $\Omega$ is the angular speed of rotation of the beam, $\mathbf{f}$ is the vector $\{F_1, F_2, F_3, \ldots, F_{n-1}\}^T$ of axial forces on each of the $(n-1)$ rigid links i.e. $\mathbf{F} = diag(\mathbf{f})$, $\mathbf{r}$ is the distance vector defined as $\{R+x_1, R+x_2, R+x_3, \ldots, R+x_{n-1}\}^T$. All the difference equations (2.50), (2.43), (2.13), (2.15) and (2.17) can be combined and simplified to obtain

$$\mathbf{M}\ddot{\mathbf{u}} + [\mathbf{E}\mathbf{L}^{-1}\mathbf{E}\hat{\mathbf{K}}\mathbf{E}^T\mathbf{L}^{-1}\mathbf{E}^T + \mathbf{E}\mathbf{L}^{-1}\mathbf{F}\mathbf{E}^T]\mathbf{u} = \frac{\tau_{n-1}}{l_{n-1}}\mathbf{E}\mathbf{e}_{n-1} + \Phi_n \mathbf{e}_{n-1} \tag{2.51}$$

thus yielding the expression for stiffness matrix $\mathbf{K}$ as

$$\mathbf{K} = \mathbf{E}\mathbf{L}^{-1}\mathbf{E}\hat{\mathbf{K}}\mathbf{E}^T\mathbf{L}^{-1}\mathbf{E}^T + \mathbf{E}\mathbf{L}^{-1}\mathbf{F}\mathbf{E}^T \tag{2.52}$$

Using equation (2.8), MRSM of the discrete model of a rotating beam is formulated as

$$\mathbf{A} = \mathbf{D}^{-1}\left[\mathbf{E}\mathbf{L}^{-1}\mathbf{E}\hat{\mathbf{K}}\mathbf{E}^T\mathbf{L}^{-1}\mathbf{E}^T + \mathbf{E}\mathbf{L}^{-1}\mathbf{F}\mathbf{E}^T\right]\mathbf{D}^{-1} \tag{2.53}$$

The $\mathbf{M}$ matrix is diagonal and unconditionally positive definite. The $\mathbf{K}$ matrix is symmetric, pentadiagonal and positive definite, since axial forces developed on account of beam rotation are always tensile. Matrix $\mathbf{A}$ too is symmetric and pentadiagonal. It is positive definite since $\mathbf{K}$ is positive definite.

## 2.8 DISCRETIZING A BEAM

In this section, we study the approach to be followed to assign numerical values for the discrete elements of each discrete model. For a general cantilever beam, the following expressions are used.

$$l_r = \frac{L}{n-1}; \qquad m_r = \rho A_r l_r; \qquad m_{n-1} = 0.5 \rho A_{n-1} l_r$$

(2.54)

$$k_r = \frac{EI_{r-1}}{l_r}; \qquad k_1 = \frac{2EI_0}{l_r}; \qquad W_r = m_r g; \qquad T_r = m_r \Omega^2 (R + x_r);$$

Corresponding non-dimensional expressions are,

$$\bar{l}_r = \frac{l_r}{L}; \qquad \bar{m}_r = \frac{m_r}{\rho A_0 L}; \qquad \bar{k}_r = \frac{k_r L}{EI_0}; \qquad \bar{T}_r = \frac{T_r L^2}{EI_0}; \qquad \bar{W}_r = \frac{W_r L^2}{EI_0} \quad (2.55)$$

Non-dimensional gravity parameter, $\gamma$, tip force parameter, $\psi$ and angular speed parameter, $\lambda$, are defined as

$$\gamma = \frac{\rho A_0 g L^3}{EI_0}; \qquad \psi = \frac{F_{end} L^2}{EI_0}; \qquad \lambda = \Omega L^2 \sqrt{\frac{\rho A_0}{EI_0}} \qquad (2.56)$$

so that $\bar{W}_r = \bar{m}_r \gamma$ and $\bar{T}_r = \bar{m}_r \lambda^2 (\bar{R} + \bar{x}_r)$, where $\bar{R} = \frac{R}{L}$ and $\bar{x}_r = \frac{x_r}{L}$. Additionally, the non-dimensional distance along length of the beam is represented as $\bar{x} = \frac{x}{L}$.

### 2.8.1 UNIFORM BEAMS

In the discrete model of such beams, stiffness/mass of every discrete rotational spring/discrete mass, except the one at clamped/free end, would be equal. The following expressions are used.

$$\bar{m}_r = \frac{1}{n-1}; \qquad \bar{l}_r = \frac{1}{n-1}; \qquad \bar{k}_{r+1} = (n-1); \qquad \bar{m}_{n-1} = 0.5 \bar{m}_r; \qquad \bar{k}_1 = 2\bar{k}_{r+1}$$

(2.57)

### 2.8.2 RECTANGULAR BEAMS WITH LINEARLY TAPERING DEPTH AND WIDTH

The mass and stiffness of discrete elements are calculated using the following expressions.

$$\bar{m}_r = \frac{c_m}{n-1} \left( 1 - \frac{r\delta_1}{n-1} \right) \times \left( 1 - \frac{r\delta_2}{n-1} \right); \qquad \bar{l}_r = \frac{1}{n-1}$$

(2.58)

$$\bar{k}_r = c_k \times (n-1) \times \left[ 1 - \frac{(r-1)\delta_1}{n-1} \right]^3 \times \left[ 1 - \frac{(r-1)\delta_2}{n-1} \right];$$

$c_m = 1$ for $r = \{1, 2, \ldots, n-2\}$ and $c_m = 0.5$ for $r = n-1$
$c_k = 1$ for $r = \{2, 3, \ldots, n-1\}$ and $c_k = 2$ for $r = 1$.

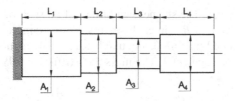

**Figure 2.10**
Schematic of a three-step cantilever beam

$\delta_1$ and $\delta_2$ denote the taper ratios defined as $\delta_1 = 1 - \dfrac{h_f}{h_c}$ and $\delta_2 = 1 - \dfrac{w_f}{w_c}$ where $h$ and $w$ denote depth and width, respectively, of the column cross-section and subscripts $f$ and $c$ denote the free end and clamped end, respectively. The above expressions can be used in the case of columns with circular cross-sections of linearly tapering diameter by setting $\delta_1 = 1 - \dfrac{\phi_f}{\phi_c} = \delta_2$ where $\phi$ denotes diameter.

### 2.8.3 STEPPED BEAMS

We consider beams of total length $L$ with many segments, as shown in Fig. 2.10. Here, each segment has a uniform cross section. We identify three types of stepped beams .
*Type A*: Rectangular beams of constant depth and with step changes in width.
*Type B*: Rectangular beams of constant width and with step changes in depth.
*Type C*: Beams with step changes in depth and width, i.e. with similar cross-sections in each segment - for example, a beam of circular cross-section with step changes in diameter [147].

In the discrete model of such a beam, every portion, $BS_r$ (other than $BS_0$ or $BS_{n-1}$), within a given $(i^{th})$ stepped-segment would have the same mass and stiffness defined as:

$$\bar{m}_r = \frac{(d_i)^p}{n-1}; \quad \bar{l}_r = \frac{1}{n-1}; \quad \bar{k}_{r+1} = (n-1)d_i{}^q; \quad \bar{m}_{n-1} = 0.5\bar{m}_{n-2}; \quad \bar{k}_1 = 2\bar{k}_2$$

$$(2.59)$$

where $d_i$ denotes the ratio of 'active' dimension of the current $(i^{th})$ segment to that of the segment with the clamped end. Thus, for the segment with the clamped end, $d_i = d_1 = 1$. 'Active' dimension, denoted as $A_i$ in Fig. 2.10, is a term used in [147] and refers to width, depth and diameter for Type A, Type B and Type C, respectively. With reference to Fig. 2.10, $d_i = \dfrac{A_i}{A_1}$. Exponents '$p$' and '$q$' are defined as follows:

for *Type A*, $p = 1$ and $q = 1$.
for *Type B*, $p = 1$ and $q = 3$.
for *Type C*, $p = 2$ and $q = 4$.

Note that we use the same length for all rigid links in the model. Consequently, certain portions, $BS_r$, can overlap the discontinuity in the beam and be part of two adjacent stepped-segments. The stiffness and mass of all such $BS_r$ are estimated depending on the proportion of $BS_r$ in each segment. For example, consider a portion $BS_b$, adjacent to the second step in a beam, that has a part of length (non-dimensional) $\bar{l}_a$ in segment 2 and $\bar{l}_b$ in segment 3. The mass and stiffness of $BS_b$ are calculated as:

$$m_b = \bar{l}_a \times (d_2)^p + \bar{l}_b \times (d_3)^p \quad \text{and} \quad k_{b+1} = \left[ \frac{\bar{l}_a}{(d_2)^q} + \frac{\bar{l}_b}{(d_3)^q} \right]^{-1} \quad (2.60)$$

### 2.8.4   BEAMS WITH POLYNOMIAL VARIATION IN MASS AND STIFFNESS

Beams that have a polynomial variation in mass and stiffness distributions are typical of wind turbine and helicopter rotor blades . Along the length of the beam, let the variations in mass per unit length, $m(x)$ and flexural stiffness, $EI(x)$ be given by

$$m(x) = m(0)(1 + \delta \bar{x}) \quad (2.61)$$

$$EI(x) = EI(0) \left( 1 + \beta_1 \bar{x} + \beta_2 (\bar{x})^2 + \beta_3 (\bar{x})^3 + \beta_4 (\bar{x})^4 \right) \quad (2.62)$$

where $\delta \neq 1$ is a taper parameter for mass distribution and $\beta_i, (i = 1,2,3,4)$ are taper parameters for stiffness distribution. Mass and stiffness of elements in the discrete model of such beams are estimated using the following expressions.

$$\bar{m}_r = \frac{c_m}{n-1} \left( 1 + \frac{r\delta}{n-1} \right); \qquad \bar{l}_r = \frac{1}{n-1}$$

$$\bar{k}_r = c_k(n-1) \left( 1 + \beta_1 \bar{x}_r + \beta_2 \bar{x}_r^2 + \beta_3 \bar{x}_r^3 + \beta_4 \bar{x}_r^4 \right) \quad (2.63)$$

where $\bar{x}_r$ is calculated as $\dfrac{r-1}{n-1}$,   $c_m = 1$ for $r = \{1,2,\dots,n-2\}$ and $c_m = 0.5$ for $r = n-1$,   $c_k = 1$ for $r = \{2,3,\dots,n-1\}$ and $c_k = 2$ for $r = 1$.

Likewise, cantilever beams with any complex cross-sectional geometry can be modeled by choosing appropriate numerical values for the discrete elements in the model.

### 2.9   NON-DIMENSIONAL FREQUENCIES FROM MRSM

In sections 2.5, 2.6 and 2.7, the approach followed to formulate MRSM of different types of ALCBs was described. Eigenvalues of $\mathbf{A}$ give the square of natural frequencies $\left( \omega_r^2 \right)$ of transverse vibration of the corresponding ALCB. Matrices $\mathbf{M}$, $\mathbf{L}$ and $\hat{\mathbf{K}}$, used in the computation of $\mathbf{A}$, may be made non-dimensional by using dimensionless parameters of mass ($\bar{m}_r$), length ($\bar{l}_r$) and stiffness ($\bar{k}_r$), respectively, as diagonal elements of these matrices.

In the case of *beams with a constant tip-force*, the axial force $F$ may be non-dimensionalized as $\overline{F} = \dfrac{FL^2}{EI_0}$. However, most published literature on *uniform* beams with tip-force use the non-dimensional parameter $\xi = \dfrac{F}{F_{crit}}$ where $F_{crit}$ denotes the critical axial-load that causes the beam to buckle. For a uniform cantilever beam, $F_{crit} = \dfrac{\pi^2 E I_0}{4 L^2}$ and hence $\overline{F} = \dfrac{\xi \pi^2}{4}$. Results of our studies on *uniform* beams with tip-force are presented in terms of $\xi$ and for *stepped* beams, we use the non-dimensional parameter $\overline{F}$.

For *gravity-loaded* beams with a tip-force, matrix $\mathbf{F}$ may be made non-dimensional by using the expression,

$$\mathbf{F} = diag(\bar{\mathbf{f}}); \qquad \bar{\mathbf{f}} = \mathbf{E}^{-1} [\mathbf{M}\boldsymbol{\gamma} + \psi \mathbf{e}_{n-1}] \tag{2.64}$$

where $\boldsymbol{\gamma}$ is the vector $\gamma \times \{1, 1, 1, \ldots, 1\}$ with $(n-1)$ elements. It may be noted that definition of $\psi$ is identical to that of $\overline{F}$.

For *rotating beams*, matrix $\mathbf{F}$ may be made non-dimensional by using the expression,

$$\mathbf{F} = diag(\bar{\mathbf{f}}); \qquad \bar{\mathbf{f}} = \lambda^2 \mathbf{E}^{-1} \mathbf{M} \bar{\mathbf{r}} \text{ and } \bar{\mathbf{r}} = \dfrac{1}{L} \times \mathbf{r} \tag{2.65}$$

When MRSM $\mathbf{A}$ is formulated using non-dimensional matrices $\mathbf{M}$, $\mathbf{L}$, $\hat{\mathbf{K}}$, $\mathbf{F}$ and parameter $\overline{F}$, square roots of its eigenvalues give dimensionless natural frequency parameters $\mu_r$, defined as

$$\mu_r = \omega_r L^2 \sqrt{\dfrac{\rho A_0}{E I_0}} \tag{2.66}$$

## 2.10 CRITICAL LOADING PARAMETERS

The discrete model can also be used for determining critical combinations of loading parameters that cause the onset of dynamic instability in a beam. For example, in a gravity-loaded vertical beam/column with a tip-force, critical combinations of $\psi$ and $\gamma$ can be determined for which the fundamental frequency parameter, $\mu_1$, goes to zero. This is the condition for onset of dynamic instability [148]. An iterative algorithm similar to the bisection method of root-searching is used. Given a value of $\gamma$, critical value of $\psi$, denoted as $\psi_{crit}$, is determined as follows.

i. Assume trial values for $\psi$ in equation (2.64), formulate MRSM $\mathbf{A}$ [using equation (2.48)] for each value of $\psi$ and determine its first eigen value, $\eta_1$. Note $\mu_1 = \sqrt{\eta_1}$.

ii. Determine an interval, $(\psi_1, \psi_2)$, of values for $\psi$ in which $\eta_1$ changes sign.

iii. Set $\psi = 0.5 \times (\psi_1 + \psi_2)$, formulate $\mathbf{A}$ with this value of $\psi$ and determine its $\eta_1$.

iv. If $0 < \eta_1 < 10^{-6}$, then $\psi_{crit} = \psi$

v. Else if $\eta_1 < 0$, set $\psi_2 = \psi$ and go to step (iii).

vi. Else if $\eta_1 > 10^{-6}$, set $\psi_1 = \psi$ and go to step (iii).

A similar procedure can be used to determine $\gamma_{crit}$ for a given $\psi$.

## 2.11   OBTAINING A FAMILY OF UCBS

In section 2.4.2, the matrix factorization technique introduced by Gladwell [77] was described using which a matrix $\mathbf{A'}$ that has properties characteristic of a realistic MRSM could be decomposed into a three-parameter, isospectral family of mass, length and stiffness matrices corresponding to discrete models of UCBs. In sections 2.5, 2.6 and 2.7, the approach to be followed to formulate MRSM of different types of ALCBs was described. These two procedures can be coupled together to obtain a three-parameter family of UCBs isospectral to a given ALCB. The following procedure is used.

i. Compute matrix $\mathbf{A}$, the MRSM of the given ALCB using methods presented in sections 2.5, 2.6 or 2.7.
ii. Set $\mathbf{A'} = \mathbf{A}$ so that $\mathbf{A'}$ will have the same eigenvalues as $\mathbf{A}$.
iii. Select a value for parameter $g_2$ and decompose $\mathbf{A'}$ into its factors $\mathbf{M'}, \mathbf{L'}$ and $\hat{\mathbf{K}}'$ using equations (2.30), (2.35) and (2.36).

These factors define the discrete model of a UCB. As described in section 2.4.2, parameters $\alpha$, $\beta$ and $g_2$ define characteristics of the isospectral models. Thus, we get a three-parameter *family of discrete models*, each corresponding to a UCB isospectral to the given CFRB.

### 2.11.1   CONSTRUCTING GEOMETRY OF A UCB FROM DISCRETE MODEL

We describe a procedure to construct a realistic UCB (i.e to determine variations in cross-sectional dimensions along the length of the beam) from its discrete model description. Numerical values of all discrete masses, lengths of all rigid links and stiffness of all rotational springs shown in Fig. 2.3 are required for this construction. The discrete model description of the isospectral UCB, however, does not contain information about mass $m_0$ and $k_{segment}$ part of stiffness $k_n$. We assume approximate values for these quantities. We also assume that over each segment $BS_r$, the non-rotating beam has a uniform rectangular cross section. Along the length of the beam (i.e. over different segments $BS_r$), width and depth of beam cross-sections vary. Also, the UCB has constant density, $\rho$, and Young's Modulus, $E$, equal to those of the given ALCB. Thus, once we obtain the mass, length and stiffness matrices of the discrete model of an isospectral UCB using methods described in section 2.11, we use the procedure described below to construct realistic beams. It may be noted that diagonal elements of $\mathbf{L'}$ are not all equal.

i. Extract diagonal elements of $\mathbf{L'}$ matrix and store in a vector $\mathbf{t}$.
ii. Create a new vector, $\bar{\mathbf{s}}$ with elements defined as

$$\bar{\mathbf{s}}(1) = 0.5 \times \mathbf{t}(1); \qquad \bar{\mathbf{s}}(n) = 0.5 \times \mathbf{t}(n-1)$$
$$\bar{\mathbf{s}}(r) = 0.5 \times [\mathbf{t}(r-1) + \mathbf{t}(r)] \text{ for } r = 2, 3, \ldots, (n-1).$$

iii. Extract the diagonal elements of $\mathbf{M'}$ matrix and store in a vector $\bar{\mathbf{m}}$. Append the assumed value of $\bar{m}_0$ (non-dimensional) at the beginning of $\bar{\mathbf{m}}$.

iv. Extract the diagonal elements of $\hat{\mathbf{K}}'$ matrix and store in a vector $\overline{\mathbf{k}}$. Append the assumed value of $\overline{k}_{segment}$ (non-dimensional) part of $\overline{k}_n$ at the end of $\overline{\mathbf{k}}$.

v. Compute non-dimensional width, $\overline{b}_r$, of the beam cross-section (normalized with respect to width of the given ALCB at the clamped end) for every $BS_r$ of the isospectral beam ($r = 1, 2, \ldots, n$) using the expression

$$\overline{b}_r = \frac{1}{[\overline{\mathbf{s}}(r)]^2} \sqrt{\frac{\overline{\mathbf{m}}(r)^3}{\overline{\mathbf{k}}(r)}} \qquad (2.67)$$

vi. Compute non-dimensional depth, $\overline{h}_r$, of the beam cross-section (normalized with respect to depth of the given ALCB at the clamped end) for every $BS_r$ of the isospectral beam ($r = 1, 2, \ldots, n$) using the expression

$$\overline{h}_r = \overline{\mathbf{s}}(r)\sqrt{\frac{\overline{\mathbf{k}}(r)}{\overline{\mathbf{m}}(r)}} \qquad (2.68)$$

Thus we see that given an ALCB with any specified variation in cross-section dimensions, a family of UCBs with rectangular cross-sections, varying in width and depth may be constructed.

In the next three chapters we present results of numerical studies on different types of ALCBs where we examine the practical utility of the discrete model to estimate frequency parameters and critical loads of ALCBs. We also examine whether the UCBs constructed using the methods presented here are practically realizable and whether these are indeed isospectral to the ALCB.

## 2.12  SUMMARY

Analysis of transverse vibration of slender, beam-type components under the influence of axially-distributed forces has considerable practical significance and is investigated in this chapter. It was found that natural frequencies of such beams depend on the nature of axial-force distribution. The objective was to develop the mathematical framework of finding isospectral systems using discrete models of axially loaded beams. Two or more systems are said to be isospectral if they have the same vibration spectrum. Finding UCBs isospectral to ALCBs has considerable practical significance from the perspective of modal experiments that involve rotating beams or beams with a constant tip-force. Moreover, it can aid in the design of space-based components with data obtained from ground-based tests. Discrete models for analysis of longitudinal and transverse vibrations of bars and beams are constructed based on dimensions and material-properties of these structures. MRSM are formulated using methods presented in this chapter. Eigenvalues of MRSM give the square of vibration frequencies. Given a matrix $\mathbf{A}'$ that has properties characteristic of a realistic MRSM, a three-parameter family of *isospectral* discrete models corresponding to UCBs with $\mathbf{A}'$ as the MRSM could be constructed using methods presents in section 2.4.2

# 3 Beams with Tip Force - Discrete Models

The partial differential equation governing transverse vibration of a *uniform* beam under the influence of a constant axial force, $F$ applied at the tip is given as

$$\frac{\partial^4 Y(x,t)}{\partial x^4} - \frac{F}{EI}\frac{\partial^2 Y(x,t)}{\partial x^2} + \frac{\rho A}{EI}\frac{\partial^2 Y(x,t)}{\partial t^2} = 0 \qquad (3.1)$$

where $Y(x,t)$ represents the transverse displacement of the beam as function of distance, $x$ along the beam and time, $t$. Young's modulus and density of the beam material are denoted as $E$ and $\rho$, respectively, while $I$ and $A$ denote moment of inertia and area of the beam cross-section. Axial force $F$ is positive when the applied tip-force is tensile. For a compressive tip-force, $F$ is negative. When the beam oscillates in one of its natural modes, the method of separation of variables can be used to write $Y(x,t) = y(x)\sin\omega t$. On substituting this expression in equation (3.1) we get the mode-shape differential equation as

$$\frac{d^4 y(x)}{dx^4} - \frac{F}{EI}\frac{d^2 y(x)}{dx^2} - \frac{\rho A}{EI}\omega^2 y(x) = 0 \qquad (3.2)$$

Exact solutions exist for this equation. These exact solutions, when combined with the boundary conditions, produce a set of homogeneous equations. The condition for existence of non-trivial solutions to these homogeneous equations yields the frequency equations. Shaker [179] presented the characteristic frequency equations. These are transcendental equations, first derived by Amba-Rao [2] who investigated the effect of end-conditions on the transverse vibration frequencies of uniform straight columns. He derived and numerically solved the transcendental equations for the fundamental frequencies of transverse vibration of compressively-loaded columns for four sets of boundary conditions. Galef [69] developed a simplified approximation for the results proposed in [2] and obtained a simple relation between the natural frequencies of a beam with and without axial loading. Chi et al. [39] investigated the flexural and torsional vibration of axially-loaded uniform beams for the general case of ends elastically restrained against rotation so that the usual end-conditions could be treated as particular cases. Raju and Rao [161] invoked the Rayleigh-Ritz formulation and included the effect of shear and rotatory inertia on the free vibration characteristics of a simply supported slender and short uniform beam at higher modes. Bokaian investigated the influence of constant axial loads, compressive [29] and tensile [30], on mode shapes and natural frequencies. He presented characteristic transcendental equations for ten different boundary conditions. Numerical solutions of these equations give the natural frequencies of transverse vibration. It was found that the variation of normalized natural frequency with normalized axial loads is identical for certain combinations of end-conditions.

In the case of *non-uniform* beams under the influence of a constant axial tip-force, the governing differential equation has the following form.

$$\frac{d^2}{dx^2}\left[EI(x)\frac{d^2y(x)}{dx^2}\right] - F\frac{d^2y(x)}{dx^2} - \rho A\omega^2 y(x) = 0 \tag{3.3}$$

This equation does not have a simple closed-form solution. Numerous authors have investigated different approximation techniques and examined the effect of an axial-load on natural frequencies of different types of non-uniform beams. Sato [174] obtained, numerically using the Ritz method, the fundamental frequencies of linearly tapered beams with ends supported rigidly against transverse displacement and restrained elastically against rotation. Williams and Banerjee [12, 206] studied the flexural vibration of beams with linear or parabolic taper and presented curves for determining the first five natural frequencies of such beams for eleven combinations of boundary conditions and three types of taper. Raju and Rao [162] used the Rayleigh-Ritz method and evaluated the effects of cross-section taper, shear deformation and rotary inertia on the free-vibration behavior of simply supported beams. Yeh and Liu [218] used the Galerkin method to study the free vibrations of non-uniform beams with ends on general rotational and translational constraints. Chen [37] introduced the shooting method to solve the governing differential equation and estimated the natural vibration frequencies of compression bars with two hinged-ends, although the method is applicable to other boundary conditions.

More recently, Auciello [7] introduced two different approaches to determine the free vibration frequencies of tapered beams and beams with discontinuous cross-section, in the presence of axial loads. The first method based on the Rayleigh-Ritz approach gives the upper bound values of the frequencies. In the second method, the structure is reduced to rigid elements connected together by elastic hinges, and lower bounds to the true frequencies are obtained. Amalgamating the results from these two approaches, a narrow range of values to which the actual frequencies belong, could be obtained. Naguleswaran [147] analyzed the transverse vibration of beams with up to three-step changes in cross-section and in which the axial force in each portion is a constant, but varying from one portion to another. Frequency equations for 16 combinations of classical boundary conditions were expressed as fourth-order determinants equated to zero. The critical combinations of axial forces that can cause the beam to buckle, were also obtained. Grossi and Quintana [94] presented an interesting method based on the weak solution of the eigenvalue problem and determined, using the method of separation of variables, the exact frequencies and mode shapes of a beam with several complicating effects. This method could be used to analyze transverse vibration of non-homogeneous beams subjected to constant or distributed axial forces, with an arbitrarily-located internal hinge and elastic support, and with ends elastically restrained against rotation and translation.

Results of *experimental studies* on beams with a constant tip-force [134, 158] are already discussed in the previous chapter.

All the aforementioned studies primarily address the *forward problem*. Publications on isospectral systems for Euler-Bernoulli beams under the influence of a constant tip-force are limited. Recently, Kambampati and Ganguli [113] presented the results of studies on systems isospectral to *uniform* beams and piano strings under the influence of a constant tip-force. The Barcilon-Gottlieb transformation was used to transform the governing differential equation of a non-uniform beam in transverse vibration. Coefficients in this transformed equation were then equated with those of the governing equation of an axially-loaded uniform beam to obtain the mass and stiffness distribution of isospectral beams. The technique presented was mathematically laborious as it involved finding solutions to coupled fourth-order ordinary differential equations (ODEs). Two special cases were considered and the ODEs were solved analytically. The authors also mentioned a practical application of the idea presented - i.e. frequencies of uniform beams under axial loads could be experimentally determined by testing their unloaded non-uniform isospectral analogues. However, extending the idea presented in [113] to the analysis of axially-loaded (constant tip-force) *non-homogeneous* beams, especially beams with discontinuous cross-section (stepped beams), is not straightforward.

The discrete system model presented in section 2.5 was used to numerically study transverse vibration of example cantilever beams with a tip-force (CBTF). Computer codes using conventional MATLAB subroutines were executed for this purpose. Results of these numerical studies are presented in this chapter. All results are presented in terms of dimensionless values.

## 3.1 CONVERGENCE STUDIES

This section presents results of a convergence study carried out to estimate the optimum number of elements required in the discrete model for accurate estimation of transverse vibration frequencies of cantilever beams under the influence of a constant tip-force.

We estimate variation in values of first three frequency parameters obtained from the discrete system model of a *uniform* CBTF as the number of elements in the model is varied. Two load cases are considered:

Case 1: Applied compressive axial load = 0.6 times $F_{crit}$ i.e. $\xi = -0.6$.
Case 2: Applied tensile axial load = 1.5 times $|F_{crit}|$ i.e. $\xi = 1.5$.

Results from the discrete model for these cases are compared with numerical solutions of governing transcendental equations for cantilever beams, presented in [29, 30]. De Rosa and Lippiello [50] suggested using 300 elements in a discrete model for analyzing non-uniform beams although they did not present results of a convergence study. We carry out the convergence study with $n$ varying from 10 to 300. Variations in frequency parameters, $\mu_1$, $\mu_2$, $\mu_3$, with respect to number of elements in the model for the two cases identified above are shown in Fig. 3.1. It is observed that frequencies computed from the discrete model asymptotically approach the theoretical values as the number of elements in the model is increased. The discrete results converge satisfactorily to theoretical values when $n = 300$. For all subsequent studies, we shall use $n = 300$.

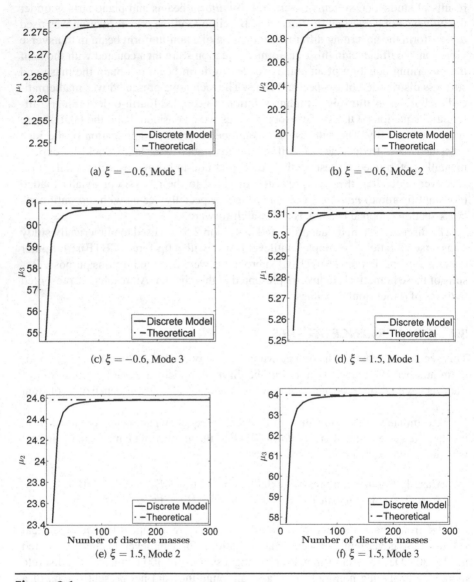

**Figure 3.1**
Frequency parameters Vs number of discrete masses for two cases
*Case 1*: compressive axial load ($\xi = -0.6$);   *Case 2*: tensile axial load ($\xi = 1.5$)

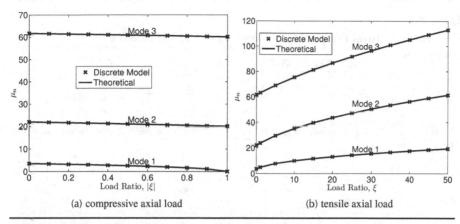

(a) compressive axial load                    (b) tensile axial load

**Figure 3.2**
Variation of first three frequency parameters with axial load ratio: Results from discrete model compared with theoretical results

## 3.2 NATURAL FREQUENCY PARAMETERS OF CBTF (CANTILEVER BEAM WITH TIP FORCE)

We now calculate the natural frequency parameters of a uniform beam and three types of stepped beams for different values of the applied axial load and compare these results with the published results.

### 3.2.1 UNIFORM CBTFS

Figure 3.2 shows variation of frequency parameters with respect to applied axial a) compressive and b) tensile loads. Curves representing theoretical values are obtained from numerical solutions of transcendental equations presented in [29, 30]. It can be observed that the values obtained from the 300-element discrete model match well with the theoretical results. Table 3.1 gives a comparison between the two results. The error between the two results is within 0.01% for both tensile and compressive axial loads.

### 3.2.2 STEPPED CBTFS

To validate the efficacy of the discrete technique for vibration-analysis of non-homogeneous beams, we apply it to analyze the transverse vibration of stepped beams, a typical example of non-homogeneous beams. First three frequency parameters $\alpha_{11,1}$, $\alpha_{11,2}$ and $\alpha_{11,3}$ of each of the three types of stepped beams, defined as *Type A*, *Type B* and *Type C* in section 2.8.3, are presented in [147, Table 1] for three different values of $\overline{F}$ (which is same in all segments) and 16 different combinations of end-conditions. We compute these parameters using the discrete technique and compare with corresponding values presented in [147] for the case of clamped-free beams. All three beams considered, have a three-step

## Table 3.1
### First Three Natural Frequency Parameters of Uniform CBTFs

| Compressive Axial Loads | | | | | | | | | | | |
|---|---|---|---|---|---|---|---|---|---|---|---|
| $\xi \longrightarrow$ | 0 | −0.1 | −0.2 | −0.3 | −0.4 | −0.5 | −0.6 | −0.7 | −0.8 | −0.9 | −1 |
| $\mu_1$   [29] | 3.5160 | 3.3478 | 3.1682 | 2.9750 | 2.7652 | 2.5346 | 2.2764 | 1.9799 | 1.6237 | 1.1533 | 0.0000 |
|     Discrete | 3.5160 | 3.3478 | 3.1682 | 2.9750 | 2.7652 | 2.5345 | 2.2764 | 1.9798 | 1.6236 | 1.1533 | 0.0000 |
| $\mu_2$   [29] | 22.0345 | 21.8521 | 21.6681 | 21.4822 | 21.2946 | 21.1052 | 20.9139 | 20.7207 | 20.5256 | 20.3285 | 20.1294 |
|     Discrete | 22.0334 | 21.8511 | 21.6670 | 21.4812 | 21.2936 | 21.1041 | 20.9128 | 20.7197 | 20.5246 | 20.3275 | 20.1284 |
| $\mu_3$   [29] | 61.6972 | 61.5425 | 61.3873 | 61.2317 | 61.0758 | 60.9194 | 60.7627 | 60.6055 | 60.4480 | 60.2900 | 60.1317 |
|     Discrete | 61.6910 | 61.5362 | 61.3811 | 61.2255 | 61.0696 | 60.9132 | 60.7565 | 60.5994 | 60.4418 | 60.2839 | 60.1255 |
| **Tensile Axial Loads** | | | | | | | | | | | |
| $\xi \longrightarrow$ | 1 | 5 | 10 | 15 | 20 | 25 | 30 | 35 | 40 | 45 | 50 |
| $\mu_1$   [30] | 4.8148 | 7.7007 | 9.9131 | 11.6100 | 13.0444 | 14.3122 | 15.4616 | 16.5213 | 17.5096 | 18.4396 | 19.3204 |
|     Discrete | 4.8147 | 7.7006 | 9.9130 | 11.6098 | 13.0442 | 14.3119 | 15.4614 | 16.5210 | 17.5093 | 18.4392 | 19.3201 |
| $\mu_2$   [30] | 23.7701 | 29.5202 | 35.0896 | 39.6633 | 43.6249 | 47.1682 | 50.4044 | 53.4033 | 56.2120 | 58.8638 | 61.3831 |
|     Discrete | 23.7690 | 29.5190 | 35.0882 | 39.6617 | 43.6232 | 47.1664 | 50.4025 | 53.4013 | 56.2099 | 58.8615 | 61.3808 |
| $\mu_3$   [30] | 63.2236 | 68.9704 | 75.4647 | 81.3482 | 86.7479 | 91.7581 | 96.4485 | 100.8721 | 105.0693 | 109.0718 | 112.9046 |
|     Discrete | 63.2172 | 68.9638 | 75.4578 | 81.3409 | 86.7404 | 91.7503 | 96.4405 | 100.8639 | 105.0609 | 109.0631 | 112.8957 |

change in cross-section. Geometry parameters of each of the three beams are $\begin{bmatrix} d_1 & d_2 & d_3 & d_4 \end{bmatrix} = \begin{bmatrix} 1.0 & 0.8 & 0.65 & 0.45 \end{bmatrix}$ and $\begin{bmatrix} R_1 & R_2 & R_3 & R_4 \end{bmatrix} = \begin{bmatrix} 0.25 & 0.30 & 0.25 & 0.2 \end{bmatrix}$ where definition of $d_i$ is given in section 2.8.3 and $R_i = \dfrac{L_i}{L}$ (Fig. 2.10). It may be noted that the definition of $\alpha_{11,1}$, $\alpha_{11,2}$ and $\alpha_{11,3}$ in [147] correspond to $\sqrt{\mu_1}$, $\sqrt{\mu_2}$ and $\sqrt{\mu_3}$, respectively. Table 3.2 gives a comparison between the two results and it can be inferred that there is good agreement.

Results from these studies indicate that the discrete modeling technique discussed here is indeed a valid approximation for the study of transverse vibration of straight CBTFs of any cross-sectional geometry.

## 3.3   FAMILY OF ISOSPECTRAL UCBS

We now determine the discretized distribution of mass and stiffness on models of a family of UCBs isospectral to the given CBTFs. We also observe how parameter $g_2$ affects these distributions. Scaling parameters, $\alpha$ and $\beta$ are fixed based on the condition that the total length and total mass of the isospectral beam are, respectively, equal to those of the given CBTF. We present results for a uniform beam and a Type A stepped beam with geometry parameters identical to those considered in section 3.2.2. For the uniform beam, we consider the load cases defined in section 3.1 and for the stepped beam, we consider $\overline{F} = 1.5$.

### 3.3.1   UNIFORM CBTFS

Equation (2.28) provides the expression for parameter $g_2$. In section 2.4.2, we defined the permissible range of values for $v$ as $0 < v < 1$. Different isospectral models are obtained when $v$ takes different values within this permissible range. Mass and stiffness distributions shown in Fig. 3.3 all correspond to discrete models of a family of UCBs, each isospectral to a uniform CBTF with $\xi = -0.6$. The fact that all these

**Table 3.2**

**First Three Natural Frequency Parameters of Axially-loaded Stepped Cantilever Beams**

| $\overline{F}$ | | Type A Beam | | | Type B Beam | | | Type C Beam | | |
|---|---|---|---|---|---|---|---|---|---|---|
| | | $\sqrt{\mu_1}$ | $\sqrt{\mu_2}$ | $\sqrt{\mu_3}$ | $\sqrt{\mu_1}$ | $\sqrt{\mu_2}$ | $\sqrt{\mu_3}$ | $\sqrt{\mu_1}$ | $\sqrt{\mu_2}$ | $\sqrt{\mu_3}$ |
| 5 | [147] | 2.76884 | 5.40033 | 8.25798 | 2.73517 | 5.06316 | 7.32240 | 3.09942 | 5.50602 | 7.72680 |
| | Discrete | 2.76883 | 5.40021 | 8.25758 | 2.73516 | 5.06306 | 7.32203 | 3.09940 | 5.50591 | 7.72643 |
| 0 | [147] | 2.14409 | 4.90763 | 7.95377 | 2.03860 | 4.30933 | 6.71184 | 2.29222 | 4.49412 | 6.75383 |
| | Discrete | 2.14408 | 4.90751 | 7.95337 | 2.03859 | 4.30921 | 6.71145 | 2.29220 | 4.49399 | 6.75343 |
| -2 | [147] | 0.99470 | 4.64476 | 7.81977 | *uns* | 3.67837 | 6.37372 | *uns* | 2.97856 | 6.00673 |
| | Discrete | 0.99467 | 4.64464 | 7.81936 | *uns* | 3.67823 | 6.37333 | *uns* | 2.97843 | 6.00635 |

*uns* indicates unstable mode

models are isospectral is verified by formulating the MRSM of each discrete model from its mass, length and stiffness matrices using equation (2.20). The eigenvalues of each of these MRSMs match exactly with those of the discrete model MRSM of the given CBTF obtained using equation (2.41).

Although all the models are isospectral, mass distribution on the models are highly non-uniform when $v$ takes values greater than 0.01 (Fig. 3.3a). As the value of $v$ is reduced, uniformity in the mass and stiffness distributions improves ($v = 10^{-3}$ in Fig. 3.3b). However, when $v$ is reduced further, the distributions again become non-uniform ($v = 10^{-4}$ in Fig. 3.3b). Thus, it is inferred that the value of $v$ (or $g_2$) can be tailored using a trial-and-error approach ($v$ between $10^{-4}$ and $10^{-2}$ for this case) to obtain practically realistic discrete models i.e. models with a fairly uniform distribution for mass and stiffness. For example, we can even determine a value of $v$ (and $g_2$), for which we get a purely uniform distribution of point masses along $\overline{x}$ (Fig. 3.3c). Results of a similar study for the case of a beam with tensile axial load ($\xi = 1.5$) are shown in Fig. 3.4. Note that once an appropriate value of $g_2$ is selected, computation time for obtaining an isospectral discrete model using conventional MATLAB subroutines is less than 1 second.

It can be deduced that parameter $g_2$ has a significant role in determining *distributions* of mass and stiffness along the length of isospectral UCBs. Although $g_2$, theoretically, can have a wide range of permissible values, the range of values that yield practically useful discrete models, is narrow. Moreover, the discrete technique is well-suited for computer implementation.

### 3.3.2 STEPPED CBTFS

A similar study was conducted on all the three types of stepped beams. It is noticed that, for stepped beams too, parameter $g_2$ can be adjusted to obtain more uniform distributions of mass and stiffness. Moreover, the range of values of $g_2$, that yields

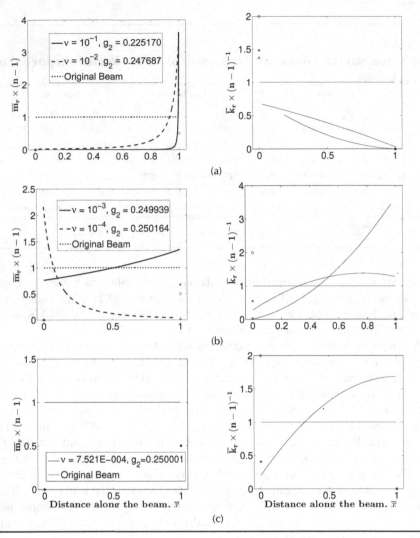

**Figure 3.3**

Mass and stiffness distributions on discrete models of UCBs isospectral to a uniform CBTF with *compressive* axial load ($\xi = -0.6$)

---

points represent values at beam ends

more uniform distributions, is narrow. Typical distributions of mass and stiffness obtained with optimum value of $g_2$ for the case of a Type A beam with $\overline{F} = 1.5$ (tensile axial load) and geometry parameters identical to those considered in section 3.2.2, are shown in Fig. 3.5. It can be seen that the isospectral UCB too has a three-step change in the mass and stiffness distributions. While the given stepped

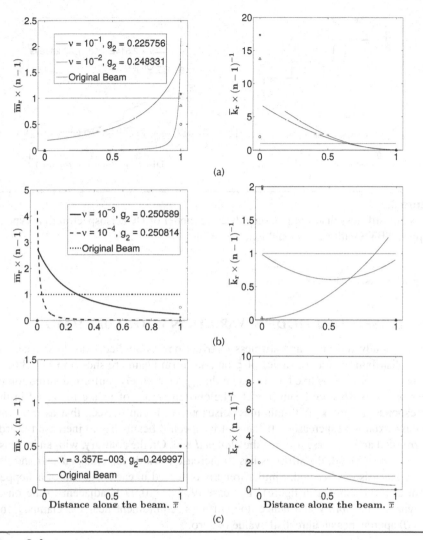

**Figure 3.4**
Mass and stiffness distributions on discrete models of UCBs isospectral to a uniform CBTF with *tensile* axial load ($\xi = 1.5$)

points represent values at beam ends

CBTF has a uniform distribution of both mass and stiffness in each of the segments, the isospectral UCB has a uniform distribution of only mass. Stiffness distribution is non-uniform.

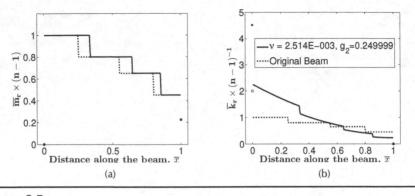

**Figure 3.5**

Mass and stiffness distributions on a discrete model of UCB isospectral to a Type A stepped CBTF with *tensile* axial load ($\overline{F} = 1.5$)

---

points represent values at beam ends

## 3.4   ISOSPECTRAL MODELS: VARIATION WITH AXIAL-LOAD

We now study the mass and stiffness distributions as applied axial load changes. These distributions for six values of $\xi$ on a uniform beam are shown in Fig. 3.6. In all these cases, $g_2$ was fixed at the value that gives a purely uniform distribution of mass on the isospectral beam model. The chosen values of $g_2$ are indicated in the corresponding figures. With this mass distribution, it can be seen that as the *compressive* axial load increases, stiffness of isospectral beams has to increase towards the *free* end and decrease towards the *clamped* end. On the contrary, with an increase in axial *tensile* load, the stiffness has to increase towards the *clamped* end and decrease towards the *free* end. Similar trend is observed in each segment of a stepped beam also (not shown in figure). The case of $\xi = -0.99$ is adjacent to the onset of dynamic instability and hence, the stiffness requirement near the clamped end ($\overline{x} = 0$) approaches an unrealistic value of zero.

## 3.5   GEOMETRY OF ISOSPECTRAL BEAMS

We apply the method described in section 2.11.1 to construct the cross-sectional geometry of beams with variable *rectangular* cross-section and which are isospectral to a given CBTF, shown schematically in Fig. 3.7a. Mass $\overline{m}_0$ is assumed as $0.5 \times \mathbf{M}'(1,1)$ and $\overline{k}_{segment}$ part of $\overline{k}_n$ is assumed to be twice the stiffness of the adjacent spring [i.e. $2 \times \mathbf{K}'(n-1, n-1)$]. Fig. 3.7b and Fig. 3.7c show variations in width and depth of beams isospectral to this CBTF for compressive axial loads, $\xi = -0.2$ and $\xi = -0.4$, respectively. We apply the mass and stiffness distributions shown in Fig. 3.6a. Similar variations for the case when the same beam is loaded with tensile loads, $\xi = 0.4$ and $\xi = 1$, are shown in Fig. 3.8. Here, mass and stiffness distributions correspond to those shown in Fig. 3.6b.

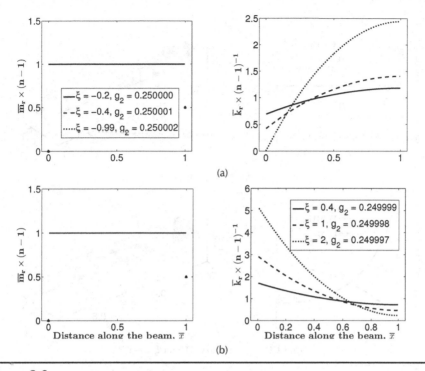

(a)

(b)

**Figure 3.6**

Mass and stiffness distributions on discrete models of UCBs isospectral to a uniform CBTF for different values of $\xi$

For clarity, stiffness values at beam-ends are not shown

We also contrive the geometry of UCBs isospectral to a Type A stepped CBTF with an axial-load of $\overline{F} = 1.5$. The given stepped CBTF is shown in Fig. 3.9a and the isospectral UCB is shown in Fig. 3.9b. This variation in width and depth correspond to the mass and stiffness distributions shown in Fig. 3.5. It can be perused that the isospectral beam also has a three-step change in width while its depth has a continuous distribution.

## 3.6 NATURAL FREQUENCIES OF UCBS: FEM RESULTS

Frequency parameters of UCBs constructed in the previous sections are calculated using FEM. The procedure adopted for FE formulation is a very general method comprehensively described in standard textbooks [45]. Since dimensions of the constructed beams are known only at discrete points, the FEM model also uses 300 elements (2D beam) with one node at each end. Each node has two degrees of freedom - lateral translation and rotation. Each element has a rectangular cross-section with uniform width and depth of dimensions as obtained in the geometric

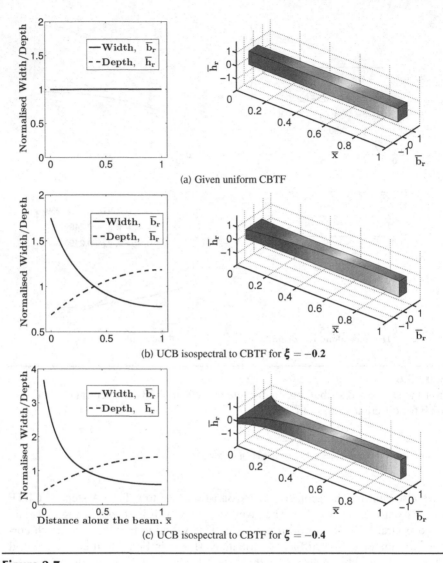

(a) Given uniform CBTF

(b) UCB isospectral to CBTF for $\xi = -0.2$

(c) UCB isospectral to CBTF for $\xi = -0.4$

**Figure 3.7**
Geometry of UCBs isospectral to a given *uniform* CBTF for different values of applied *compressive* axial load, $\xi$

model described previously. Within each element, lateral displacement is assumed to have a cubic variation and Hermite shape functions $(N_1, N_2, N_3, N_4)$, defined below, are used for interpolation. These functions have $C^1$ continuity.

$$N_1(x) = 1 - \frac{3x^2}{l_e^2} + \frac{2x^3}{l_e^3}; \qquad N_2(x) = x - \frac{2x^2}{l_e} + \frac{x^3}{l_e^2}$$

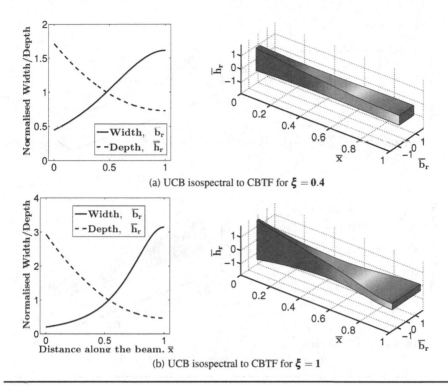

(a) UCB isospectral to CBTF for $\xi = 0.4$

(b) UCB isospectral to CBTF for $\xi = 1$

**Figure 3.8**
Geometry of UCBs isospectral to a given *uniform* CBTF for different values of applied *tensile* axial load, $\xi$

$$N_3(x) = \frac{3x^2}{l_e^2} - \frac{2x^3}{l_e^3}; \qquad N_4(x) = -\frac{x^2}{l_e} + \frac{x^3}{l_e^2}$$

Here $l_e$ denotes the total length of the element under consideration, value of which is obtained from the corresponding entry in $\bar{s}$ vector. Element stiffness and mass matrices (square matrices of order $4 \times 4$) are formulated as

$$\mathbf{k}_{ele} = \int_0^{l_e} \mathbf{B}^T EI_e \mathbf{B}\, dx; \qquad \mathbf{m}_{ele} = \int_0^{l_e} \rho A_e \mathbf{N}^T \mathbf{N}\, dx \qquad (3.4)$$

where $E$ and $\rho$ denote the Young's Modulus and density of the beam material, $I_e$ and $A_e$ denote the moment of inertia and cross-sectional area, respectively of the element, $\mathbf{N}$ and $\mathbf{B}$ are the row vectors defined as $\mathbf{N} = \begin{bmatrix} N_1 & N_2 & N_3 & N_4 \end{bmatrix}$ and $\mathbf{B} = \dfrac{d^2\mathbf{N}}{dx^2}$. The stiffness and mass matrices of all 300 elements in the model are formulated and assembled into global matrices, $\mathbf{K}_{global}$ and $\mathbf{M}_{global}$. Boundary conditions of a cantilever beam are imposed by eliminating the first two rows and columns of the global matrices to give $\mathbf{K}_{reduced}$ and $\mathbf{M}_{reduced}$. Square root of eigenvalues of the matrix $[\mathbf{M}_{reduced}]^{-1} \times [\mathbf{K}_{reduced}]$ gives the natural frequencies of the beam.

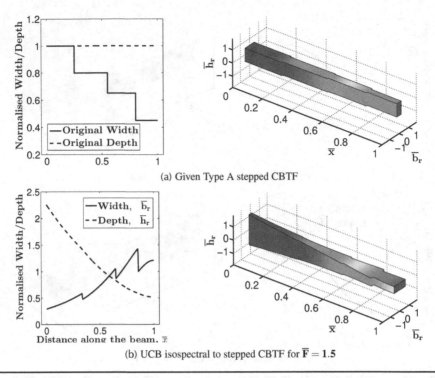

(a) Given Type A stepped CBTF

(b) UCB isospectral to stepped CBTF for $\overline{F} = 1.5$

**Figure 3.9**
Geometry of a UCB isospectral to a given Type A stepped CBTF for $\overline{F} = 1.5$

The results obtained from the FE formulation, expressed in non-dimensional form, are shown in Table 3.3. Natural frequency parameters of the discrete model, from which geometries of the UCBs were deciphered, are also shown. These frequency parameters are computed as the square roots of first three eigenvalues of MRSMs formulated from equation (2.20) using corresponding $\mathbf{M}'$, $\mathbf{L}'$ and $\hat{\mathbf{K}}'$ matrices. For the case of the Type A stepped UCB, frequency parameters of the original CBTF, obtained from its discrete model, are also shown. In all cases, it is observed that the frequencies computed using FEM match with the discrete model frequencies within 0.02% error. The values in this table, for the case of uniform CBTFs, may also be compared with corresponding values in Table 3.1. It is found that the natural frequencies of UCBs computed using FEM match well with theoretical frequencies of uniform CBTFs. The error between the two is also within 0.02%. Thus, discrete models of UCBs obtained for different values of $\xi$ (or $\overline{F}$) are *indeed isospectral* to the given CBTF with an end-axial load corresponding to $\xi$ (or $\overline{F}$).

From the numerical studies outlined in this chapter, it is concluded that the discrete modeling technique introduced here is a versatile methodology that is useful for the *forward* problem and also for determining isospectral beams associated with CBTFs.

## Table 3.3
## First Three Natural Frequency Parameters of UCBs:
## FEM Results Vs Discrete Model Results

| | Uniform Beam | | | | | | | | Type 1 Stepped Beam | | |
| | $\xi = -0.2$ | | $\xi = -0.4$ | | $\xi = 0.4$ | | $\xi = 1$ | | $\bar{F} = 1.5$ | | |
| | (a) | FEM | (a) | FEM | (a) | FEM | (a) | FEM | (a) | FEM | (b) |
|---|---|---|---|---|---|---|---|---|---|---|---|
| $\mu_1$ | 3.1682 | 3.1681 | 2.7652 | 2.7657 | 4.1032 | 4.1027 | 4.8147 | 4.8150 | 5.8089 | 5.8092 | 5.8089 |
| $\mu_2$ | 21.6670 | 21.6681 | 21.2936 | 21.2948 | 22.7462 | 22.7471 | 23.7690 | 23.7700 | 25.7624 | 25.7634 | 25.7624 |
| $\mu_3$ | 61.3811 | 61.3874 | 61.0696 | 61.0761 | 62.3061 | 62.3121 | 63.2172 | 63.2231 | 64.7868 | 64.7932 | 64.7868 |

(a)     from discrete model of UCB
(b)     from discrete model of given CBTF

# 4 Gravity Loaded Beams - Discrete Models

In this chapter, we present results of numerical studies on gravity-loaded clamped-free beams with a tip-force, hereinafter referred to by the acronym GLCFTF. The discrete model presented in section 2.6 is applied to calculate the natural frequencies of a uniform and three types of tapered beams for arbitrary combinations of loading parameters. Furthermore, we estimate the critical axial-loads that cause onset of dynamic instability in a typical gravity-loaded tapered beam, using the iterative algorithm presented in section 2.10. These results are compared with published results. We then construct weightless and unloaded clamped-free beams (WUCF) isospectral to given GLCFTFs. Natural frequencies of these WUCFs are determined using FEM and compared with those of the corresponding GLCFTFs. The material in this chapter is adapted from [165].

## 4.1 CONVERGENCE STUDIES

We investigate the variation in the first three natural frequency parameters of a uniform GLCFTF as the number of discrete elements in the model is varied. These results are compared with values presented in [148] where Naguleswaran studied the transverse vibration of uniform Euler-Bernoulli beams under a linearly varying axial force distribution which has two components - a constant component, $\tau(0)$, due to a constant tip force and a varying component, $\gamma$, due to an axial force uniformly distributed along the length of the beam. He discovered three frequency parameters, $\alpha_1$, $\alpha_2$ and $\alpha_3$, of lateral vibration under different combinations of $\tau(0)$ and $\gamma$ for sixteen different boundary conditions. For a *uniform* GLCFTF, the axial force distribution along its length has a linear variation identical to the distribution addressed in [148] for the case of a free-clamped (*fr/cl*) beam. Thus, results presented in [148] for a *fr/cl* beam can be used directly for validating results obtained from the discrete model presented here. Note that the definitions of parameters $\tau(0)$ and $\gamma$ in [148] are, respectively identical to $\psi$ and $\gamma$ described here. Frequency parameters $\alpha_1$, $\alpha_2$ and $\alpha_3$ in [148] correspond to $\sqrt{\mu_1}$, $\sqrt{\mu_2}$ and $\sqrt{\mu_3}$.

A convergence study is performed with $n$ varying from 10 to 300 for two cases:

Case 1: A hanging beam with a compressive tip force ($\gamma = 4$ and $\psi = -2$).
Case 2: A standing beam with a tensile tip force ($\gamma = -3$ and $\psi = 10$).

Figure 4.1 shows the variation in $\sqrt{\mu_1}$, $\sqrt{\mu_2}$ and $\sqrt{\mu_3}$ with change in $n$ for the two cases described above. It can be observed that the frequencies computed from the discrete model asymptotically approach values in [148] as the number of elements in the model is increased. Discrete results converge satisfactorily when 300 elements are used in the model. For subsequent studies, $n = 300$ is used.

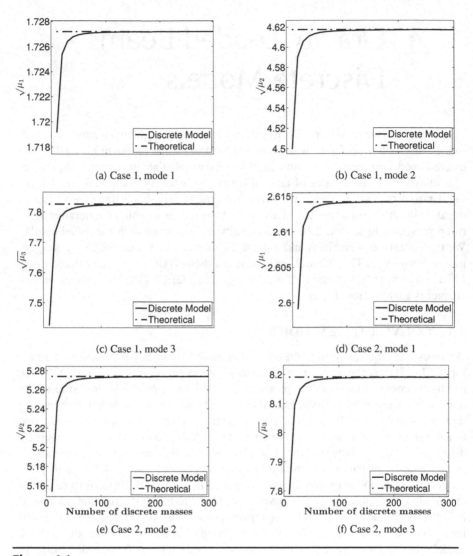

(a) Case 1, mode 1

(b) Case 1, mode 2

(c) Case 1, mode 3

(d) Case 2, mode 1

(e) Case 2, mode 2

(f) Case 2, mode 3

## Figure 4.1

Frequency parameters Vs number of discrete masses for two cases - Case 1: Hanging beam with compressive tip force ($\gamma = 4$, $\psi = -2$), Case 2: Standing beam with tensile tip force ($\gamma = -3$, $\psi = 10$)

**Table 4.1**

**First Three Frequency Parameters of Uniform GLCFTFs**

| $\psi$ | | $\gamma = 100$ | | | $\gamma = 4$ | | | $\gamma = -3$ | | |
|---|---|---|---|---|---|---|---|---|---|---|
| | | $\sqrt{\mu_1}$ | $\sqrt{\mu_2}$ | $\sqrt{\mu_3}$ | $\sqrt{\mu_1}$ | $\sqrt{\mu_2}$ | $\sqrt{\mu_3}$ | $\sqrt{\mu_1}$ | $\sqrt{\mu_2}$ | $\sqrt{\mu_3}$ |
| 10 | [148] | 3.7806 | 6.3957 | 9.1596 | 2.7547 | 5.3778 | 8.2704 | 2.6142 | 5.2739 | 8.1912 |
| | Discrete | 3.7806 | 6.3956 | 9.1592 | 2.7547 | 5.3777 | 8.2700 | 2.6142 | 5.2738 | 8.1908 |
| 0 | [148] | 3.5873 | 6.0423 | 8.8871 | 2.0777 | 4.7755 | 7.9057 | 1.6627 | 4.6300 | 7.8158 |
| | Discrete | 3.5873 | 6.0422 | 8.8867 | 2.0777 | 4.7754 | 7.9053 | 1.6627 | 4.6299 | 7.8154 |
| -2 | [148] | 3.5394 | 5.9599 | 8.8292 | 1.7272 | 4.6176 | 7.8261 | *uns* | 4.4575 | 7.7337 |
| | Discrete | 3.5394 | 5.9598 | 8.8287 | 1.7272 | 4.6175 | 7.8257 | *uns* | 4.4574 | 7.7333 |

*uns* indicates an unstable mode

## 4.2 NATURAL FREQUENCY PARAMETERS OF GLCFTFS (GRAVITY-LOADED CLAMPED-FREE BEAMS WITH TIP-FORCE)

The adequacy of a 300-element discrete model to estimate natural frequency parameters of homogeneous and non-homogeneous GLCFTFs is now demonstrated, with an accuracy deemed satisfactory from an engineering perspective. The frequency parameters of a uniform beam and three types of tapered beams for arbitrarily-chosen combinations of $\gamma$ and $\psi$ are obtained from the discrete model and compared with published results.

### 4.2.1 UNIFORM GLCFTFS

Table 4.1 lists values of $\sqrt{\mu_1}$, $\sqrt{\mu_2}$ and $\sqrt{\mu_3}$ calculated using the discrete model of a uniform GLCFTF for different combinations $\psi$ and $\gamma$ and compares these predictions with results in [148]. The discrete model results match well with values reported in [148] and the relative error is within 0.01% for all cases.

### 4.2.2 TAPERED GLCFTFS

Three classes of tapered beams are considered - (i) rectangular beams with linearly tapering depth and constant width (ii) rectangular beams with constant depth and linearly tapering width, and (iii) circular beams with linearly tapering diameter. Sudheer et al. [189] presented the first three frequency parameters for all three classes of gravity-loaded standing beams (without a tip force i.e. $\psi = 0$) for different combinations of taper ratios, $\delta_1$, $\delta_2$, and gravity parameter, $\gamma$. These values, along with those computed from the discrete model, are listed in Tables 4.2, 4.3 and 4.4. (Note that the parameter $\beta$ in [189] is identical to $\gamma$ used here but with an opposite sign convention. Moreover, $c$ in [189] denotes taper ratio of the column) There is good agreement between the two results (except for the value marked with $^{\$\$}$ in Table 4.4 which indicates a possible typographical error in [189]).

**Table 4.2**

**Frequency Parameters ($\mu_1$, $\mu_2$, $\mu_3$) of Rectangular GLCFs ($\psi = 0$) with Linearly Tapering Depth and Constant Width**

| | | | Taper Ratio, $\delta_1$ | | | |
|---|---|---|---|---|---|---|
| | 0.1 | | 0.3 | | 0.7 | |
| $\gamma$ | a) | b) | a) | b) | a) | b) |
| | 3.558702 | 3.558672 | 3.666749 | 3.666726 | 4.081714 | 4.081712 |
| 0 | 21.338102 | 21.337111 | 19.880606 | 19.879744 | 16.625269 | 16.624664 |
| | 58.979904 | 58.974014 | 53.322198 | 53.317000 | 40.587991 | 40.584030 |
| | 2.948465 | 2.948438 | 3.062087 | 3.062067 | 3.497114 | 3.497116 |
| -2.5 | 20.836806 | 20.835826 | 19.368556 | 19.367706 | 16.085256 | 16.084660 |
| | 58.469500 | 58.463630 | 52.805953 | 52.800778 | 40.051452 | 40.047506 |
| | 0.846644 | 0.846607 | 1.093815 | 1.093801 | 1.815472 | 1.815490 |
| -7.5 | 19.794401 | 19.793443 | 18.300150 | 18.299324 | 14.946583 | 14.946008 |
| | 57.432830 | 57.426999 | 51.755963 | 51.750833 | 38.954952 | 38.951034 |

a) from Table 5 of Ref [189]        b) from discrete model

**Table 4.3**

**Frequency Parameters ($\mu_1$, $\mu_2$, $\mu_3$) of Rectangular GLCFs ($\psi = 0$) with Constant Depth and Linearly Tapering Width**

| | | | Taper Ratio, $\delta_2$ | | | |
|---|---|---|---|---|---|---|
| | 0.1 | | 0.3 | | 0.7 | |
| $\gamma$ | a) | b) | a) | b) | a) | b) |
| | 3.631027 | 3.630995 | 3.916033 | 3.916004 | 4.931642 | 4.931626 |
| 0 | 22.254029 | 22.252979 | 22.785958 | 22.784928 | 24.687279 | 24.686347 |
| | 61.909628 | 61.903405 | 62.436120 | 62.429964 | 64.526628 | 64.520782 |
| | 3.037632 | 3.037603 | 3.363823 | 3.363797 | 4.479281 | 4.479269 |
| -2.5 | 21.770841 | 21.769802 | 22.332727 | 22.331707 | 24.315225 | 24.314301 |
| | 61.416302 | 61.410099 | 61.975560 | 61.969422 | 64.152600 | 64.146768 |
| | 1.132558 | 1.132522 | 1.802415 | 1.802393 | 3.395995 | 3.395993 |
| -7.5 | 20.769043 | 20.768023 | 21.396124 | 21.395124 | 23.552729 | 23.551821 |
| | 60.415453 | 60.409286 | 61.042332 | 61.036228 | 63.397027 | 63.391224 |

a) from Table 6 of Ref [189]        b) from discrete model

**Table 4.4**

**Frequency Parameters ($\mu_1$, $\mu_2$, $\mu_3$) of Circular GLCFs ($\psi = 0$) with Linearly Tapering Diameter**

| | | Taper Ratio, $\delta$ | | | | |
|---|---|---|---|---|---|---|
| | | 0.1 | | 0.3 | | 0.7 |
| $\gamma$ | a) | b) | a) | b) | a) | b) |
| | 3.673701 | 3.673672 | 4.066932 | 4.066914 | 5.509268 | 5.509286 |
| **0** | 21.550253 | 21.549269 | 20.555506 | 20.554671 | 18.641218 | 18.640709 |
| | 59.188637 | 59.182772 | 54.015186 | 54.010072 | 42.810666 | 42.806928 |
| | 3.082123 | 3.082098 | 3.518063 | 3.518048 | 5.053637 | 5.053659 |
| **-2.5** | 21.062033 | 21.061061 | 20.084902 | 20.084078 | 18.212036 | 18.211526 |
| | 58.692572 | 58.686727 | 53.543037 | 53.537944 | 42.385379 | 42.381629 |
| | 1.213739 | 1.213713 | 2.003562 | 2.003557 | 3.983635 | 3.983666 |
| **-7.5** | 20.048376 | 20.047425 | 19.107981 | 19.107179 | 17.321630 | 17.321121 |
| | 57.685602 | 57.679796 | 57.584453[SS] | 52.579480 | 41.521269 | 41.517495 |

a) from Table 7 of Ref [189]     b) from discrete model

**Table 4.5**

**Critical Tip Force Parameter, $\psi_{crit}$, of Rectangular GLCFTFs with Linearly Tapered Depth and Width ($\delta_1 = \delta_2 = 0.6$)**

| $\dfrac{4\gamma}{\pi^2} \longrightarrow$ | Hanging Beams | | | | | | Standing Beams | | | | |
|---|---|---|---|---|---|---|---|---|---|---|---|
| | 20 | 16 | 12 | 8 | 4 | 0 | -4 | -8 | -12 | -16 | -20 |
| $\psi_{crit}$ Ref [44] | -2.118 | -1.913 | -1.686 | -1.431 | -1.131 | -0.757 | -0.247 | 0.503 | 1.589 | 3.000 | 4.659 |
| $\psi_{crit}$ Discrete | -2.118 | -1.912 | -1.685 | -1.430 | -1.131 | -0.757 | -0.247 | 0.503 | 1.590 | 3.001 | 4.659 |

## 4.3 ESTIMATION OF CRITICAL LOADING PARAMETERS

In section 2.10, a procedure to use the discrete model of a GLCFTF and estimate critical combinations of $\psi$ and $\gamma$ that cause onset of dynamic instability is described. Such critical loads for both uniform and tapered beams were calculated by Wei et al. [204] and also by Cifuentes and Kapania [44]. Table 4.5 compares the values obtained from the discrete model and those presented in [44]. Tapered GLCFTFs of rectangular cross-section with a linear taper in both width and depth ($\delta_1 = \delta_2 = 0.6$) are considered. Here too, a good agreement between the two results is observed. (Note that the symbols $\lambda_p$ and $\lambda_q$ are used in [44] in place of $\psi$ and $\gamma$ used here. Furthermore, sign convention for $\lambda_p$ and $\lambda_q$ is opposite to that used for $\psi$ and $\gamma$.)

Results from these studies indicate that the discrete modeling technique discussed here is a valid approximation for analysis of transverse vibration and stability of straight, GLCFTFs with any given cross-sectional geometry.

## 4.4 FAMILY OF ISOSPECTRAL WUCFS

Discretized distributions of mass and stiffness on models of a family of WUCFs isospectral to given GLCFTFs are now determined. We also study how parameter

$g_2$ affect these distributions. As done earlier, scaling parameters, $\alpha$ and $\beta$ are fixed based on the condition that the total length and total mass of the isospectral beam are, respectively, equal to those of the given GLCFTF.

### 4.4.1   UNIFORM GLCFTFS

For the case of a uniform GLCFTF we carry out this study for the two cases defined in section 4.1. Figures 4.2 and 4.3 present results of the study for Case 1 and Case 2, respectively. It shows the stiffness and mass distributions on three discrete models of WUCFs, all isospectral to the given GLCFTF, but obtained with different values of $g_2$. The parameter $g_2$ has a significant impact in determining these distributions. The range of values of $g_2$ ensuring practically useful discrete models is narrow. As done in the case of CBTFs, here too we can follow a trial and error approach and obtain distributions that correspond to practically realistic discrete models. In Fig. 4.2c, it is observed that for case 1, a value of $g_2$ that gives a purely uniform mass distribution on the isospectral model, also gives a realistic distribution of stiffness. However, for case 2 it was observed that value of $g_2$ that gives uniform mass distribution, produced a highly non-uniform stiffness distribution (not shown in figure). A different value of $g_2$ could be estimated (by trial and error) that gives realistic distributions for mass and stiffness. This distribution is shown in Fig. 4.3c.

### 4.4.2   TAPERED GLCFTFS

A similar study was carried out on all three types of tapered GLCFTFs. Typical distributions of discretized mass and stiffness on a WUCF beam, obtained with optimum value of $g_2$ and isospectral to a GLCFTF ($\gamma = -3$ and $\psi = 1$) with constant depth and linear taper in width ($\delta_1 = 0$, $\delta_2 = 0.7$), are shown in Fig. 4.4.

## 4.5   ISOSPECTRAL MODELS: VARIATION WITH LOAD PARAMETERS

We now investigate the variations in distributions of mass and stiffness on isospectral discrete models as loading parameters, $\gamma$ and $\psi$, on a uniform GLCFTF, change. These distributions are shown in Fig. 4.5. In all these cases, the parameter $g_2$ was fixed at the value that gives a purely uniform distribution of mass on the isospectral model. Mass and stiffness distributions represented in Fig. 4.5a correspond to the case of zero tip force. In this case, it can be observed that WUCF beams isospectral to *hanging* beams (positive values of $\gamma$) should have higher stiffness near the clamped end and lower stiffness near the free end compared to that of the original column (which has a normalized stiffness value = 1 throughout the length). On the other hand, stiffness distribution on WUCFs isospectral to *standing* beams (negative values of $\gamma$) follows an opposite trend. It can also be seen that as the *magnitude* of gravity parameter, $\gamma$, increases, *uniformity* in distribution of stiffness along the length of the isospectral WUCF decreases.

Figure 4.5b indicates that a tensile tip force increases (decreases) the stiffness requirement near clamped (free) end of the isospectral WUCF while from Fig. 4.5c,

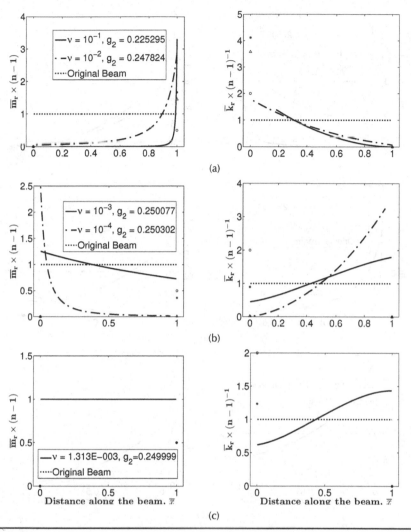

**Figure 4.2**
Mass and stiffness distributions on discrete models of WUCFs isospectral to a uniform *hanging GLCFTF with compressive* tip-force ($\gamma = 4, \psi = -2$)

points represent values at beam ends

it can be surmised that a compressive tip force increases (decreases) the stiffness requirement near free (clamped) end. In Fig. 4.5a and Fig. 4.5c, for the case of $\gamma = -6$, stiffness requirement near clamped end of the isospectral WUCF demands an unrealistic value close to zero. This is because loading parameter combinations for

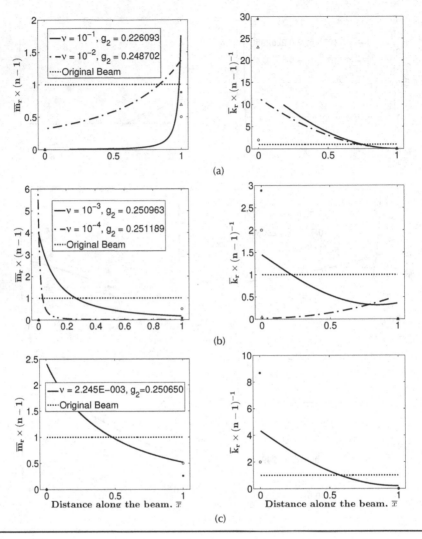

**Figure 4.3**
Mass and stiffness distributions on discrete models of WUCFs isospectral to a uniform *standing GLCFTF with tensile* tip-force ($\gamma = -3, \psi = 10$)

---

points represent values at beam ends

---

these cases are close to the critical values that cause onset of buckling . Thus, we observe that the method of finding isospectral systems described here is applicable only when the loading parameter combination is below the corresponding critical values.

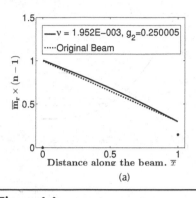
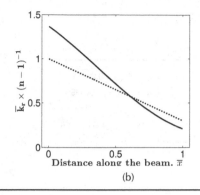

(a)                                                          (b)

**Figure 4.4**

Mass and stiffness distributions on discrete model of a WUCF isospectral to a tapered GLCFTF ($\delta_1 = 0$, $\delta_2 = 0.7$) for $\gamma = -3$ and $\psi = 1$

For clarity, stiffness values at beam-ends are not shown

## 4.6   CONSTRUCTING GEOMETRY OF ISOSPECTRAL WUCFS

The method described in section 2.11.1 is now used to construct the cross-sectional geometry of WUCFs isospectral to uniform and tapered GLCFTFs for example combinations of loading parameters. Mass $\overline{m}_0$ is assumed as $0.5 \times \mathbf{M}'(1,1)$ and $\overline{k}_{segment}$ part of $\overline{k}_n$ is assumed to be twice the stiffness of the adjacent spring [i.e. $2 \times \mathbf{K}'(n-1, n-1)$]. It is assumed that isospectral WUCFs have variable *rectangular* cross-section.

### 4.6.1   WUCFS ISOSPECTRAL TO UNIFORM GLCFTFS

Four combinations of $(\gamma, \psi) = (1, 2), (-1, 2), (1, -0.5)$ and $(-1, -0.5)$, on a *uniform* GLCFTF are considered and for each combination, we find cross-sectional geometry of an isospectral WUCF. The discrete model description of isospectral beams corresponding to these values of loading parameters are illustrated by the mass and stiffness distributions shown in Figs. 4.5b and 4.5c. The given uniform GLCFTF column is schematically represented in Fig. 4.6a. Variation in cross-sectional dimensions, along length of isospectral beams for the given combinations of loading parameters are shown in Figs. 4.6b, 4.6c and 4.7.

### 4.6.2   WUCF ISOSPECTRAL TO TAPERED GLCFTF

A non-homogeneous beam with a linear taper in width and constant depth ($\delta_1 = 0$, $\delta_2 = 0.7$) is shown schematically in Fig. 4.8a. Geometry of a WUCF isospectral to this non-homogeneous beam with loading parameters $\gamma = -3$ and $\psi = 1$ are shown in Fig. 4.8b. This variation in width and depth correspond to mass and stiffness distributions shown in Fig. 4.4.

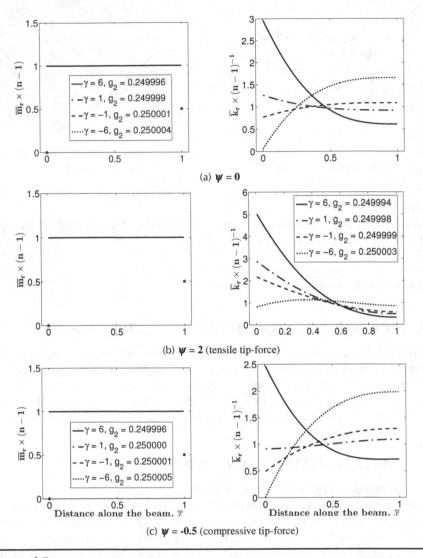

(a) $\psi = 0$

(b) $\psi = 2$ (tensile tip-force)

(c) $\psi = -0.5$ (compressive tip-force)

**Figure 4.5**
Mass and stiffness distributions on discrete models of WUCFs isospectral to a uniform GLCFTF for different values of $\gamma$ and $\psi$

For clarity, stiffness values at beam-ends are not shown

## 4.7  NATURAL FREQUENCIES OF WUCFS: FEM RESULTS

Natural frequency parameters of WUCFs constructed in sections 4.6.1 and 4.6.2, are calculated using FEM. The procedure undertaken for FE formulation is the same as that described in section 3.6. These frequency parameters are expressed

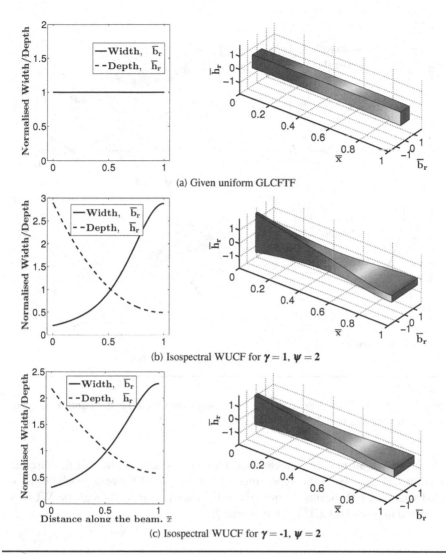

(a) Given uniform GLCFTF

(b) Isospectral WUCF for $\gamma = 1$, $\psi = 2$

(c) Isospectral WUCF for $\gamma = -1$, $\psi = 2$

## Figure 4.6

Geometry of WUCFs isospectral to a given *uniform* GLCFTF for *tensile* tip force parameter and two different values of $\gamma$

in non-dimensional form and compared with frequency parameters of the discrete model of the given GLCFTFs in Table 4.6. The table also presents the frequency parameters of discrete models of WUCFs from which the geometries were extracted. It is observed that the frequency parameters calculated using FEM match well with the discrete model frequencies within 0.02% error. It can thus be inferred that the WUCFs obtained for different combinations of loading parameters, $\gamma$ and $\psi$, are *indeed isospectral* to the corresponding GLCFTF.

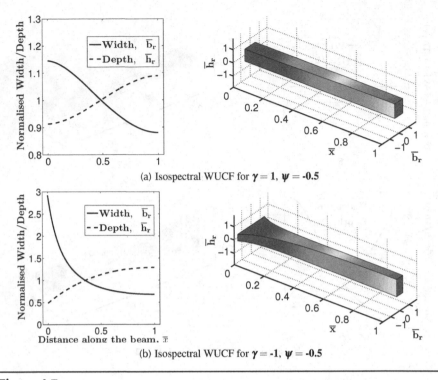

(a) Isospectral WUCF for $\gamma = 1$, $\psi = -0.5$

(b) Isospectral WUCF for $\gamma = -1$, $\psi = -0.5$

**Figure 4.7**
Geometry of WUCFs isospectral to a given *uniform* GLCFTF for *compressive* tip force parameter and two different values of $\gamma$

Results from all the above mentioned numerical studies show that the discrete modeling technique is a valid approximation for analysis of *forward* problems associated with GLCFTF beams. Using this methodology, practically realistic WUCFs isospectral to given GLCFTFs can be created.

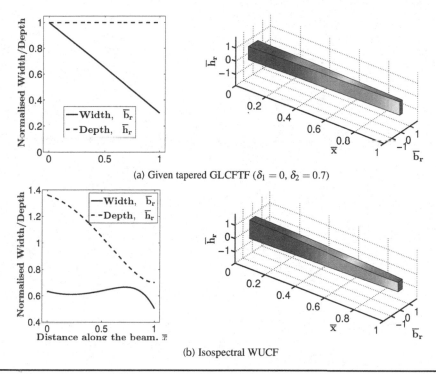

(a) Given tapered GLCFTF ($\delta_1 = 0$, $\delta_2 = 0.7$)

(b) Isospectral WUCF

**Figure 4.8**

Geometry of a WUCF isospectral to a given tapered GLCFTF, $\gamma = -3$, $\psi = 1$

**Table 4.6**

**Comparison of Natural Frequency Parameters: FEM Vs Discrete Model Results**

| | WUCF Isospectral to Uniform GLCFTF | | | | | | | | | | | |
|---|---|---|---|---|---|---|---|---|---|---|---|---|
| | $\psi = 2, \gamma = 1$ | | | $\psi = 2, \gamma = -1$ | | | $\psi = -0.5, \gamma = 1$ | | | $\psi = -0.5, \gamma = -1$ | | |
| | a) | b) | c) | a) | b) | c) | a) | b) | c) | a) | b) | c) |
| $\mu_1$ | 4.7726 | 4.7723 | 4.7723 | 4.4340 | 4.4345 | 4.4345 | 3.4030 | 3.4033 | 3.4033 | 2.9034 | 2.9031 | 2.9031 |
| $\mu_2$ | 23.6380 | 23.6371 | 23.6371 | 23.2659 | 23.2650 | 23.2650 | 21.8613 | 21.8602 | 21.8602 | 21.4632 | 21.4621 | 21.4621 |
| $\mu_3$ | 63.1352 | 63.1294 | 63.1294 | 62.7381 | 62.7320 | 62.7320 | 61.5859 | 61.5797 | 61.5797 | 61.1798 | 61.1734 | 61.1734 |
| | WUCF Isospectral to Tapered GLCFTF $\gamma = -3$ and $\psi = 1$ | | | | | | | | | | | |
| | $\mu_1$ | | | $\mu_2$ | | | $\mu_3$ | | | – | | |
| | a) | b) | c) | a) | b) | c) | a) | b) | c) | – | | |
| | 5.4768 | 5.4770 | 5.4770 | 25.6242 | 25.6233 | 25.6233 | 65.3318 | 65.3259 | 65.3259 | – | | |
| a) from FEM | b) from discrete model of GLCFTF column | | | | c) from discrete model of WUCF column | | | | | | | |

# 5 Rotating Beams - Discrete Models

Results from numerical studies on clamped-free rotating beams, hereinafter referred to as CFRBs, are provided in this chapter. A uniform beam, a tapered beam (which can be considered as a good model of a helicopter rotor blade) and a beam with general polynomial variation in mass and stiffness (which can be considered as a good model of a wind turbine rotor blade) are analyzed. Discrete models of these beams are generated using methods discussed in sections 2.7 and 2.8. Natural frequency parameters are estimated and compared with published results. Isospectral cantilever non-rotating beams (NRBs) are constructed. Natural frequencies of the NRBs are estimated using FEM and compared with those of corresponding CFRBs. The material in this chapter is adapted from Ref. [166].

## 5.1 CONVERGENCE STUDIES

We evaluate the variation in the first four natural frequency parameters ($\mu_r$, $r = 1$ to 4) of a uniform CFRB with zero root offset ($\overline{R} = 0$), as the number of discrete elements in the model is varied. These results are compared with values presented in [207] where a similar study was conducted using the Frobenius method . The number of discrete elements in the model, $n$, is varied from 10 to 300. The rotation speed of the beam is selected as $\lambda = 5$. Fig. 5.1 shows the variation in $\mu_r$ due to change in $n$. It is observed that frequencies obtained from the discrete model asymptotically approach the values given in [207] as the number of elements in the model is increased. The results from the discrete model converge satisfactorily when we use 300 elements in our model. For subsequent studies we use $n = 300$.

## 5.2 NATURAL FREQUENCY PARAMETERS OF CFRBS (CLAMPED FREE ROTATING BEAM)

The frequency parameters of the uniform and non-uniform CFRBs, are estimated from the 300-element discrete model for different values of $\lambda$. All values are compared with published results.

### 5.2.1 UNIFORM CFRBS

Tables 5.1 and 5.2 list values of frequency parameters, $\mu_r$, of a uniform CFRB (with and without offset, respectively), calculated from the discrete model for different values of $\lambda$ and compares these values with the results in Ref. [207]. It can be deduced that the discrete model results compare well with the values reported in [207] and the relative error is within 0.02% for all cases.

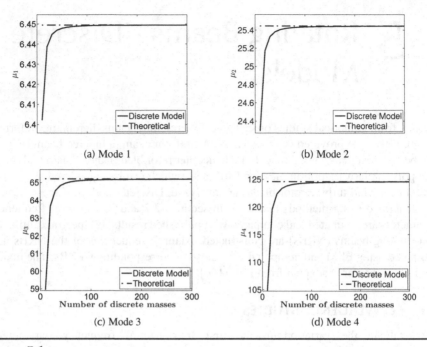

**Figure 5.1**
Uniform CFRB with $\lambda = 5$ and $\overline{R} = 0$: Frequency parameters Vs number of discrete masses

**Table 5.1**

## Frequency Parameters of Uniform CFRBs with Zero Root Offset ($\overline{R} = 0$)

|        | Mode 1 | | Mode 2 | | Mode 3 | | Mode 4 | |
| $\lambda$ | [207] | Discrete | [207] | Discrete | [207] | Discrete | [207] | Discrete |
|---|---|---|---|---|---|---|---|---|
| 0 | 3.5160 | 3.5160 | 22.0345 | 22.0334 | 61.6972 | 61.6910 | 120.902 | 120.881 |
| 1 | 3.6817 | 3.6816 | 22.1810 | 22.1800 | 61.8418 | 61.8355 | 121.051 | 121.030 |
| 2 | 4.1373 | 4.1373 | 22.6149 | 22.6139 | 62.2732 | 62.2669 | 121.497 | 121.476 |
| 3 | 4.7973 | 4.7972 | 23.3203 | 23.3192 | 62.9850 | 62.9787 | 122.236 | 122.214 |
| 4 | 5.5850 | 5.5850 | 24.2734 | 24.2723 | 63.9668 | 63.9604 | 123.261 | 123.240 |
| 5 | 6.4495 | 6.4495 | 25.4461 | 25.4450 | 65.2050 | 65.1987 | 124.566 | 124.545 |
| 6 | 7.3604 | 7.3603 | 26.8091 | 26.8079 | 66.6840 | 66.6775 | 126.140 | 126.119 |
| 7 | 8.2996 | 8.2996 | 28.3341 | 28.3329 | 68.3860 | 68.3794 | 127.972 | 127.951 |
| 8 | 9.2568 | 9.2568 | 29.9954 | 29.9941 | 70.2930 | 70.2863 | 130.049 | 130.028 |
| 9 | 10.2257 | 10.2256 | 31.7705 | 31.7692 | 72.3867 | 72.3799 | 132.358 | 132.336 |
| 10 | 11.2023 | 11.2022 | 33.6404 | 33.6390 | 74.6493 | 74.6424 | 134.884 | 134.862 |
| 11 | 12.1843 | 12.1842 | 35.5890 | 35.5876 | 77.0638 | 77.0568 | 137.614 | 137.592 |
| 12 | 13.1702 | 13.1700 | 37.6031 | 37.6016 | 79.6145 | 79.6073 | 140.534 | 140.512 |

**Table 5.2**

**Frequency Parameters of Uniform CFRBs with Root Offset ($\bar{R}$)**

| | $\bar{R}=0.1$ | | | | $\bar{R}=1.0$ | | | |
|---|---|---|---|---|---|---|---|---|
| | Mode 1 | | Mode 2 | | Mode 1 | | Mode 2 | |
| $\lambda$ | [207] | Discrete | [207] | Discrete | [207] | Discrete | [207] | Discrete |
| 0 | 3.5160 | 3.5160 | 22.0345 | 22.0334 | 3.5160 | 3.5160 | 22.0345 | 22.0334 |
| 1 | 3.7029 | 3.7029 | 22.2005 | 22.1994 | 3.8888 | 3.8888 | 22.3750 | 22.3739 |
| 2 | 4.2123 | 4.2122 | 22.6912 | 22.6902 | 4.8337 | 4.8337 | 23.3660 | 23.3650 |
| 3 | 4.9412 | 4.9411 | 23.4863 | 23.4852 | 6.0818 | 6.0817 | 24.9277 | 24.9266 |
| 4 | 5.8026 | 5.8025 | 24.5559 | 24.5548 | 7.4751 | 7.4750 | 26.9573 | 26.9561 |
| 5 | 6.7414 | 6.7414 | 25.8655 | 25.8643 | 8.9404 | 8.9403 | 29.3528 | 29.3516 |
| 6 | 7.7260 | 7.7260 | 27.3797 | 27.3785 | 10.4439 | 10.4438 | 32.0272 | 32.0259 |
| 7 | 8.7384 | 8.7383 | 29.0653 | 29.0641 | 11.9691 | 11.9690 | 34.9116 | 34.9101 |
| 8 | 9.7682 | 9.7681 | 30.8930 | 30.8917 | 13.5074 | 13.5072 | 37.9538 | 37.9522 |
| 9 | 10.8092 | 10.8091 | 32.8377 | 32.8363 | 15.0541 | 15.0539 | 41.1154 | 41.1137 |
| 10 | 11.8578 | 11.8577 | 34.8784 | 34.8770 | 16.6064 | 16.6062 | 44.3682 | 44.3664 |
| 11 | 12.9116 | 12.9115 | 36.9981 | 36.9966 | 18.1625 | 18.1623 | 47.6916 | 47.6895 |
| 12 | 13.9692 | 13.9690 | 39.1829 | 39.1813 | 19.7215 | 19.7213 | 51.0701 | 51.0679 |

## 5.2.2   NON-UNIFORM CFRBS

Two classes of non-uniform CFRBs are now considered.

*Type 1*: Beams with mass distribution given by $m(x) = m(0)(1 - 0.5\bar{x})$ and stiffness distribution given by $EI(x) = EI(0)(1 - 0.5\bar{x})^3$. Such a beam is a good mathematical model for a helicopter rotor blade [124].

*Type 2*: Beams with mass distribution given by $m(x) = m(0)(1 - 0.8\bar{x})$ and stiffness distribution given by $EI(x) = EI(0)(1 - 0.95\bar{x})$. Such a beam is a good mathematical model for a wind turbine rotor blade [124].

The vibration frequencies of these beams are reported in several publications [11, 104, 124, 129, 199, 207]. Tables 5.3 and 5.4 provide a comparison of the frequency parameter values ($\mu_1$, $\mu_2$ and $\mu_3$) of such beams obtained from the discrete model with those reported in [199]. There is good agreement between the two results.

The results from these numerical simulations indicate that the discrete model is suitable for analysis of transverse vibration of straight, CFRBs with any given cross-sectional geometry.

## 5.3   FAMILY OF NRBS ISOSPECTRAL TO UNIFORM CFRBS

We now construct a family of discrete models corresponding to NRBs, isospectral to a given uniform CFRB with $\bar{R} = 0$ rotating at $\lambda = 3$. We evaluate the effect of parameter $g_2$ in defining characteristics of isospectral models. As done before, scaling parameters, $\alpha$ and $\beta$, are fixed based on the condition that the total length and the total mass of the isospectral NRB are respectively equal to those of the given CFRB.

The mass and stiffness distributions illustrated in Fig. 5.2 all correspond to discrete models of NRBs, each isospectral to the given CFRB ($\lambda = 3$ and $\bar{R} = 0$). Here, it is observed that the distributions of mass and stiffness on the isospectral models are highly non-uniform when $\nu$ takes values greater than $10^{-2}$ (Fig. 5.2a). As

**Table 5.3**
**Frequency Parameters of Type 1 Non-uniform CFRBs with Zero Root Offset**
$(\overline{R} = 0)$

| | Mode 1 | | Mode 2 | | Mode 3 | |
|---|---|---|---|---|---|---|
| $\lambda$ | [199] | Discrete | [199] | Discrete | [199] | Discrete |
| 0 | 3.8238 | 3.8238 | 18.3173 | 18.3165 | 47.2648 | 47.2603 |
| 1 | 3.9866 | 3.9866 | 18.4740 | 18.4733 | 47.4173 | 47.4127 |
| 2 | 4.4368 | 4.4368 | 18.9366 | 18.9359 | 47.8716 | 47.8671 |
| 3 | 5.0927 | 5.0926 | 19.6839 | 19.6831 | 48.6190 | 48.6144 |
| 4 | 5.8788 | 5.8787 | 20.6852 | 20.6844 | 49.6456 | 49.6410 |
| 5 | 6.7434 | 6.7434 | 21.9053 | 21.9045 | 50.9338 | 50.9291 |
| 6 | 7.6551 | 7.6551 | 23.3093 | 23.3084 | 52.4633 | 52.4585 |
| 7 | 8.5956 | 8.5955 | 24.8647 | 24.8638 | 54.2124 | 54.2076 |
| 8 | 9.5540 | 9.5539 | 26.5437 | 26.5427 | 56.1595 | 56.1545 |
| 9 | 10.5239 | 10.5238 | 28.3227 | 28.3217 | 58.2833 | 58.2782 |
| 10 | 11.5015 | 11.5015 | 30.1827 | 30.1817 | 60.5639 | 60.5586 |
| 11 | 12.4845 | 12.4844 | 32.1085 | 32.1074 | 62.9829 | 62.9775 |
| 12 | 13.4711 | 13.4710 | 34.0877 | 34.0865 | 65.5237 | 65.5181 |

**Table 5.4**
**Frequency Parameters of Type 2 Non-uniform CFRBs with Zero Root Offset**
$(\overline{R} = 0)$

| | Mode 1 | | Mode 2 | | Mode 3 | |
|---|---|---|---|---|---|---|
| $\lambda$ | [199] | Discrete | [199] | Discrete | [199] | Discrete |
| 0 | 5.2738 | 5.2738 | 24.0041 | 24.0033 | 59.9708 | 59.9650 |
| 1 | 5.3903 | 5.3903 | 24.1069 | 24.1062 | 60.0703 | 60.0645 |
| 2 | 5.7249 | 5.7249 | 24.4129 | 24.4122 | 60.3676 | 60.3617 |
| 3 | 6.2402 | 6.2402 | 24.9148 | 24.9141 | 60.8598 | 60.8539 |
| 4 | 6.8928 | 6.8928 | 25.6013 | 25.6005 | 61.5420 | 61.5360 |
| 5 | 7.6443 | 7.6443 | 26.4580 | 26.4573 | 62.4078 | 62.4016 |
| 6 | 8.4653 | 8.4653 | 27.4692 | 27.4685 | 63.4494 | 63.4430 |
| 7 | 9.3347 | 9.3347 | 28.6184 | 28.6176 | 64.6579 | 64.6513 |
| 8 | 10.2379 | 10.2379 | 29.8893 | 29.8885 | 66.0238 | 66.0168 |
| 9 | 11.1651 | 11.1650 | 31.2667 | 31.2659 | 67.5370 | 67.5296 |
| 10 | 12.1092 | 12.1091 | 32.7367 | 32.7359 | 69.1875 | 69.1796 |
| 11 | 13.0657 | 13.0656 | 34.2868 | 34.2860 | 70.9653 | 70.9566 |
| 12 | 14.0313 | 14.0312 | 35.9060 | 35.9053 | 72.8604 | 72.8507 |

$v$ is decreased, uniformity in the distributions improves ($v = 10^{-2}$ compared with $v = 10^{-1}$). For $v = 10^{-3}$ (Fig. 5.2b), the mass and stiffness distributions are fairly uniform. However, as $v$ is decreased further, the distributions again become non-uniform ($v = 10^{-4}$). Thus, it is gleaned that values of $v$ (or $g_2$) can be tailored between $10^{-3}$ and $10^{-2}$ using a trial and error approach to obtain practically realistic discrete models. The value of $v$ (and $g_2$), which yields a purely uniform distribution of point masses along $\overline{x}$ is shown in Fig. 5.2c.

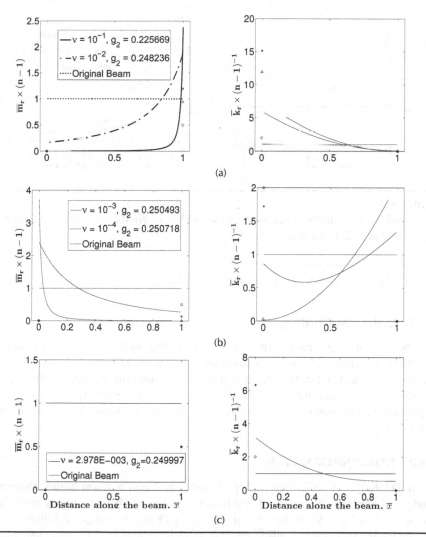

**Figure 5.2**
Mass and stiffness distributions on discrete models of non-rotating beams isospectral
to a uniform CFRB ($\lambda = 3, \overline{R} = 0$)

---

points represent values at beam ends

## 5.4 ISOSPECTRAL MODELS: VARIATION WITH ROTATION SPEED

We now peruse the variations in distributions of mass and stiffness on isospectral
discrete models as rotation speed, $\lambda$, of the CFRB changes.

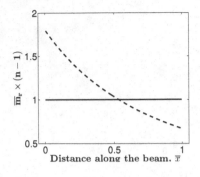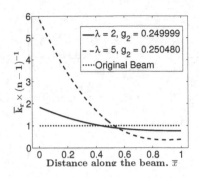

**Figure 5.3**

Mass and stiffness distributions on discrete models of NRBs isospectral to uniform CFRBs for $\lambda = 2, 5$ and $\overline{R} = 0$

---

For clarity, values at beam-ends are not shown

### 5.4.1   UNIFORM CFRB

For the case of a uniform CFRB, the mass and stiffness distributions on isospectral NRBs are shown in Fig. 5.3 for two different values of $\lambda$. In both these cases, parameter $g_2$ was fixed at the value that gives a fairly uniform mass and stiffness distributions on the isospectral model. It can be deduced that as $\lambda$ increases, the stiffness requirement on the non-rotating beam increases towards $\overline{x} = 0$ and decreases towards $\overline{x} = 1$.

### 5.4.2   NON-UNIFORM CFRBS

A similar study was performed on Type 1 and Type 2 non-uniform beams. Discrete models of a family of isospectral non-rotating beams can be constructed with different values of $g_2$. As in the case of uniform CFRBs, the range of values of $g_2$ that guarantee practically realistic discrete models, is narrow. The mass and stiffness distributions on typical non-rotating beams isospectral to Type 1 and Type 2 non-uniform CFRBs for two different rotation speeds are provided in Fig. 5.4.

## 5.5   GEOMETRY OF NRBS

We now deduce the geometry of NRBs whose mass and stiffness distributions were obtained in section 5.4. The NRBs are assumed to have variable rectangular cross-sections. Mass $\overline{m}_0$ is assumed as $0.5 \times \mathbf{M}'(1, 1)$ and $\overline{k}_{segment}$ part of $\overline{k}_n$ is assumed to be twice the stiffness of the adjacent spring [i.e. $2 \times \mathbf{K}'(n - 1, n - 1)$].

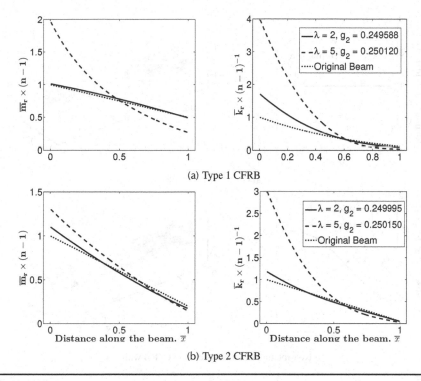

(a) Type 1 CFRB

(b) Type 2 CFRB

## Figure 5.4

Mass and stiffness distributions on discrete models of NRBs isospectral to Type 1 and Type 2 non-uniform CFRBs for $\lambda = 2, 5$ and $\overline{R} = 0$

For clarity, values at beam-ends are not shown

### 5.5.1 NRBS ISOSPECTRAL TO UNIFORM CFRBS

We construct two realistic NRBs - one isospectral to a uniform CFRB rotating with $\lambda = 2$ and the other isospectral to the same CFRB rotating with $\lambda = 5$. Discrete model description of these isospectral non-rotating beams are represented by mass and stiffness distributions shown in Fig. 5.3. The given uniform CFRB is schematically shown in Fig. 5.5a. Variation in cross-sectional width and depth, along the length of isospectral NRBs for the given values of $\lambda$ are shown in Fig. 5.5b and Fig. 5.5c.

### 5.5.2 NRBS ISOSPECTRAL TO NON-UNIFORM CFRBS

Realistic non-rotating beams, isospectral to Type 1 and Type 2 non-uniform CFRBs, are also created using the methods presented here. Fig. 5.6a illustrates a Type 1 non-uniform CFRB. Fig. 5.6b and Fig. 5.6c show non-rotating beams isospectral to this Type 1 CFRB rotating at $\lambda = 2$ and $\lambda = 5$, respectively. A similar set of beams

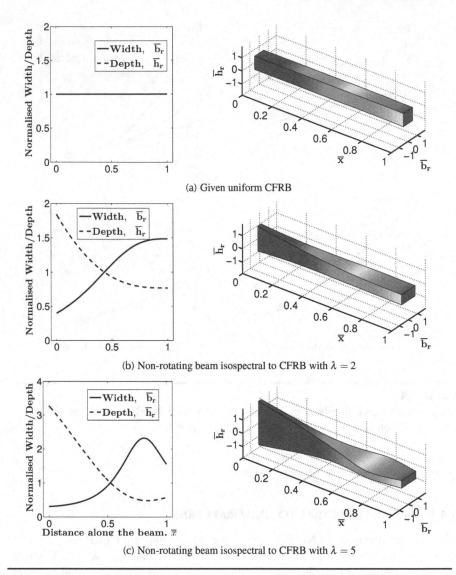

(a) Given uniform CFRB

(b) Non-rotating beam isospectral to CFRB with $\lambda = 2$

(c) Non-rotating beam isospectral to CFRB with $\lambda = 5$

### Figure 5.5
Geometry of non-rotating beams isospectral to a given uniform CFRB

corresponding to Type 2 non-uniform CFRBs are depicted in Fig. 5.7. The isospectral beams in Fig. 5.6 and Fig. 5.7 are constructed based on mass and stiffness distributions obtained in Fig. 5.4.

We thus demonstrate that the discrete approach presented in this chapter is a powerful and versatile tool which can be used to construct physically realizable, non-rotating beams isospectral to CFRBs with any complex cross-sectional geometry. Using the methods presented in [114, 116], non-rotating *cantilever* beams isospectral

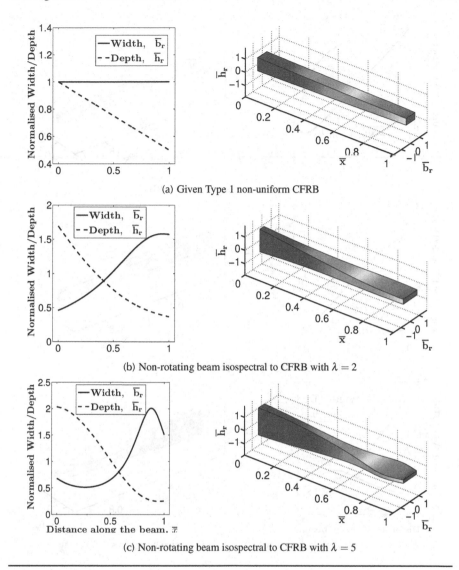

(a) Given Type 1 non-uniform CFRB

(b) Non-rotating beam isospectral to CFRB with $\lambda = 2$

(c) Non-rotating beam isospectral to CFRB with $\lambda = 5$

**Figure 5.6**
Geometry of non-rotating beams isospectral to a given Type 1 non-uniform CFRB

to CFRBs rotating at a high rotational speed ($\lambda = 5$) could not be determined. Instead, a non-rotating beam with torsional and translational springs at the free end was presented as the isospectral counterpart. Nevertheless, using the methods presented here, we will determine a practically realizable isospectral non-rotating cantilever beam for high values of $\lambda$.

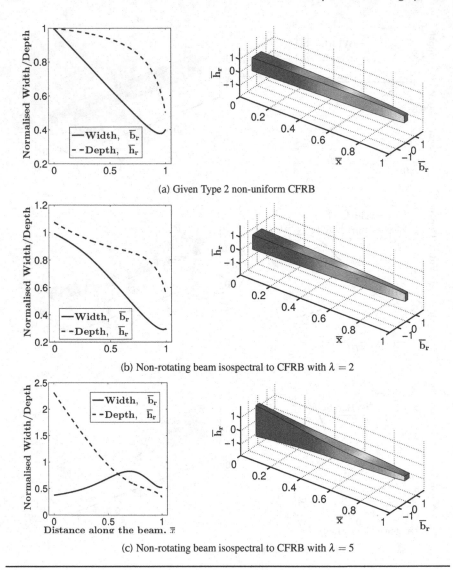

(a) Given Type 2 non-uniform CFRB

(b) Non-rotating beam isospectral to CFRB with $\lambda = 2$

(c) Non-rotating beam isospectral to CFRB with $\lambda = 5$

## Figure 5.7
Geometry of non-rotating beams isospectral to a given Type 2 non-uniform CFRB

### 5.5.3   NATURAL FREQUENCIES OF NRBS: FEM RESULTS

The natural frequency parameters of NRBs constructed in sections 5.5.1 and 5.5.2, are calculated using FEM. The procedure adopted for FE formulation is identical to that described in section 3.6. In Table. 5.5, the frequencies obtained from FE formulation are put in non-dimensional form and compared with frequency parameters $(\mu_1, \mu_2, \mu_3)$ of the discrete model of the NRB from which geometries of the beams

**Table 5.5**

**Natural Frequency Parameters of Isospectral NRBs: FEM Vs Discrete Model Results**

| | Uniform Beam | | | | Type 1 Non-Uniform Beam | | | | Type 2 Non-uniform Beam | | | |
|---|---|---|---|---|---|---|---|---|---|---|---|---|
| | $\lambda = 2$ | | $\lambda = 5$ | | $\lambda = 2$ | | $\lambda = 5$ | | $\lambda = 2$ | | $\lambda = 5$ | |
| | (a) | FEM | (a) | FEM | (a) | FEM | (a) | FEM | (a) | FEM | (a) | FEM |
| $\mu_1$ | 4.1373 | 4.1368 | 6.4495 | 6.4491 | 4.4368 | 4.4369 | 6.7434 | 6.7433 | 5.7249 | 5.7249 | 7.6443 | 7.6443 |
| $\mu_2$ | 22.6139 | 22.6147 | 25.4450 | 25.4456 | 18.9359 | 18.9366 | 21.9045 | 21.9051 | 24.4122 | 24.4130 | 26.4573 | 26.4579 |
| $\mu_3$ | 62.2669 | 62.2729 | 65.1987 | 65.2043 | 47.8671 | 47.8715 | 50.9291 | 50.9331 | 60.3617 | 60.3668 | 62.4016 | 62.4063 |

Values in columns marked (a) are obtained from discrete model of isospectral NRB

were extracted. The values in this table are also compared with corresponding values in Tables 5.1, 5.3 and 5.4. It is observed that the frequency parameters computed using FEM concur with theoretical and discrete model frequencies. Thus, the NRBs constructed using the procedures described here *are indeed isospectral* to the corresponding CFRB.

Based on the results of these numerical studies in this chapter, it is established that the discrete modeling methodology is capable of estimating natural frequencies of CFRBs with any complex cross-sectional geometry. This methodology can also determine the geometry of physically realizable NRBs that are isospectral to given CFRBs.

# 6 Isospectral Beams - Electromagnetism Optimization

The previous chapters have indicated that considerable research has been conducted into the development of discrete isospectral systems [77, 80, 84, 91, 187]. While the first chapter focussed on discrete spring-mass systems, further chapter addressed the problem of determining isospectral beams by converting the beams into discrete lumped mass system. From this chapter onwards, we address the problem of beams more directly. As a transition chapter, here we address the beam isospectral system using the finite element method for discretization. The finite element method is a modern numerical method which is ubiquitous in structural vibration analysis.

Subramanian and Raman [187] established the mathematical relation, in terms of non-linear algebraic equations, between a linearly tapered beam and a cubic tapered beam. The cubic tapered beam is isospectral to the linearly tapered beam. Gladwell [77] proposed two procedures for finding families of isospectral cantilever beams for a given cantilever beam. In one procedure, a shifted **QR** factorization of the mass-reduced stiffness matrix is proposed and that can yield several members of an isospectral family of beams. In another procedure, a family of isospectral mass-reduced stiffness matrix from isospectral flow equation, is used [77]. Gottlieb [91] discovered seven different classes of analytically solvable inhomogeneous Euler-Bernoulli beams with continuous density and flexural rigidity functions which are isospectral to a homogeneous beam for clamped-clamped boundary condition. In another publication, Gladwell [80] showed a way to obtain closed form expression for isospectral families of beams under specific boundary conditions. This analysis exploited the fact that a certain family of Euler-Bernoulli beams are equivalent to a certain family of strings, under suitable boundary conditions. Those Euler-Bernoulli beams belong to a special class for which the product between the bending stiffness and the linear mass density is constant.

Determination of isospectral vibrating beams can expand various avenues in research. For instance, determination of isospectral beams can be useful in structural damage identification from frequency measurements. The existence of isospectral beams could reveal if structural damage can be identified uniquely from frequency measurements, as there exist multiple beams with identical frequency response. Pawar and Ganguli [157] developed a genetic fuzzy system which allows automatic rule generation for various structures, to find the location and extent of damage in beams and helicopter rotor blades. Raich and Liszkai [160] posed the problem of damage detection from noisy frequency response function information, as an unconstrained optimization problem and solved using this problem using genetic algorithm combined with a local hill-climbing procedure. A multi-stage approach, utilizing genetic algorithm, was proposed in [184] for identification and quantification of damage in structures.

In this chapter, we focus on finding non-uniform cantilever beams that are isospectral to a given uniform cantilever beam using an evolutionary computational algorithm. For a given uniform cantilever beam, there can exist multiple isospectral cantilever beams which will reveal themselves as multiple optima of the objective function. Evolutionary methods are attractive for such multimodal optimization problems [41, 96, 119, 145]. We pose the problem of finding non-uniform beams which are isospectral to a given uniform beam, as an optimization problem. The first Q natural frequencies $\mu_i$ ($i = 1, 2, ..., Q$) of the given uniform Euler-Bernoulli beam can be determined using an analytical solution. The first Q natural frequencies $\eta_i$ ($i = 1, 2, ..., Q$) of a non-uniform beam are obtained with the help of finite element modeling. The objective function in this optimization problem is the difference between the spectra of the given uniform beam and the non-uniform beam. The error function is: $\sum_{i=1}^{Q} \frac{|\eta_i - \mu_i|}{\mu_i}$. In this chapter, this error function is minimized using electromagnetism based optimization technique, to obtain non-uniform beams that are isospectral to the given uniform beam.

Evolutionary computation methods are population based optimization techniques [27, 48, 68, 178, 221, 223]. Previous works [58, 200, 201, 203] witness that evolutionary methods are good at solving structural optimization problems. In the first chapter, we have explored numerous isospectral spring-mass systems using an evolutionary optimization technique (the *firefly algorithm*). However, we did not address continuous systems such as beams. Population based optimization algorithms search from a set of solution points and they do not need any gradient information. In this chapter, we use the electromagnetism based optimization algorithm (EM-algorithm) [27] to solve the isospectral problem for a continuous system. Population based algorithms, such as genetic algorithm (GA), have been extensively applied in solving structural optimization problems. GA involves computationally expensive operators such as crossover and mutation, and sometimes produces pre-mature convergence due to slow random search. The EM algorithm is a new meta-heuristic global optimization algorithm, and it has several advantages over early population based algorithms. Birbil and Fang [27] showed that the application of EM algorithm on nonlinear test functions yields satisfactory results. The heuristic algorithm was implemented in training neural networks [202], solving combinatorial optimization problem [209] etc. In this chapter, the EM algorithm is utilized to solve a structural optimization problem which yields isospectral systems as the solution.

The organization of the chapter is as follows: In section 6.1, the formulation of the optimization problem is described. In section 6.2, the electromagnetism inspired optimization technique which is used in the solution of the optimization problem is discussed. Section 6.3 presents the numerical results for isospectral beams found through this optimization driven methodology. The section includes an application of the proposed methodology to a damage identification problem. Section 5 summarizes the chapter. The material in this chapter is adapted from Ref. [60].

## 6.1 FORMULATION OF THE OPTIMIZATION PROBLEM

In this section, the problem of finding non-uniform beams that are isospectral to a given uniform beam is posed as an optimization problem. A beam with fixed-free boundary condition, i.e. a cantilever beam, is considered for the numerical illustration.

The bending vibration of an Euler-Bernoulli beam is governed by

$$\frac{\partial^2}{\partial x^2}[EI(x)\frac{\partial^2 u}{\partial x^2}] + m(x)\frac{\partial^2 u}{\partial t^2} = 0 \qquad (6.1)$$

Here, $EI(x)$ is the element flexural stiffness, $m(x)$ is the linear mass density (mass/length) and $u(x,t)$ is the transverse displacement.

*Uniform Beam:* The natural frequencies of an uniform beam can be determined directly from the closed form solutions. Natural frequency of the $n$'th mode of the beam with fixed-free (cantilever) boundary condition, is [140]

$$\mu_n = \frac{\beta_n^2}{2\pi}\sqrt{\frac{EI}{mL^4}} \quad Hz \qquad (6.2)$$

where, $\beta_1 = 1.87510$, $\beta_2 = 4.69409$, $\beta_{n\geq 3} = (2n-1)\frac{\pi}{2}$. In the above equation (6.2), $EI$ is the constant flexural stiffness, $m$ is the constant linear mass density (mass per unit length), and $L$ is the length of the beam.

*Non-uniform beam:* The natural frequencies of a non-uniform beam can be computed by an approximate technique such as the finite element method [35], which provides an approximate discrete model of a continuous system. From the finite element model (FEM) of the beam, the global stiffness matrix **K** and the global mass matrix **M** are constructed by assembling the element level local stiffness and mass matrices, respectively. The equation of motion for free vibration of the beam is

$$\mathbf{K}\underline{\phi} = \omega^2 \mathbf{M}\underline{\phi} \qquad (6.3)$$

where, $\omega$ is the natural frequency (in rad/sec) and $\underline{\phi}$ is the vector of the nodal amplitudes of vibration. The natural frequencies of the non-uniform beam are obtained by solving the generalized eigenvalue equation (6.3). The natural frequencies $\eta_i$ (in Hz) are obtained as

$$\eta_i = \frac{\omega_i}{2\pi} \quad Hz \qquad (6.4)$$

Copious details of finite element modeling of beams are available in [35, 157]. As the number of elements in the finite element model increases, the accuracy of the results also improves.

The methodology proposed in the present chapter, is presented figuratively by a block diagram (Fig. 6.1). Moreover, we introduce a numerical example to facilitate understanding of the optimization problem, in the following.

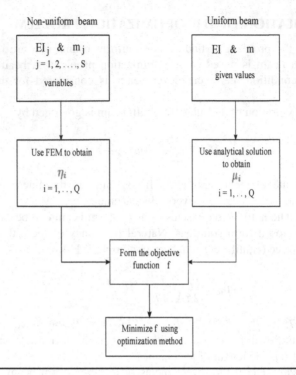

**Figure 6.1**
Block diagram of the problem formulation

*Illustrative Numerical Example*: Consider a uniform cantilever beam with $EI = 400$ N.msq, and $m = 1.8816$ kg/m, as shown in Fig. 6.2. The first six natural frequencies of this uniform beam are obtained using equation (6.2). The first six natural frequencies are: $\mu_1 = 22.6639$, $\mu_2 = 142.0322$, $\mu_3 = 397.6944$, $\mu_4 = 779.3224$, $\mu_5 = 1288.2761$, and $\mu_6 = 1924.4621$ Hz.

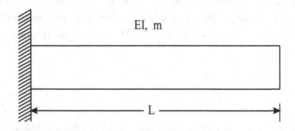

**Figure 6.2**
A uniform cantilever beam

Consider a non-uniform cantilever beam with 5 segments of equal length, as shown in Fig. 6.3. The flexural stiffness values in five segments are: $EI_1 = 451.1758$, $EI_2 = 403.6674$, $EI_3 = 431.0792$, $EI_4 = 426.3894$, and $EI_5 = 413.5099$. The mass per unit length values in the segments are: $m_1 = 1.9203$, $m_2 = 1.9154$, $m_3 = 2.1019$, $m_4 = 2.0406$, and $m_5 = 2.0375$. The first six natural frequencies of this non-uniform beam are obtained using equation (6.3). The first six natural frequencies are: $\eta_1 = 22.6638$, $\eta_2 = 142.0664$, $\eta_3 = 397.4179$, $\eta_4 = 779.3311$, $\eta_5 = 1288.5652$, and $\eta_6 = 1925.2065$ Hz. We observe that the first six natural frequencies of the non-uniform beam are the same (for most practical purposes) as the first six natural frequencies of the uniform beam, and therefore this non-uniform beam is isospectral to the given uniform beam.

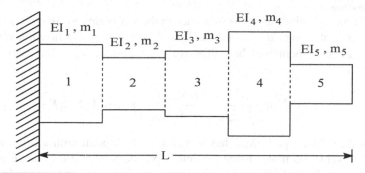

**Figure 6.3**
A non-uniform segmented cantilever beam, with five segments

From the aforementioned example, it can be concluded that the problem of finding isospectral non-uniform beam involves the determination of the flexural stiffness and mass per unit length values in each segment of the beam. This inverse vibration problem is transformed into an optimization problem and solved using electromagnetism based (EM) algorithm. In the next section, the EM algorithm is described.

## 6.2   ELECTROMAGNETISM (EM) INSPIRED OPTIMIZATION TECHNIQUE

We compendiously discuss the electromagnetism-based optimization algorithm (EM-algorithm) presented in [27, 28]. The optimization problem considered is

$$
\begin{aligned}
Minimize \quad & f(X) \\
Such\ that \quad & X \in [L,U] \\
where \quad & [L,U] := \{X \in R^d \mid L_k \leq X_k \leq U_k; \quad k = 1,2,...,d\} \quad (6.5)
\end{aligned}
$$

In the aforementioned equation, $L_k$ and $U_k$ are the upper and lower bounds on decision variables in dimension $k$.

The EM-algorithm is a population based iterative methodology that exploits the attraction-repulsion mechanism. The algorithm starts with an initial population of $N$ solutions. Each solution $X_p$ ($p = 1,2,...,N$), represents a point in $d$-dimensional search space. Each point $X_p$ is a possible solution to the optimization problem. Furthermore, each point $X_p$ is associated with an objective function value denoted as

$f(X_p)$. In each iteration, the population of points are moved towards the optimal solution. The EM-algorithm has the following steps:

Given, $N$ = number of solution points; $MAXITER$ = maximum number of iterations; $\delta$ = local search parameter $\delta \in [0,1]$; $LSITER$ = maximum number of local search iterations:

**STEP.1: Initialization:** In this step, $N$ solution points are randomly generated from the feasible space of the decision variables. The objective function $(f(X_p))$ value at each point is computed. The point that has the best objective function value is denoted as $X_{best}$.

**STEP.2: Force Calculation:** In this step, each of these $N$ points are assigned a charge $q_p$. The charge of a point $q_p$ is computed based on the objective function value of this point $X_p$ and the point with best objective function value $X_{best}$. The charge $q_p$ is computed using

$$q_p = exp\{-d\frac{f(X_p) - f(X_{best})}{\sum_{p=1}^{N}[f(X_p) - f(X_{best})]}\} \quad p = 1, 2, ..., N \tag{6.6}$$

We can observe that a point $X_{best}$ has a charge of 1. A point with a better objective function value has more charge than other points. Note that the charge associated with a point in the EM algorithm is always positive unlike the actual electrical charges. A point with a superior objective function value means more in maximization problems and less in minimization problems. In other words, a point with a lower objective function value has a higher charge, in a minimization problem.

The force of attraction or repulsion acting on a point $p$ depends on the charge $q_p$. The direction of the force acting between any two points is obtained by comparing the objective function values at those points. The total force $F_p$ exerted on the point $p$ by the other points is evaluated as:

$$F_p = \sum_{l=1}^{N} \begin{cases} (X_l - X_p)\frac{q_p q_l}{\|X_l - X_p\|^2} & if \ f(X_l) < f(X_p) \\ (X_p - X_l)\frac{q_p q_l}{\|X_l - X_p\|^2} & if \ f(X_l) \geq f(X_p) \end{cases} \tag{6.7}$$

**STEP.3: Movement of points:** In this step, all the points are moved along the force vector direction obtained in the previous step. The point $p$ moves in the direction of $F_p$ and its coordinate $X_p$ gets updated by the rule, given below:

$$X_p = X_p + \sigma\frac{F_p}{\|F_p\|}(G), \quad p = 1, 2, ..., N \ and \ p \neq best \tag{6.8}$$

Here, $\sigma$ is the random step length which is usually chosen [27] between 0 to 1. In equation (6.8), $G$ stands for allowed feasible movement which limits the movement of the point $p$ with respect to the upper bound $U_k$ or the lower bound $L_k$, for the dimension $k$, $k = 1, 2, ..., d$.

**STEP.4: Local search:** This local search step is optional. The EM algorithm works even without this step. In this local search step, the search for other better points in the neighborhood is undertaken. If local search is excluded, then the number of function evaluations is reduced. A simple local search procedure is used in [27, 28].

Another powerful local search algorithm described by [182], is incorporated into the EM algorithm in [85]. A complete description of EM algorithm is given in [27, 28].

## 6.2.1 ISOSPECTRAL BEAM DETERMINATION USING EM ALGORITHM

Typically, the number of segments in the non-uniform beam is considered as $S$. The number of decision variables (also called design variables) in the optimization problem is $(2S)$. The design variables are $EI_1, EI_2, ...., EI_S$ and $m_1, m_2, ...., m_S$. The objective function is

$$f = \sum_{i=1}^{Q} \frac{|\eta_i - \mu_i|}{\mu_i} \qquad (6.9)$$

where, $Q$ represents the number of modes of vibration. Specifically, the objective function $f$ is an error function; i.e. $f$ is the sum of absolute frequency errors for first $Q$ modes of vibration. Therefore, the optimization problem can be written as

$$\min_{EI_j, m_j} \quad f = \sum_{i=1}^{Q} \frac{|\eta_i - \mu_i|}{\mu_i} \qquad (6.10)$$
$$subject \ to \ EI_{LB} \leq EI_j \leq EI_{UB}, \ m_{LB} \leq m_j \leq m_{UB} \ (j = 1, ..., S)$$

In the above equation, $EI_{LB}$ and $EI_{UB}$ are the lower and upper bounds on flexural stiffness. The lower and upper bounds on the linear mass density are $m_{LB}$ and $m_{UB}$. The solution to this optimization problem gives the flexural stiffness and linear mass density values in each segment of the non-uniform beam which is isospectral to the given uniform beam. In order to minimize the objective function $f$, we use the electromagnetism based optimization technique.

The optimization problem is both higher dimensional and multimodal. As the number of segments increase, the dimension of the problem also increases. Moreover, the objective function does not have a direct relationship with the independent variables, and it is onerous to extract gradient information from the objective function. Therefore, we apply an evolutionary optimization approach using the population based EM algorithm instead of using traditional gradient based and initial condition dependent algorithms, to solve the above problem.

The EM algorithm starts with an initial population of $N$ non-uniform beams (solutions). We know the number of segments in the non-uniform beam is $S$. Hence, each solution $X_p$, $(p = 1, 2, ..., N)$, is a point in $2S$-dimensional search space. Each point $X_p$ contains the values of $EI_1, EI_2, ...., EI_S$ and $m_1, m_2, ...., m_S$, which is also a possible solution to the optimization problem. Each point $X_p$ is associated with an objective function value denoted as $f(X_p)$. In each iteration, the population of points is moved towards the optimal solution $x_{best}$, at which objective function value is the best (minimum for a minimization problem). The pseudocode of the algorithm is provided next.

### Electromagnetism based optimization algorithm: pseudocode

- Given: $N$ number of solutions in the population.
  S: the number of segments in the non-uniform beam.
  MAXITER: maximum number of iterations.
  LSITER: maximum number of local search iterations.
  $\delta$: the local search parameter $\delta \in [0,1]$.
  Initialize population of $N$ points $X_p$, $p = 1, 2, ...., N$.
  In each $X_p$, the decision variables $EI_1, ..., EI_S$ and $m_1, ..., m_S$ are contained.
- **While** (*counter* < *MAXITER*)
      **for** p=1:N
      Evaluate the objective function value $f(X_p)$ using equation (6.9).
      Assign charge $q_p$ to point $X_p$ using equation (6.6).
      Calculate force $F_p$ exerted on point $X_p$ using equation (6.7).
      Move point $X_p$ in $F_p$'s direction using equation (6.8).
      **end for**
- **end while**
  All points converge to the $X_{best}$ which gives the optimal objective function value.

## 6.3   NUMERICAL SIMULATIONS

In this section, we illustrate the non-uniform beams obtained for a given uniform beam, using the electromagnetism based optimization algorithm. The material properties and the dimensions of the given uniform cantilever beam is obtained from a finite element book [35]. The properties of the cantilever beam are given as follows:

- Modulus of elasticity   $= 2 \times 10^5 \ N/mm^2$
- Cross sectional area   $= 240 \ mm^2$
- Moment of inertia   $= 2000 \ mm^4$
- Density   $= 7840 \times 10^{-9} \ Kg/mm^3$
- Length   $= 600 \ mm$

From the aforementioned properties, we obtain the value of $EI = 400 \ N.m^2$, and $m = 1.8816$kg/m for the given uniform cantilever beam. The first six natural frequencies of this uniform beam are computed using equation (6.2). The first six natural frequencies are: $\mu_1 = 22.6639$, $\mu_2 = 142.0322$, $\mu_3 = 397.6944$, $\mu_4 = 779.3224$, $\mu_5 = 1288.2761$, and $\mu_6 = 1924.4621$ Hz.

In this problem, we have considered a population of 20 sample points in the search space i.e. $N = 20$, and the step size in EM algorithm $\sigma =$ random number between 0 to 1. The search space for the optimization problem is enclosed by the lower and upper bounds on the decision variables. The bound on flexural stiffness ($EI$ in each segment) is $300 \leq EI_j \leq 500 \ N.m^2$, and the bound on mass per unit length ($m$ in each segment) is $1 \leq m_j \leq 3 \ kg/m$ for $j = 1, 2, ..., S$. The number of segments $S$ is assumed. The algorithm is terminated when the objective function value becomes less than or equal to a fixed tolerance value of 0.005; and the local search is excluded, i.e. *LSITER* = 0. The first six natural frequencies are considered in the objective function; i.e. $Q = 6$.

In order to evaluate the objective function value, we need to find the first six natural frequencies ($\eta_i$, $i = 1, 2, ..., 6$) of the non-uniform beam. We apply the finite element model to determine the natural frequencies ($\eta_i$) of the non-uniform beams. The discrete non-uniform beams are divided into 100 finite elements of equal length. These non-uniform beams have jump discontinuities [216, 217] at segment interfaces.

Numerical simulations are run on a computer with 2.40 GHz Intel(R) Core™2 Quad Processor Q6600 and 1.99 GB of RAM. The results obtained show the existence of several non-uniform isospectral beams for a certain number of segments. In this algorithm, we have used the number of segments as 5, 10 and 25; i.e. $S = 5$, 10 and $S = 25$.

*Case 1. Number of segments $S = 5$*: We discover several non-uniform isospectral beams with 5 segments, which have identical (very close) first six natural frequencies as that of the uniform beam. The average number of iterations taken by the EM algorithm to fulfill the stopping criteria is 414, and the average computational time required is 522 s. The flexural stiffness and linear mass density values across the segments of five non-uniform isospectral beams, are shown in Table 6.3.

Now, we apply another evolutionary algorithm named as the Firefly Algorithm (FA) [58, 178] to solve the same problem. Chapter 1 has already showed the effectiveness of FA to explore isospectral spring-mass systems. Application of the FA to the addressed optimization problem, as shown by equation (6.10), results in several non-uniform isospectral beams. The parameters involved with the firefly algorithm are chosen as: $\beta_0 = 0.8$, $\gamma = 0.002$, $\alpha = 0.2$. The average number of iterations taken by the FA to fulfill the stopping criteria is 108, and the average computational time required is 1584 s. Table 6.1 shows a performance comparison study between the EM algorithm and the FA over 10 simulation runs, in a problem related with the application to isospectral beam problem. It can be observed from Table 6.1 that the computational time required by FA is considerably higher than the time required by EM algorithm. We also perform Wilcoxon's rank-sum test indexWilcoxon's rank-sum test among two sample sets of numeric data (containing mass and stiffness elements), obtained by FA and EM algorithm. The test results in a $p$-value of 0.6232, which indicates a failure to reject the null hypothesis at 5% significance level. Numerical results show that evolutionary algorithms are good at solving the isospectral beam problem. In the present chapter, we select the EM algorithm to proceed with and to find isospectral beams with various number of segments.

It is well understood that the performance of an evolutionary algorithm is problem dependent. Moreover, the algorithm performance depends on the parameters (such as: step size) associated with the algorithm. For the isospectral beam problem, the EM algorithm comes up with superior performance compared to the FA.

## Table 6.1

**Comparison of the Performance of EM Algorithm and FA with the Objective Function $f$ Over 10 Runs**

| Algorithm Used | Population Size | Best Value | Standard Deviation | No. of Iterations | Average Run Time (s) |
|---|---|---|---|---|---|
| EM | 20 | 0.001 | 0.05 | 2000 | 2682 |
| FA | 20 | 0.003 | $1.8 \times 10^{-8}$ | 2000 | 28845 |

*Case 2. Number of segments $S = 10$*: We obtain a number of distinct non-uniform beams (with 10 segments), which have identical first six natural frequencies as that of

the uniform beam. The average number of iterations required by the EM algorithm to fulfill the stopping criteria is 370. The flexural stiffness and the mass per unit length values in each segment of the five non-uniform beams (beams A, B, C, D, E), that are isospectral to the given beam, is shown in Table 6.5. The first six natural frequencies of the isospectral beams (beams A, B, C, D, E) are shown in Table 6.6. In Fig. 6.5a, 6.5b, the flexural stiffness and mass density distributions of one of the isospectral beams (beam E) are shown.

*Case 3. Number of segments S = 25*: We find a number of distinct non-uniform beams (with 25 segments), which have identical first six natural frequencies as that of the uniform beam. The average number of iterations required to reach the stopping criteria is 236. The flexural stiffness and the mass per unit length values in each segment of the five non-uniform beams (beams F, G, H, I, J), that are isospectral to the given beam, is shown in Table 6.7. The first six natural frequencies of the isospectral beams (beams F, G, H, I, J) are listed in Table 6.8. In Fig. 6.6a, 6.6b, the flexural stiffness and linear mass density distributions of the isospectral beams I and G are shown.

Next, the tolerance criteria is relaxed and the simulation is run several times for a higher number of iterations, to get a statistical analysis of the numerical results. We follow the above for beams with different number of segments (5, 10 and 25). Table 6.2 presents statistical measures of the results over 20 simulation runs with different initial population of 20 points. The average, best and worst values of the objective function along with the average run time are listed in this table. Figure 6.4 shows the function value as it decreases with the iterations. Table 6.4 lists the best isospectral beams found by numerical simulation, with different number of segments.

## Table 6.2
## Statistical Details of the Numerical Results Found by EM Algorithm Over 20 Runs

| No. of Segments | Best Value | Worst Value | Mean Value | Standard Deviation | No. of Iterations | Average Run Time (s) |
|---|---|---|---|---|---|---|
| 5  | $10.496 \times 10^{-4}$ | 0.1364 | 0.0824 | 0.0463 | 10,000 | 8224 |
| 10 | $6.235 \times 10^{-4}$  | 0.1087 | 0.0436 | 0.0388 | 10,000 | 8152 |
| 25 | $1.665 \times 10^{-4}$  | 0.0258 | 0.0094 | 0.0082 | 10,000 | 8165 |

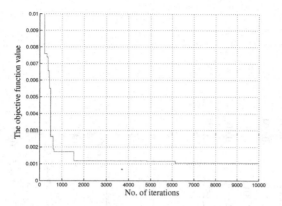

(a) Variation in objective function in case of 5 segments

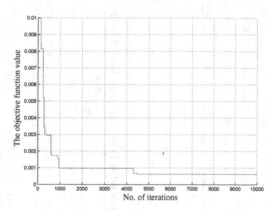

(b) Variation in objective function in case of 10 segments

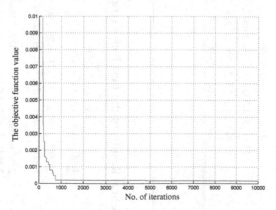

(c) Variation in objective function in case of 25 segments

**Figure 6.4**
Variation in the objective function value with iterations

**Table 6.3**

**Segment-wise Flexural Stiffness and Linear Mass Density Values of Five Isospectral Beams, Having 5 Segments**

| Isospectral Beam | Linear Mass Density (kg/m) | Flexural Stiffness (N.$m^2$) | Error Value |
|---|---|---|---|
| 1 | 1.9450, 1.8786, 2.0783, 2.0106, 1.8902 | 414.3579, 400.8896, 428.6178, 414.8999, 418.6149 | 0.0047 |
| 2 | 2.0058, 2.0032, 1.9589, 1.9336, 2.0260 | 435.7822, 403.0367, 419.4251, 423.6223, 428.0426 | 0.0047 |
| 3 | 1.9065, 1.9358, 1.8884, 2.0577, 1.8779 | 416.1168, 407.3466, 390.0328, 425.3725, 410.0013 | 0.0027 |
| 4 | 2.0199, 1.9353, 2.0499, 1.9274, 2.0119 | 430.4404, 400.3612, 430.0509, 419.7599, 432.8830 | 0.0046 |
| 5 | 1.8566, 1.8974, 2.0362, 2.0721, 1.8041 | 403.4609, 402.2277, 414.2294, 420.4756, 407.1347 | 0.0045 |

**Table 6.4**

**Numerical Results of the Best Isospectral Beams with Different Number of Segments**

| No. of Segments | Linear Mass Density (kg/m) | Flexural Stiffness (N.$m^2$) | Error Value |
|---|---|---|---|
| 5 | 1.9859, 1.8231, 1.9152, 2.0249, 1.9564. | 422.9086, 416.4253, 402.7772, 393.9051, 425.8623. | $10.496 \times 10^{-4}$ |
| 10 | 2.0587, 1.9815, 2.0733, 1.9606, 2.0016, 1.8294, 2.0851, 1.7978, 1.7919, 2.1322. | 430.7504, 394.4221, 403.2665, 438.2535, 429.0256, 403.7874, 440.4355, 433.9973, 401.0551, 417.4502. | $6.235 \times 10^{-4}$ |
| 25 | 2.0176, 1.9259, 1.9223, 2.0329, 2.1068, 2.1643, 2.0672, 1.7776, 2.0048, 1.9845, 2.1325, 1.7633, 1.5365, 1.7892, 1.7906, 1.9827, 1.6379, 1.9872, 1.8638, 1.7925, 1.7505, 1.7965, 2.2097, 1.9109, 1.8094. | 443.1705, 355.2345, 387.2830, 338.6440, 432.1637, 447.8219, 371.7510, 443.2740, 417.5579, 438.4571, 463.1498, 382.0309, 373.4506, 353.6324, 430.0756, 392.1861, 396.0922, 404.1752, 436.6537, 351.1996, 432.3028, 421.1471, 354.1478, 371.8569, 355.2794. | $1.665 \times 10^{-4}$ |

The numerical simulations yielding isospectral beams divulge the fact that the error function attains a better value (minimum) as the number of segment rises. However, the EM based methodology is able to generate isospectral beams with lower as well as higher number of segments. Although the optimal error function value is restricted to a certain level for lower number of segments, the resultant isospectral beams can be considered as closely or almost isospectral beams. From a design perspective, these almost isospectral beams can fulfil the requirements (related to dynamic performance) of a particular beam. Moreover, it can be deciphered from the computational results that the structure of the proposed methodology can be slightly modified by introducing the number of segments ($S$) as an extra design variable in the optimization problem, according to the requirement. Then, the evolutionary optimization approach can be applied to obtain appropriate values of $S$ within a certain range based on the design needs.

### 6.3.1   REDUCING COMPUTATIONAL EFFORT

In this section, a method of reducing the computational time in obtaining non-uniform isospectral beams is discussed. So far, we have used both the flexural stiffness and the linear mass density in each segment as decision variables; i.e. $EI_j$ and $m_j$ as decision variables for $j = 1, 2, ..., S$. Thus, we have $2S$ decision variables in this approach. Instead of treating both the flexural stiffness to linear mass density as decision variables, a new condition is imposed on the variables. We concentrate on finding isospectral non-uniform (segmented) beams which have the same flexural stiffness to linear mass density ratio in each segment as that of the uniform beam, i.e. $\frac{EI_j}{m_j} = \frac{EI}{m}$ for all $j = 1, 2, ..., S$. The decision variables are only $EI_j$. In this manner, the number of decision variables in the optimization problem reduced to $S$, instead of $2S$. This reduced design space leads to a faster convergence to the optimal solution. We now consider a non-uniform beam with $S = 10$ segments. The flexural stiffness and linear mass density values in each segment of the five non-uniform beams (beams K-O), that are isospectral to the given beam, are shown in Table 6.9.

**Table 6.5**

**Segment-wise Flexural Stiffness and Linear Mass Density Values of the Five Isospectral Beams (A-E), Having 10 Segments**

| Segment Order | Beam A | | Beam B | | Beam C | | Beam D | | Beam E | |
|---|---|---|---|---|---|---|---|---|---|---|
| | EI ($N.m^2$) | m ($kg/m$) | EI ($N.m^2$) | m ($kg/m$) | EI ($N.m^2$) | m ($kg/m$) | EI ($N.m^2$) | m ($kg/m$) | EI ($N.m^2$) | m ($kg/m$) |
| 1 | 466.6800 | 2.1912 | 431.3626 | 2.0441 | 461.8189 | 2.2351 | 476.6294 | 2.2375 | 470.5137 | 2.2641 |
| 2 | 480.7200 | 2.2837 | 405.4668 | 1.7186 | 481.5352 | 2.2837 | 462.7171 | 2.2671 | 463.4560 | 2.2230 |
| 3 | 471.6826 | 2.2279 | 413.8376 | 2.0025 | 481.0110 | 2.2802 | 478.9141 | 2.2240 | 468.9762 | 2.1992 |
| 4 | 472.3206 | 2.1905 | 396.3654 | 1.8229 | 471.6953 | 2.1965 | 463.2287 | 2.1790 | 460.9091 | 2.2048 |
| 5 | 457.8118 | 2.2094 | 430.5029 | 2.0941 | 474.5196 | 2.2042 | 481.5296 | 2.3005 | 481.8319 | 2.2078 |
| 6 | 474.1256 | 2.2100 | 415.2816 | 2.0289 | 466.0814 | 2.2200 | 477.5600 | 2.2350 | 472.5938 | 2.3400 |
| 7 | 465.3676 | 2.2397 | 394.7257 | 1.9779 | 486.5404 | 2.3197 | 476.3901 | 2.2618 | 477.5728 | 2.2464 |
| 8 | 471.7100 | 2.1973 | 405.1293 | 2.0184 | 474.8231 | 2.2145 | 470.3369 | 2.2426 | 467.0163 | 2.2235 |
| 9 | 483.8713 | 2.2253 | 397.9569 | 2.0612 | 489.4612 | 2.2695 | 480.2927 | 2.1347 | 480.5998 | 2.1523 |
| 10 | 462.3124 | 2.1700 | 455.5929 | 1.8249 | 473.5553 | 2.1833 | 473.8662 | 2.2284 | 472.9245 | 2.1869 |

**Table 6.6**

**First Six Natural Frequencies of the Isospectral Beams (A-E)**

| Cantilever Beams | First | Second | Third | Fourth | Fifth | Sixth | Error Function Values |
|---|---|---|---|---|---|---|---|
| | | | Natural | Frequencies | (Hz) | | |
| Uniform beam | 22.6639 | 142.0323 | 397.6945 | 779.3225 | 1288.2761 | 1924.4621 | 0 |
| Beam A | 22.7471 | 142.0021 | 397.7970 | 779.3495 | 1288.3303 | 1924.2740 | 0.0043 |
| Beam B | 22.6666 | 142.1077 | 397.5262 | 777.3985 | 1287.9588 | 1924.4405 | 0.0038 |
| Beam C | 22.6279 | 141.9985 | 397.4990 | 779.3035 | 1288.2896 | 1924.5327 | 0.0024 |
| Beam D | 22.7066 | 142.0474 | 397.6165 | 779.1940 | 1288.2009 | 1924.5850 | 0.0025 |
| Beam E | 22.6665 | 141.9430 | 397.6209 | 779.1667 | 1288.4242 | 1924.5464 | 0.0013 |

**Table 6.7**

**Segment-wise Flexural Stiffness and Linear Mass Density Values of the Five Isospectral Beams (F-J), Having 25 Segments**

| Segment Order | Beam F EI ($N.m^2$) | F m ($kg/m$) | Beam G EI ($N.m^2$) | G m ($kg/m$) | Beam H EI ($N.m^2$) | H m ($kg/m$) | Beam I EI ($N.m^2$) | I m ($kg/m$) | Beam J EI ($N.m^2$) | J m ($kg/m$) |
|---|---|---|---|---|---|---|---|---|---|---|
| 1 | 422.8932 | 2.1367 | 444.8405 | 1.7454 | 529.5811 | 2.3841 | 482.8548 | 2.1659 | 473.2358 | 2.2990 |
| 2 | 398.6172 | 2.2286 | 407.1748 | 2.0671 | 494.5318 | 2.4571 | 480.5253 | 2.2686 | 478.3578 | 2.2040 |
| 3 | 435.3841 | 1.9291 | 427.8609 | 2.0008 | 526.6734 | 2.2463 | 488.9374 | 2.1361 | 477.7246 | 2.2522 |
| 4 | 391.8827 | 1.7677 | 393.9743 | 1.9171 | 400.7671 | 2.5121 | 465.4311 | 2.2228 | 441.4936 | 2.0887 |
| 5 | 401.9510 | 2.1271 | 417.5755 | 1.9095 | 527.4048 | 2.4212 | 448.9507 | 2.2859 | 472.3579 | 2.1313 |
| 6 | 384.5815 | 1.9439 | 383.4296 | 2.0009 | 515.3034 | 2.4175 | 464.5827 | 2.1068 | 483.2288 | 2.1788 |
| 7 | 422.6929 | 1.5905 | 448.6614 | 2.1134 | 518.5865 | 2.6443 | 468.9813 | 2.2220 | 478.5233 | 2.3002 |
| 8 | 405.9930 | 2.3032 | 435.7202 | 2.0151 | 527.1900 | 2.6168 | 483.0522 | 2.2957 | 465.5558 | 2.1354 |
| 9 | 412.4082 | 1.7784 | 435.2355 | 1.9748 | 520.1474 | 2.4832 | 454.6891 | 2.3218 | 459.9995 | 2.1424 |
| 10 | 380.8566 | 1.8672 | 435.9814 | 2.0622 | 521.3526 | 2.2623 | 475.6912 | 2.1951 | 452.2255 | 2.2934 |
| 11 | 416.5963 | 2.1885 | 412.0479 | 1.9241 | 484.8709 | 2.6957 | 470.0353 | 2.1628 | 475.9829 | 2.1970 |
| 12 | 423.2727 | 1.9433 | 409.4412 | 2.0548 | 469.8895 | 2.3134 | 424.4242 | 2.2335 | 459.9942 | 2.1017 |
| 13 | 423.2972 | 2.1573 | 406.5785 | 1.9369 | 521.9348 | 2.0984 | 489.9609 | 2.1969 | 487.0791 | 2.2397 |
| 14 | 381.8142 | 1.6359 | 418.2936 | 1.8911 | 499.5288 | 2.5570 | 448.4703 | 2.0930 | 449.4700 | 2.2741 |
| 15 | 431.7912 | 1.7762 | 423.6863 | 1.9319 | 527.1230 | 2.3067 | 461.7873 | 2.2780 | 475.5690 | 2.2358 |
| 16 | 340.5189 | 1.9193 | 435.3413 | 2.1778 | 519.0907 | 2.3910 | 476.0204 | 2.2989 | 483.6245 | 2.1755 |

| Segment Order | Beam F EI (N.m²) | m (kg/m) | Beam G EI (N.m²) | m (kg/m) | Beam H EI (N.m²) | m (kg/m) | Beam I EI (N.m²) | m (kg/m) | Beam J EI (N.m²) | m (kg/m) |
|---|---|---|---|---|---|---|---|---|---|---|
| 16 | 340.5189 | 1.9193 | 435.3413 | 2.1778 | 519.0907 | 2.3910 | 476.0204 | 2.2989 | 483.6245 | 2.1755 |
| 17 | 391.6028 | 1.5859 | 392.1431 | 1.8833 | 528.7005 | 2.3775 | 479.4674 | 2.1054 | 464.6688 | 2.2666 |
| 18 | 403.2099 | 1.7956 | 455.5570 | 2.0922 | 529.6670 | 2.5803 | 473.3735 | 2.2485 | 475.9890 | 2.1540 |
| 19 | 399.4619 | 2.3626 | 371.5560 | 1.7929 | 529.7819 | 2.3316 | 481.2766 | 2.3417 | 453.7230 | 2.3258 |
| 20 | 416.5947 | 1.8772 | 436.7823 | 1.9250 | 514.6729 | 2.4531 | 482.9645 | 2.2655 | 467.6610 | 2.1800 |
| 21 | 422.5163 | 1.7641 | 430.7623 | 2.1000 | 473.3390 | 2.1259 | 480.8412 | 2.1573 | 466.7178 | 2.2536 |
| 22 | 405.4691 | 2.1240 | 418.0123 | 2.0178 | 529.4585 | 2.3123 | 484.9210 | 2.0944 | 454.2804 | 2.2619 |
| 23 | 411.0713 | 2.0016 | 416.2461 | 1.8667 | 521.2259 | 2.4320 | 451.8268 | 2.2807 | 489.6573 | 2.0587 |
| 24 | 437.7817 | 1.9843 | 375.3732 | 2.0853 | 519.3348 | 2.4797 | 470.2115 | 2.2382 | 436.5842 | 2.2974 |
| 25 | 433.0454 | 1.7385 | 388.3652 | 1.9182 | 401.1183 | 2.2540 | 472.2535 | 2.2037 | 447.1599 | 2.1845 |

¹ continuation of table 6.7

**Table 6.8**
**First Six Natural Frequencies of the Isospectral Beams (F-J)**

| Cantilever Beams | Natural Frequencies (Hz) | | | | | | Error Function Values |
|---|---|---|---|---|---|---|---|
| | First | Second | Third | Fourth | Fifth | Sixth | |
| Uniform beam | 22.6639 | 142.0323 | 397.6945 | 779.3225 | 1288.2761 | 1924.4621 | 0 |
| Beam F | 22.6648 | 141.8517 | 397.5047 | 779.8652 | 1289.0454 | 1925.6285 | 0.0037 |
| Beam G | 22.6522 | 141.8575 | 397.7015 | 778.6042 | 1286.8247 | 1924.3646 | 0.0039 |
| Beam H | 22.6312 | 142.0469 | 397.6193 | 779.4197 | 1288.1055 | 1924.4716 | 0.0020 |
| Beam I | 22.6662 | 141.9236 | 397.7594 | 779.3413 | 1288.3155 | 1924.4540 | 0.0011 |
| Beam J | 22.6404 | 142.0982 | 397.6363 | 779.3697 | 1288.1235 | 1924.4676 | 0.0018 |

**Table 6.9**
**Segment-wise Flexural Stiffness and Linear Mass Density Values of the Five Isospectral Beams (K-O)**

| Segment Number | Beam K | | Beam L | | Beam M | | Beam N | | Beam O | |
|---|---|---|---|---|---|---|---|---|---|---|
| | EI $(N.m^2)$ | m $(kg/m)$ | EI $(N.m^2)$ | m $(kg/m)$ | EI $(N.m^2)$ | m $(kg/m)$ | EI $(N.m^2)$ | m$(kg/m)$ | EI $(N.m^2)$ | m $(kg/m)$ |
| 1 | 383.8124 | 1.8054 | 392.3160 | 1.8454 | 401.7918 | 1.8900 | 391.9962 | 1.8439 | 434.9778 | 2.0461 |
| 2 | 411.5136 | 1.9358 | 385.0282 | 1.8112 | 397.7563 | 1.8710 | 405.0935 | 1.9056 | 412.8024 | 1.9418 |
| 3 | 395.9108 | 1.8624 | 424.5670 | 1.9972 | 428.6987 | 2.0166 | 418.6356 | 1.9693 | 427.8832 | 2.0128 |
| 4 | 398.5195 | 1.8746 | 368.0550 | 1.7313 | 414.4307 | 1.9495 | 372.6452 | 1.7529 | 376.3232 | 1.7702 |
| 5 | 397.2445 | 1.8686 | 410.8173 | 1.9325 | 404.2379 | 1.9015 | 451.3717 | 2.1232 | 383.3303 | 1.8032 |
| 6 | 419.8425 | 1.9749 | 403.6976 | 1.8989 | 395.4301 | 1.8601 | 403.9324 | 1.9001 | 422.2234 | 1.9861 |
| 7 | 399.1751 | 1.8777 | 391.1421 | 1.8399 | 433.5035 | 2.0392 | 411.9827 | 1.9379 | 402.4768 | 1.8932 |
| 8 | 390.4239 | 1.8365 | 402.4256 | 1.8930 | 419.4087 | 1.9729 | 392.8120 | 1.8478 | 387.3228 | 1.8219 |
| 9 | 406.1950 | 1.9107 | 395.2425 | 1.8592 | 398.2603 | 1.8734 | 412.2993 | 1.9395 | 429.8087 | 2.0218 |
| 10 | 383.0510 | 1.8019 | 388.9143 | 1.8294 | 403.5196 | 1.8982 | 387.0092 | 1.8205 | 429.1106 | 2.0185 |

**Table 6.10**

**First Six Natural Frequencies of the Isospectral Beams (K-O)**

| Cantilever Beams | Natural Frequencies (Hz) | | | | | | Error Function Values |
|---|---|---|---|---|---|---|---|
| | First | Second | Third | Fourth | Fifth | Sixth | |
| Uniform beam | 22.6639 | 142.0323 | 397.6945 | 779.3225 | 1288.2761 | 1924.4621 | 0 |
| Beam K | 22.7041 | 142.0985 | 397.4910 | 778.7396 | 1287.6794 | 1923.9177 | 0.0042 |
| Beam L | 22.6537 | 142.0686 | 397.6167 | 779.5259 | 1288.9129 | 1925.5171 | 0.0022 |
| Beam M | 22.6343 | 141.9816 | 397.4492 | 779.8103 | 1287.6017 | 1924.5967 | 0.0035 |
| Beam N | 22.6954 | 142.0032 | 397.6026 | 779.3773 | 1288.2630 | 1925.4360 | 0.0024 |
| Beam O | 22.6561 | 142.1853 | 397.6356 | 779.3405 | 1288.8863 | 1924.0767 | 0.0023 |

## 6.3.2 APPLICATION TO DAMAGE DETECTION

Earlier results discussed in this chapter have shown that there can exist various mass and stiffness distributions corresponding to a single set of natural frequencies. It is therefore difficult to uniquely identify a cantilever beam using only the first few frequencies. Therefore, detection of structural damage by using only frequencies as measurements is very difficult for cantilever beams.

With the help of proposed methodology, we want to find beams which are isospectral to a damaged beam. We incorporate a localized damage at a section of 10 percent of the uniform beam. From continuum damage mechanics, it is known that structural damage causes a localized stiffness reduction [157, 167]. The flexural stiffness of 10 adjacent elements near the center (51-60) is reduced by 25 percent, keeping the mass as constant.

In this optimization problem, we allow the flexural stiffness to vary across the segments of the beam, while holding the mass uniform throughout the beam. The decision variables in the optimization problem, are $EI_1, EI_2, ..., EI_s$ where $S$ is the number of the segments. The number of the decision variables becomes identical to the number of the segments. The stopping criteria in simulations is kept unchanged, the tolerance value is set to 0.005. The simulation results show several beams, having 10 segments, which are isospectral (beams with approximately equal set of first six natural frequencies) to the damaged beam. The flexural stiffness values of the beams isospectral to the damaged beam, are provided in Table 6.11.

Numerical simulations show that structural damage detection using the first few frequencies is infeasible. Thus, the damage identification problem for a cantilever beam cannot be solved accurately by only frequency measurements. This is an important finding with ramifications in the area of structural health monitoring [70, 156].

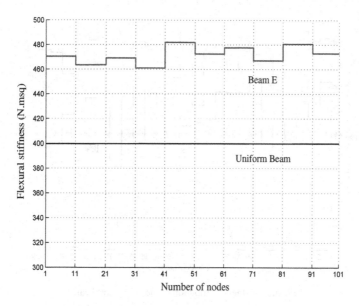

(a) Flexural stiffness distribution of the isospectral beam

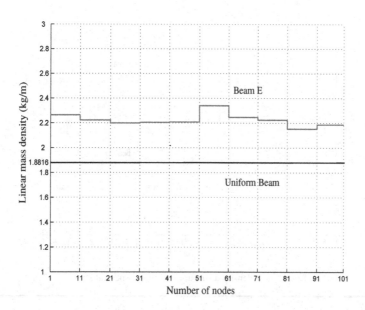

(b) Linear mass density distribution of the isospectral beam

**Figure 6.5**
Flexural stiffness and linear mass density of the beam (Beam E in Table 6.5), isospectral to an uniform beam

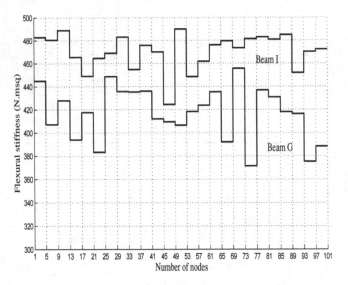

(a) Flexural stiffness distribution of the isospectral beam

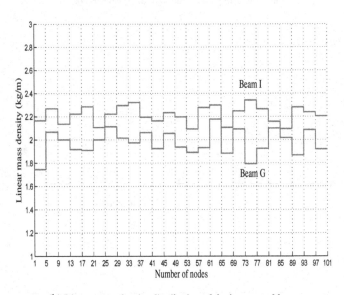

(b) Linear mass density distribution of the isospectral beam

## Figure 6.6
Flexural stiffness and linear mass density of the beam (Beam I in Table 6.7), isospectral to an uniform beam

**Table 6.11**

**Flexural Stiffness Values of the Beams (P-T), Isospectral to a Damaged Beam**

| Cantilever Beams | Flexural stiffness Values | | | | | | | | | |
|---|---|---|---|---|---|---|---|---|---|---|
| | $EI_1$ | $EI_2$ | $EI_3$ | $EI_4$ | $EI_5$ | $EI_6$ | $EI_7$ | $EI_8$ | $EI_9$ | $EI_{10}$ |
| Damaged beam | 400 | 400 | 400 | 400 | 300 | 400 | 400 | 400 | 400 | 400 |
| Beam P | 388.3593 | 375.0694 | 420.4895 | 433.2842 | 361.6750 | 354.4456 | 388.1367 | 386.7776 | 390.2441 | 400.1017 |
| Beam Q | 389.7734 | 400.3618 | 393.4299 | 416.3972 | 350.2228 | 350.9435 | 389.1919 | 416.9342 | 386.8858 | 398.9161 |
| Beam R | 376.8440 | 395.5652 | 425.9517 | 420.0728 | 343.6349 | 370.7063 | 402.5571 | 385.5708 | 386.5570 | 382.3662 |
| Beam S | 366.0215 | 378.8812 | 469.9064 | 441.2317 | 365.8260 | 365.8891 | 410.3469 | 360.1100 | 382.7252 | 363.8036 |
| Beam T | 369.4561 | 369.2166 | 448.1080 | 483.1202 | 367.3786 | 366.2588 | 368.5821 | 376.2482 | 379.7320 | 378.8380 |

**Table 6.12**

**First Six Natural Frequencies of the Isospectral Beams (P-T)**

| Cantilever Beams | Natural frequencies (Hz) | | | | | | Error function Values |
|---|---|---|---|---|---|---|---|
| | First | Second | Third | Fourth | Fifth | Sixth | |
| Damaged beam | 22.4244 | 138.0842 | 395.7077 | 763.0873 | 1271.0979 | 1898.2184 | 0 |
| Beam P | 22.4209 | 138.2922 | 395.5289 | 763.5667 | 1268.8883 | 1898.0972 | 0.0045 |
| Beam Q | 22.4257 | 138.0801 | 395.6078 | 764.2999 | 1268.6005 | 1897.1541 | 0.0045 |
| Beam R | 22.4240 | 138.1122 | 395.1459 | 763.2933 | 1269.6784 | 1898.9003 | 0.0034 |
| Beam S | 22.4479 | 138.3475 | 395.7578 | 763.2179 | 1270.7240 | 1900.5510 | 0.0048 |
| Beam T | 22.4161 | 138.0447 | 395.4256 | 763.0313 | 1267.2941 | 1898.7524 | 0.0047 |

### 6.3.3 DISCUSSION

In this chapter, we formulate the problem of finding segmented (stepped) beams isospectral to a given uniform beam, as a numerical optimization problem. The optimization problem is higher dimensional and multimodal. The problem is solved using an evolutionary optimization technique. The proposed methodology results in numerous isospectral beams with various number of segments. Each isospectral beam corresponds to a local optimum of the objective function. The best solution found after a finite number of simulation runs is nearest to the global optimum.

Cantilever beams are ubiquitous in engineering. Uniform cantilever beam have applications in MEMS devices [159], modeling of carbon nanotubes [64], and are used in space structures. The numerical results in this chapter show that unique system identification of these beams from the spectra is not possible. From a design perspective, there exist dynamically similar beams which can be used in place of uniform beams. Cantilever beams, uniform [171] and segmented [128], have applications in vibration energy harvesting. Non-uniform isospectral beams (segmented) can be substituted for a uniform beam by profering better strain or fatigue characteristics while incorporating the same natural frequencies. Moreover, the isospectral beams can be used to verify the convergence of the finite element codes as the non-uniform beams should yield frequencies almost identical to those obtained from closed form solution of the uniform beam.

### 6.4 SUMMARY

The problem of finding non-uniform beams, isospectral to a given uniform beam, is posed as an optimization problem in this chapter. The objective is to minimize a non-negative error function which calculates the difference between the spectra of the given uniform beam and the determined non-uniform beam. The decision variables in this approach, are the flexural stiffnesses and the linear-mass densities of all the segments of a beam. The flexural stiffness and linear mass density are permitted to vary independently across the segments of a beam. Electromagnetism based optimization algorithm is applied to minimize the error function, and results in several optimum points, based on a stopping criteria. Each optimum point indicates an isospectral non-uniform beam. The optimization approach, used in the proposed methodology, performs well and determines the isospectral beams for a given uniform beam. Numerical results of the isospectral beams having 10 segments, and the isospectral beams having 25 segments, are presented in this chapter. Several isospectral beams are found thereby demonstrating the non-uniqueness of the identification problem for a uniform cantilever beam. Moreover, the isospectral beams can be useful for structural design and for verifying the convergence of finite element codes. The proposed methodology is also applied to a structural damage identification problem. Numerical simulations indicate that structural damage is onerous to detect accurately using only frequency measurements. The results in this chapter also cement the fact that the EM based evolutionary algorithm is a powerful candidate for the solution of multimodal optimization problems resulting from inverse problems in vibration.

# 7 Isospectral Beams - Closed Form Solutions

In this chapter, we study isospectral systems with the aim of discovering new closed form solutions. For many structural dynamics systems, such as the *uniform* Euler-Bernoulli beam, an exact solution of the governing differential can be obtained for free vibration. Consequently, this exact solution can be used together with boundary conditions to find closed form expressions for the natural frequencies of the structure. The closed form solutions allow for easy calculation of the frequencies of the structure, and also enable one to check the scientific veracity and implementation of numerical methods for natural frequency calculation, such as those based on the theories of Rayleigh-Ritz and Galerkin method, and the finite element method. This chapter will present cases where we find a beam which has a closed form solution which is isospectral to a beam which does not have a closed form solution. In this indirect approach, we discover the closed form solutions of non-uniform and rotating beams for which closed form solutions for the natural frequencies are not available.

Recall that an analysis of isospectral Euler-Bernoulli beams with continuous density and rigidity functions was conducted by Gottlieb in [86]. In this study [86], Gottlieb corrected the Barcilon's transformation [17], and presented seven classes of beams (with continuous mass and stiffness functions) that are isospectral to a given uniform beam. Subramanian and Raman [188] presented a more generalized transformation of the dependent variable, as compared this with the Barcilon-Gottlieb transformation [86]. Ghanbari [72] factored the fourth order beam operator into two second order differential operators, and found twelve classes of isospectral beams, for four different boundary conditions. Gladwell and Morassi [80], obtained closed form expressions for beams isospectral to a given one, where the boundary conditions are any combination of pinned and sliding. A specific class of beams where the product of stiffness and mass/length is a constant, are considered. The analysis in [80] is based on the fact that a special class of beams is equivalent to a string, and uses a Darboux lemma after reduction of the string equation to Sturm-Liouville canonical form. Gottlieb [86, 90] enumerated the procedure for obtaining real densities of circular membranes, that are isospectral to a given uniform circular membrane under fixed and free boundary conditions. Gottlieb found that membranes isospectral to radial-density membranes no longer possess radial symmetry, and discussed some examples in [92]. However, axially loaded beams are not discussed in these studies but will be considered in this chapter.

Axially loaded beam structures are widely used in many aerospace, civil and mechanical structures. The transverse vibration of axially loaded uniform beams is discussed in textbooks [137]. Bokian [29, 30] calculated the frequencies of a uniform beam under constant compressive and tensile axial loads for ten combinations of classical boundary conditions. Another example of axially loaded structures are piano strings which can be modeled as stiff-strings which have tension and a flexural stiffness. Fletcher [65] developed the equations that govern the vibration of a stiff-string for the pinned-pinned and clamped-clamped boundary conditions. Zhang et al. [222] developed a double elastic beam model for the transverse vibrations of carbon-nano tubes under compressive axial loads. The natural frequency of such structures is sensitive to the applied axial load. This frequency shift of the carbon nanotube resonator is the basic principle of sensing.

Solutions to inverse problems and isospectral problems are used extensively for damage detection in structures. A damage or a crack at any location causes a decrease in the stiffness at that location, which produces a change in the modal characteristics of the structure [155]. Researchers have used this discrepancy between the modal data, acquired through experiments and that through analytical models, to detect damage location in structures. Farrar et al. [63] conducted a vibration based damage detection study on cylindrical concrete columns that were quasi-statically loaded to failure in an incremental manner.

The material in this chapter is adapted from Refs. [113] and [115].

## 7.1   ISOSPECTRAL NON-UNIFORM BEAMS AND UNIFORM AXIALLY LOADED BEAMS

In this section, we determine the mass and stiffness profiles of non-uniform beams under no axial forces, which are isospectral to a given uniform beam under an axial load. The Barcilon-Gottlieb transformation is modified so that the non-uniform beam equation is transformed into the axially loaded uniform beam equation. However, this transformation requires the solution of a pair of coupled ODEs. These coupled ODEs are solved for two specific cases and closed form solutions for the mass and stiffness variations of a non-uniform beam are obtained, which is isospectral to a given uniform beam under an axial load. Both the given beam and the isospectral beam to be determined are assumed to be under the clamped-clamped, and the hinged-hinged boundary conditions. Finally, the analysis is applied to physically realizable beams with a rectangular cross section.

This method of using the Barcilon-Gottlieb transformation was studied extensively for finding isospectral systems for rotating beams, such as helicopter and wind-turbine blades, in [115, 116, 117]. Three different isospectral possibilities between rotating and non-rotating beams are explored in these three papers. In these studies, rotating beams are modeled as beams under radially varying axial loads (centrifugal forces). However, due to the complicated nature of the centrifugal stiffening term in the governing equation, these isospectral mass and stiffness functions derived are not closed-form expressions, but are presented as implicit functions. The isospectral properties of these functions are verified by performing a detailed finite element analysis. The present section specifically deals with non-rotating beams. We only present the cases where the isospectral mass and stiffness functions have closed form expressions. The advantage of such closed form expressions is that the isospectral properties can be readily verified by substituting these functions in the governing differential equations.

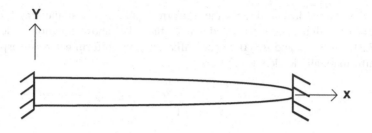

**Figure 7.1**
A schematic of a non-uniform beam under no axial force

## 7.1.1 MATHEMATICAL FORMULATION

The mathematical analysis of the problem is presented in this section. The non-dimensional form of the governing equation for the transverse vibrations of a clamped-clamped non-uniform beam under no axial forces, as shown in Fig. 7.1, is given by [139]

$$\frac{d^2}{dx^2}\left[f(x)\frac{d^2Y}{dx^2}\right] = \eta^2 m(x)Y, \quad 0 < x < x_0 \tag{7.1}$$

where $Y$ is the transverse displacement, $x_0$ is the length of the beam, and $f(x)$, $m(x)$ and $\eta$ are the stiffness, mass and frequency of the non-uniform beam. Using the Barcilon-Gottlieb transformation [17, 86], Eq. (7.1) can be transformed into

$$\frac{d^4U}{dz^4} + \frac{d}{dz}\left[A\frac{dU}{dz}\right] + BU = \eta^2 U \tag{7.2}$$

where $p$ and $q$ are the auxiliary variables given by

$$p = \left(\frac{m}{f}\right)^{\frac{1}{4}} ; \quad q = \left(\frac{1}{m^3 f}\right)^{\frac{1}{8}} \tag{7.3}$$

This transformation transforms the the variables $x$ and $Y$ into $z$ and $U$, respectively, as

$$z = \int_0^x p\,dx \quad \Leftrightarrow \quad x = \int_0^z \frac{dz}{p} \tag{7.4}$$

$$Y = qU \quad \Leftrightarrow \quad U = Y/q \tag{7.5}$$

The clamped boundary condition is preserved under the Barcilon-Gottlieb transformation, and under some special conditions, the hinged boundary condition is also preserved. The coefficients $A$ and $B$ in Eq. (7.2), in terms of $p$ and $q$ are given in [86]. Similarly, the non-dimensional governing equation for the transverse vibrations of an axially loaded clamped-clamped uniform beam, as shown in Fig. 7.2 is given by [118]

$$\frac{d^4U}{dz^4} + \frac{d}{dz}\left[-h\frac{dU}{dz}\right] = \eta^2 U \quad , \quad 0 < z < z_0 \tag{7.6}$$

where $h$ is the axial load acting on the uniform beam, and $z_0$ is the length of the beam, $z_0 = 1$ for all the cases discussed in this study. The above equation is identical to Eq. (7.2) if $A = -h$ and $B = 0$. Specifically, the non-uniform beam is isospectral to the uniform axially loaded beam if

$$A = -h \tag{7.7}$$
$$B = 0 \tag{7.8}$$

The above equations are coupled fourth fourth order ODEs and are onerous to solve. However, for two special cases: (i) $pq_z = k_0$, a constant and (ii) $q = q_0$ a constant, we show that the coupled ODEs can be solved analytically.

### 7.1.1.1  Case-1: $pq_z = k_0$, a constant

When $pq_z = k_0$, then $B = 0$ is automatically satisfied (Eq. (2.22a) in [86]). Setting $A = -h$ yields

$$A = 4\frac{q_{zz}}{q} - 6\frac{q_z^2}{q^2} - 2\frac{q_z p_z}{qp} + \frac{p_{zz}}{p} - 2\frac{p_z^2}{p^2} = -h \tag{7.9}$$

Substituting $p = k_0/q_z$ in the above equation, yields

$$-\frac{6q_z^2}{q^2} + \frac{6q_{zz}}{q} - \frac{q_{zzz}}{q_z} = -h \tag{7.10}$$

which simplifies to

$$\hat{q}_{zzz} = h\hat{q}_z \tag{7.11}$$

where

$$\hat{q} = 1/q \tag{7.12}$$

Eq. (7.11) is a linear differential equation in $\hat{q}$ with constant coefficients, and its solutions depend on whether the axial load ($h$) is compressive or tensile.

**Tensile axial load, $h > 0$**
Solving Eq. (7.11), for $h > 0$, yields

$$\hat{q} = c_1 e^{\sqrt{h}z} + c_2 e^{-\sqrt{h}z} + c_3 \tag{7.13}$$

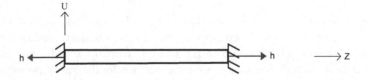

**Figure 7.2**
A schematic of an axially loaded uniform beam

$$\Rightarrow q = (c_1 e^{\sqrt{h}z} + c_2 e^{-\sqrt{h}z} + c_3)^{-1} \tag{7.14}$$

$$\Rightarrow q_z = \frac{-\sqrt{h}(c_1 e^{\sqrt{h}z} - c_2 e^{-\sqrt{h}z})}{(c_1 e^{\sqrt{h}z} + c_2 e^{-\sqrt{h}z} + c_3)^2} \tag{7.15}$$

Since $pq_z = k_0$,

$$p = \frac{-k_0(c_1 e^{\sqrt{h}z} + c_2 e^{-\sqrt{h}z} + c_3)^2}{\sqrt{h}(c_1 e^{\sqrt{h}z} - c_2 e^{-\sqrt{h}z})} \tag{7.16}$$

Using Eq. (7.4), the relationship between $x$ and $z$ can be calculated as

$$x = \frac{1}{k_0}\left(\frac{1}{c_1 e^{\sqrt{h}z} + c_2 e^{-\sqrt{h}z} + c_3} - \alpha\right) \tag{7.17}$$

where

$$\alpha = \frac{1}{c_1 + c_2 + c_3} \tag{7.18}$$

which implies

$$c_1 e^{\sqrt{h}z} + c_2 e^{-\sqrt{h}z} + c_3 = \frac{1}{k_0 x + \alpha} \tag{7.19}$$

the mass $m(x)$ and stiffness $f(x)$ functions, can be calculated by using $m = pq^{-2}$ and $f = p^{-3}q^{-2}$, and are given by

$$m(x) = \frac{p}{q^2} = -k_0\frac{(c_1 e^{\sqrt{h}z} + c_2 e^{-\sqrt{h}z} + c_3)^4}{\sqrt{h}(c_1 e^{\sqrt{h}z} - c_2 e^{-\sqrt{h}z})} \tag{7.20}$$

$$= -k_0\frac{(k_0 x + \alpha)^{-4}}{\sqrt{h}\left(\sqrt{(k_0 x + \alpha)^{-1} - c_3)^2 - 4c_1 c_2}\right)}$$

$$f(x) = \frac{1}{p^3 q^2} = \frac{(\sqrt{h}(c_1 e^{\sqrt{h}z} - c_2 e^{-\sqrt{h}z}))^3}{-k_0^3(c_1 e^{\sqrt{h}z} + c_2 e^{-\sqrt{h}z} + c_3)^4} \tag{7.21}$$

$$= \frac{(\sqrt{h}(\sqrt{(k_0 x + \alpha)^{-1} - c_3)^2 - 4c_1 c_2}))^3}{(-k_0^3(k_0 x + \alpha)^{-4})}$$

If $k_0 = -1, h = 1, c_1 = -c_2 = 1/2, c_3 = 1$, then $\alpha = 1, x_0 = 0.54$, and the expressions for $m(x)$ and $f(x)$ simplify to

$$m(x) = (1-x)^{-3}(2x^2 - 2x + 1)^{-1/2} \tag{7.22}$$

$$f(x) = \left((1-x)\sqrt{2x^2 - 2x + 1}\right)^3 \tag{7.23}$$

The $i^{th}$ mode shape of the non-uniform beam, $Y_i(x)$ can be calculated using the relation $Y_i(x) = qU_i(z)$, where $U_i(z)$ is the $i^{th}$ mode shape of the axially loaded uniform beam, as shown below.

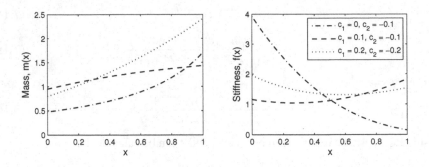

**Figure 7.3**
Mass and stiffness functions of a non-uniform beam isospectral to an axially loaded uniform beam (tensile load $h = 1$ and $pq_z = k_0 = -1$)

The $i^{th}$ mode-shape ($U_i(z)$) of a uniform axially loaded non-rotating beam, under an axial load $h$, is given as [30]:

$$U_i(z) = c_1 \sinh(\bar{\alpha}z) + c_2 \cosh(\bar{\alpha}z) + c_3 \sin(\bar{\beta}z) + c_4 \cos(\bar{\beta}z) \tag{7.24}$$

where

$$\bar{\alpha} = \sqrt{\frac{h}{2} + \sqrt{\eta_i^2 + \frac{h^2}{4}}} \tag{7.25}$$

$$\bar{\beta} = \sqrt{-\frac{h}{2} + \sqrt{\eta_i^2 + \frac{h^2}{4}}} \tag{7.26}$$

$$c_1 = 1 \tag{7.27}$$

$$c_2 = -\frac{\bar{\alpha}^2 \sinh(\bar{\alpha}) + \bar{\alpha}\bar{\beta} \sin(\bar{\beta})}{\bar{\alpha}^2 \cosh(\bar{\alpha}) + \bar{\beta}^2 \cos(\bar{\beta})} \tag{7.28}$$

$$c_3 = -\frac{\bar{\alpha}}{\bar{\beta}} \tag{7.29}$$

$$c_4 = \frac{\bar{\alpha}^2 \sinh(\bar{\alpha}) + \bar{\alpha}\bar{\beta} \sin(\bar{\beta})}{\bar{\alpha}^2 \cosh(\bar{\alpha}) + \bar{\beta}^2 \cos(\bar{\beta})} \tag{7.30}$$

and $\eta_i$ is the $i^{th}$ non-dimensional natural frequency. Therefore, substituting the derived $m(x), f(x)$ and $Y_i(x)$ functions in Eq. (7.1), we can verify the frequency equivalence of the non-uniform beam and the given axially loaded uniform beam.

In Fig. 7.3, the mass and stiffness functions of the non-uniform beams are shown, for different values of $c_1, c_2$, with $c_3$ chosen such that $x_0 = 1$, which are isospectral to a uniform beam under an axial load $h = 1$.

**Compressive axial load, $h < 0$**
    Solving $A = -h$ for $h < 0$ yields

$$\hat{q} = c_1 \cos\left(\sqrt{-hz} + \phi\right) + c_2 \tag{7.31}$$

$$\Rightarrow q = \left(c_1 \cos\left(\sqrt{-h}z + \phi\right) + c_2\right)^{-1} \tag{7.32}$$

$$\Rightarrow q_z = \frac{\sqrt{-h}c_1 \sin\left(\sqrt{-h}z + \phi\right)}{\left(c_1 \cos\left(\sqrt{-h}z + \phi\right) + c_2\right)^2} \tag{7.33}$$

Using the relation $pq_z = k_0$, yields

$$p = k_0 \frac{\left(c_1 \cos\left(\sqrt{-h}z + \phi\right) + c_2\right)^2}{\sqrt{-h}(c_1 \sin\left(\sqrt{-h}z\right) + \phi)} \tag{7.34}$$

Using (Eq. (7.4)), $x$ is calculated as

$$x = \frac{1}{k_0} \left( \frac{1}{c_1 \cos\left(\sqrt{-h}z + \phi\right) + c_2} - \alpha \right) \tag{7.35}$$

where

$$\alpha = \frac{1}{c_1 \cos(\phi) + c_2} \tag{7.36}$$

which implies

$$c_1 \cos\left(\sqrt{-h}z + \phi\right) + c_2 = \frac{1}{k_0 x + \alpha} \tag{7.37}$$

the mass $m(x)$ and stiffness $f(x)$ functions, can be calculated by using $m = pq^{-2}$ and $f = p^{-3}q^{-2}$,

$$m(x) = \frac{p}{q^2} = k_0 \frac{\left(c_1 \cos\left(\sqrt{-h}z + \phi\right) + c_2\right)^4}{\sqrt{-h}(c_1 \sin\left(\sqrt{-h}z\right) + \phi)} \tag{7.38}$$

$$= k_0 \frac{(k_0 x + \alpha)^{-4}}{\sqrt{-h}(\sqrt{c_1^2 - ((k_0 x + \alpha)^{-1} - c_2)^2})}$$

$$f(x) = \frac{1}{p^3 q^2} = \frac{\left(\sqrt{-h}(c_1 \sin\left(\sqrt{-h}z\right)) + \phi)\right)^3}{k_0^3 (c_1 \cos\left(\sqrt{-h}z + \phi\right) + c_2)^4} \tag{7.39}$$

$$= \frac{\left(\sqrt{-h}(\sqrt{c_1^2 - ((k_0 x + \alpha)^{-1} - c_2)^2})\right)^3}{(k_0^3 ((k_0 x + \alpha)^{-4}))}$$

If $k_0 = -1, h = -1, c_2 = 1, c_1 = -1, \phi = \pi/2$ then $\alpha = 1, x_0 = 0.46$, and the mass and stiffness functions simplify to

$$m(x) = (1-x)^{-3}\sqrt{1-2x} \tag{7.40}$$

$$f(x) = \left(\frac{1-x}{\sqrt{1-2x}}\right)^3 \tag{7.41}$$

In Fig. 7.4, the mass and stiffness functions of beams are shown, for different values of $c_1, c_2$, with $c_3$ chosen such that $x_0 = 1$, which are isospectral to a uniform beam under an axial load $h = -1$.

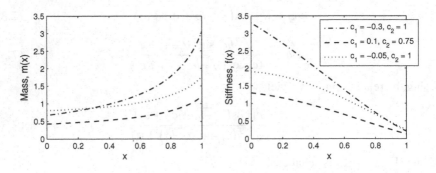

**Figure 7.4**
Mass and stiffness functions of a non-uniform beam isospectral to an axially loaded
uniform beam (compressive load $h = -1$, $pq_z = k_0 = -1$ and $c_3 = 1$)

### 7.1.1.2   Case-2: $q = q_0$, a constant

When $q = q_0$, then $B = 0$ (Section 3(c) in [86]) is automatically satisfied. Simplifying
$A = -h$ yields

$$-h = \frac{p_{zz}}{p} - \frac{2p_z^2}{p^2} \tag{7.42}$$

which simplifies to

$$\frac{d^2}{dz^2}\left[\frac{1}{p}\right] = \frac{h}{p} \tag{7.43}$$

**Axial load is tensile** ($h > 0$) Solving Eq. (7.43) for $h > 0$ yields

$$p(z) = \frac{1}{k_1 e^{\sqrt{h}z} + k_2 e^{-\sqrt{h}z}} \tag{7.44}$$

using Eq. (7.4), $x$ is calculated as

$$x = \frac{k_1 e^{\sqrt{h}z} - k_2 e^{-\sqrt{h}z} + k_2 - k_1}{\sqrt{h}} \tag{7.45}$$

$$\Rightarrow k_1 e^{\sqrt{h}z} - k_2 e^{-\sqrt{h}z} = \sqrt{h}x + k_1 - k_2 \tag{7.46}$$

Finally, the mass $m(x)$ and stiffness $f(x)$ functions calculated using $m = pq^{-2}$ and
$f = p^{-3}q^{-2}$, are given as

$$m(x) = \frac{q_0^{-2}}{k_1 e^{\sqrt{h}z} + k_2 e^{-\sqrt{h}z}} \tag{7.47}$$

$$= \frac{q_0^{-2}}{\sqrt{(\sqrt{h}x + k_1 - k_2)^2 + 4k_1 k_2}}$$

$$f(x) = q_0^{-2}(k_1 e^{\sqrt{h}z} + k_2 e^{-\sqrt{h}z})^3 \tag{7.48}$$

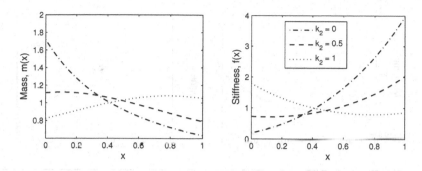

**Figure 7.5**
Mass and stiffness functions of a non-uniform beam isospectral to an axially loaded uniform beam (tensile load $h = 1$), and $q(z) = q_0 = 1$

$$= q_0^{-2}(\sqrt{(\sqrt{h}x + k_1 - k_2)^2 + 4k_1 k_2})^3$$

If $q_0 = 1, h = 1, k_1 = k_2 = 1/2$ then the expressions for $m(x)$ and $f(x)$ become

$$m(x) = (1 + x^2)^{-1/2} \tag{7.49}$$

$$f(x) = (1 + x^2)^{3/2} \tag{7.50}$$

If $x_0 = 1$, then the relationship between $k_1$ and $k_2$ is calculated by substituting $x = x_0 = 1$ and $z = z_0 = 1$ in Eq. (7.46), and is given by

$$k_1 = \frac{\sqrt{h} + k_2(e^{-\sqrt{h}} - 1)}{e^{\sqrt{h}} - 1} \tag{7.51}$$

In Fig. 7.5, the mass and stiffness functions of beam are plotted, for different values of $k_2$, such that $x_0 = 1$, $q_0 = 1$ which are isospectral to a uniform beam under an axial load $h = 1$.

**Axial load is compressive** ($h < 0$) Solving Eq. (7.43) for $h < 0$ yields

$$p(z) = \frac{1}{k_1 \cos(\sqrt{-h}z + \phi)} \tag{7.52}$$

which implies from Eq. (7.4)

$$x = \frac{k_1 \sin(\sqrt{-h}z + \phi) - k_1 \sin(\phi)}{\sqrt{-h}} \tag{7.53}$$

$$\Rightarrow k_1 \sin(\sqrt{-h}z + \phi) = \sqrt{-h}x + k_1 \sin(\phi) \tag{7.54}$$

Finally, the mass $m(x)$ and stiffness $f(x)$ functions, calculated using $m = pq^{-2}$ and $f = p^{-3}q^{-2}$, are obtained as

$$m(x) = pq_0^{-2} = q_0^{-2}\left(\frac{1}{k_1 \cos(\sqrt{-h}z + \phi)}\right) \tag{7.55}$$

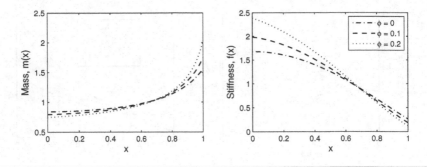

**Figure 7.6**
Mass and stiffness functions of a non-uniform beam isospectral to an axially loaded
uniform beam (compressive load $h = -1$) and $q(z) = q_0 = 1$.

$$= q_0^{-2} \left( k_1^2 - (\sqrt{-h}x + k_1 \sin(\phi))^2 \right)^{-1/2}$$

$$f(x) = q_0^{-2} \left( k_1 \cos \left( \sqrt{-h}z + \phi \right) \right)^3 \tag{7.56}$$

$$= q_0^{-2} \left( k_1^2 - (\sqrt{-h}x + k_1 \sin(\phi))^2 \right)^{3/2}$$

If $q_0 = 1, h = -1, \phi = 0, k_1 = 1$, then the expressions for $m(x)$ and $f(x)$ reduces to

$$m(x) = (1 - x^2)^{-1/2} \tag{7.57}$$

$$f(x) = (1 - x^2)^{3/2} \tag{7.58}$$

In Fig. 7.6, we plot the mass and stiffness functions of beams for different values
of $\phi$; $k_1$ is chosen such that $x_0 = 1$, which are isospectral to a uniform beam under
an axial load $h = -1$.

### 7.1.1.3 Rectangular beams

In this section, the analysis in the preceding section is applied to beams with a rectangular cross-section. The non-dimensional breadth ($\hat{b}$) and height ($\hat{h}$) profiles of the
cross sections are related to the mass and stiffness of the beams by the following
relation.

$$m = \hat{b}\hat{h} \quad ; \quad f = \hat{b}\hat{h}^3 \tag{7.59}$$

$$\Rightarrow \hat{b} = \sqrt{\frac{m^3}{f}} \quad ; \quad \hat{h} = \sqrt{\frac{f}{m}} \tag{7.60}$$

Therefore, using the $m(x)$ and $f(x)$ functions derived in the previous section, $\hat{b}$ and
$\hat{h}$ profiles of the rectangular beams can be derived. These $\hat{b}$ and $\hat{h}$ profiles of non-
uniform beams, which are isospectral to an axially loaded uniform beam, for different
axial loads, are shown in Fig. 7.7. These results indicate the tension or compression
in the axially loaded uniform beam is physically manifested in the non-uniform beam

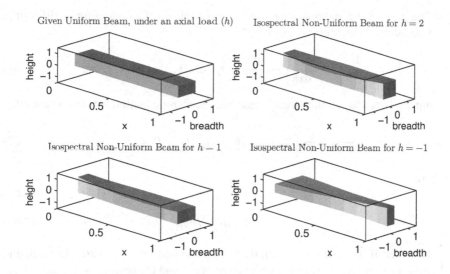

**Figure 7.7**
Breadth and height distributions of non-uniform beams, isospectral to an axially loaded uniform beam for different axial loads

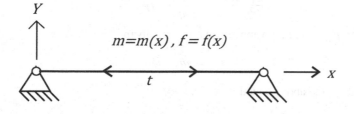

**Figure 7.8**
A schematic of a non-uniform hinged-hinged stiff-string under tension

for the natural frequencies to remain identical. A practical application of this idea is that frequencies of uniform beams under axial loads could be *experimentally determined* by testing their unloaded non-uniform isospectral analogues.

### 7.1.2 OTHER BOUNDARY CONDITIONS: HINGED-HINGED PIANO STRINGS

In this section, stiff strings under hinged-hinged boundary conditions are analyzed. Piano strings are sometimes modeled as uniform stiff-strings under a tension, hinged at both ends. We seek to find non-uniform stiff strings under a tension $t$ (Fig. 7.8), which are isospectral to a given uniform piano string under a tension $h$ (Fig. 7.9). Note that piano strings are typically uniform with a tension. However, it would be interesting to study if the tension in the string can be adjusted by tailoring the mass and stiffness distributions. Such non-uniform stiff strings could be manufactured using CNC machining techniques.

The governing equation for the transverse vibrations of a non-uniform stiff-string, in a non-dimensional form is given by [37]

$$\frac{d^2}{dx^2}\left[f(x)\frac{d^2Y}{dx^2}\right] - t\frac{d^2Y}{dx^2} = \eta^2 m(x)Y, \quad 0 < x < x_0 \tag{7.61}$$

Using the Barcilon-Gottlieb transformation, the above equation is transformed into

$$\frac{d^4U}{dz^4} + \frac{d}{dz}\left[\tilde{A}\frac{dU}{dz}\right] + \tilde{B}U = \eta^2 U \quad , \quad 0 < z < z_0 \tag{7.62}$$

where

$$\tilde{A} = A - tpq^2 \tag{7.63}$$

$$\tilde{B} = B - q\frac{d}{dz}[tpq_z] \tag{7.64}$$

Note that if $\tilde{A} = -h$ and $\tilde{B} = 0$, then Eq. (7.6) and Eq. (7.61) are identical, and then the non-uniform stiff string under a tension $t$, and the uniform stiff piano string under a tension $h$, are isospectral to each other.

In this section , we solve $\tilde{A} = -h$ and $\tilde{B} = 0$, for the case where $pq_z = k_0$, a constant. When $pq_z = k_0$, then the equations $\tilde{B} = 0$ is already satisfied, and simplifying $\tilde{A} = -h$ yields

$$-\frac{6q_z^2}{q^2} + \frac{6q_{zz}}{q} - \frac{q_{zzz}}{q_z} - tpq^2 = -h \tag{7.65}$$

Substituting $pq_z = k_0$ in the above equation, results in

$$\hat{q}_{zzz} = h\hat{q}_z + tk_0 \tag{7.66}$$

The solution to the above equation is given by

$$\hat{q} = c_1 e^{\sqrt{h}z} + c_2 e^{-\sqrt{h}z} + c_3 - tk_0 z/h \tag{7.67}$$

where $c_1, c_2, c_3$, and $c_4$ are the constants of integration. Choosing $c_1 = c_2 = 0$, yields

$$q = 1/\hat{q} = \frac{1}{c_3 - tk_0 z/h} \tag{7.68}$$

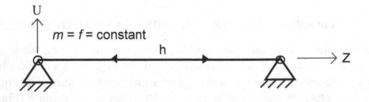

**Figure 7.9**
A schematic of a uniform hinged-hinged piano string under tension

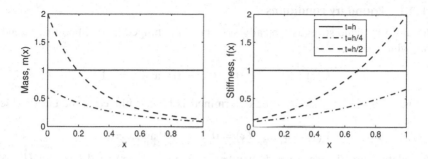

**Figure 7.10**
Mass and stiffness functions of non-uniform stiff strings under a tension ($t$), isospectral to the given piano string under a tension $h$

$$\Rightarrow q_z = \frac{tk_0}{h(c_3 - tk_0z/h)^2} \tag{7.69}$$

Since $pq_z = k_0$, we get

$$p = \frac{h(c_3 - k_0z/h)^2}{t} \tag{7.70}$$

$x$ can be calculated from Eq. (7.4), and is given by

$$x = \frac{1}{k_0}\left(\frac{1}{c_3 - k_0tz/h} - \frac{1}{c_3}\right) \tag{7.71}$$

which implies

$$k_0x + \frac{1}{c_3} = \left(c_3 - \frac{tk_0z}{h}\right)^{-1} \tag{7.72}$$

The mass and stiffness functions are given as:

$$m(x) = pq^{-2} = \frac{h}{t\,(k_0x + 1/c_3)^4} \tag{7.73}$$

$$f(x) = p^{-3}q^{-2} = \frac{t^3\,(k_0x + 1/c_3)}{h^3} \tag{7.74}$$

These mass and the stiffness functions are plotted in Fig. 7.10, for $t = h/4$ and $t = h/2$. The results shown in Fig. 7.10 show that that the tension required in a stiff-string can be adjusted by changing the mass and stiffness functions of the stiff-string appropriately, in order to retain its natural frequencies. Since the natural frequencies are the source of the piano notes, the musical characteristics of the instrument remains unaltered. For a typical uniform stiff piano string ($t = h$), non-uniform stiff-strings with the same frequency with $t = h/4$ and $t = h/2$ can be constructed, as shown in Fig. 7.10.

### 7.1.2.1 Boundary conditions

The uniform piano string is typically subject to the hinged-hinged boundary condition, which is given as

$$U(z) = U''(z) = 0 \quad \text{at} \quad z = 0 \quad \text{and} \quad z = 1 \tag{7.75}$$

The non-uniform stiff-string to be determined is hinged-hinged at the ends, which implies

$$Y(x) = 0 \; ; \; Y''(x) = 0 \text{ at } x = 0 \text{ and } x = x_0 \tag{7.76}$$

The relationships between the derivatives of $Y(Y'(x), Y''(x))$ and $U(U'(z), U''(z))$, can be obtained by differentiating Eq. (7.5), and are given by

$$Y'(x) = \frac{dY}{dx} = p\frac{d(qU)}{dz} = pq_zU + pqU_z \tag{7.77}$$

$$Y''(x) = p((p_zq_z + pq_{zz})U + (2pq_z + p_zq)U_z + pqU_{zz}) \tag{7.78}$$

Substituting the above expressions of $Y'(x)$ and $Y''(x)$ in Eq. (7.76), yields

$$p((p_zq_z + pq_{zz})U + (2pq_z + p_zq)U_z + pqU_{zz}) = \quad \text{at } z = 0 \text{ and } z = 1 \tag{7.79}$$

Since $U(z) = U''(z) = 0$ at $z = 0$ and $z = 1$, the above equations reduce to

$$(2pq_z + p_zq)U_z = 0 \quad \text{at } z = 0 \text{ and } z = 1 \tag{7.80}$$

$$\Rightarrow 2pq_z + p_zq = 0 \text{ at } z = 0 \text{ and } z = 1 \tag{7.81}$$

Therefore, by choosing $c_1 = c_2 = 0$, Eq. (7.81) is satisfied at $z = 0$ and $z = 1$. Therefore, the mass and stiffness functions presented in Equations (7.73) and (7.74), satisfy the boundary conditions in the Barcilon-Gottlieb transformation.

### 7.1.2.2 Tensionless piano strings

In this section, we endeavor to find non-uniform tensionless strings, which are isospectral to a given uniform piano string. The governing equations for the transverse vibrations of hinged-hinged stiff-strings is the same as that of beams (discussed in the previous section), except for the boundary conditions. The non-uniform stiff-string to be determined is hinged-hinged and has elastic rotational springs at the hinges (Fig. 7.11):

$$Y(x) = 0 \; ; \; Y''(x) = \tilde{k}_1Y'(x) \text{ at } x = 0 \tag{7.82}$$

$$Y(x) = 0 \; ; \; Y''(x) + \tilde{k}_2Y'(x) = 0 \text{ at } x = x_0 \tag{7.83}$$

here $\tilde{k}_1$ and $\tilde{k}_1$ are the non-dimensional spring constants of the hinge springs at $x = 0$ and $x = x_0$, respectively. Substituting Eq. (7.77) and Eq. (7.78) in Eqs. (7.82) and (7.83), results in

$$p((p_zq_z + pq_{zz})U + (2pq_z + p_zq)U_z + pqU_{zz}) = \tilde{k}_1(pq_zU + pqU_z) \text{ at } z = 0 \tag{7.84}$$

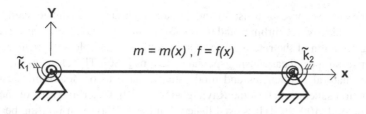

**Figure 7.11**
A schematic of a hinged-hinged non-uniform tensionless stiff-string with hinge-springs

$$p\left((p_z q_z + pq_{zz})U + (2pq_z + p_z q)U_z + pqU_{zz}\right) + \tilde{k}_2(pq_z U + pqU_z) = 0 \ \text{ at } z = 1$$
(7.85)

Since $U(z) = U''(z) = 0$ at $z = 0$ and $z = 1$, the above equations become

$$(2pq_z + p_z q - \tilde{k}_1 q)U_z = 0 \quad \text{at } z = 0$$
(7.86)

$$\Rightarrow \tilde{k}_1 = \frac{2pq_z + p_z q}{q} \ \text{ at } z = 0$$
(7.87)

Similarly, we obtain

$$(2pq_z + p_z q + \tilde{k}_2 q)U_z = 0 \quad \text{at } z = 1$$
(7.88)

$$\Rightarrow \tilde{k}_2 = -\frac{2pq_z + p_z q}{q} \ \text{ at } z = 1$$
(7.89)

Therefore, using the above values of the hinge-spring constants $(\tilde{k}_1, \tilde{k}_2)$, tensionless stiff-strings which are isospectral to a given piano string can be determined. For example, for the mass and stiffness functions of a non-uniform stiff-string derived in Eqs. (7.49) and (7.50), we can calculate the values of spring constants as $\tilde{k}_1 = 0$ and $\tilde{k}_2 = 0.49$. Therefore, the mass and stiffness functions of the non-uniform stiff-string, when combined with the hinge-springs, has the same natural frequencies as that of the given piano string.

## 7.2 ISOPECTRAL NON-UNIFORM ROTATING BEAM AND UNIFORM NON-ROTATING BEAM

In general, more than one frequency spectrum is required for a reconstruction procedure to determine the material properties of the system [138]. The isospectral systems corresponding to a vibrating string with continuous coefficients (second order governing equation) were studied by Borg [32] and with discontinuous coefficients were studied by Gottlieb [88, 89]. Gottlieb also analyzed the non-rotating beam equation and gave seven classes of non-uniform beams isospectral to the uniform beam with different boundary conditions [86]. Subramanian and Raman generalized Gottlieb's method for tapered beams [188]. However, none of these studies addressed rotating beams, which we shall do in this section.

Rotating beams serve as a useful mathematical model to simulate vibration of helicopter blades, wind turbine blades, turbo-machinery blades etc. However, the governing equation of the rotating beam does not have a simple analytical solution; except for some special cases using special functions [186]. Therefore, as mentioned before, the natural frequencies and mode shapes have to be determined using an approximation scheme such as the Rayleigh-Ritz [220], Galerkin [126] or the finite element method [100, 199]. It is established that the uniform non-rotating beam has an analytical solution for the frequencies and mode shapes. Hence, it is interesting to verify if we can find any rotating beam isospectral to the uniform non-rotating beam. This isospectral configuration would give us insight into countering the stiffening effect of the centrifugal force acting on the rotating beam. Such isospectral rotating beams can also be used to check the implementation of finite element codes as putting the mass and stiffness distributions of the isospectral rotating beams in such codes will yield the well known frequencies and mode shapes of a uniform non-rotating beam.

In this section, we discover non-uniform rotating beams (NURB) with continuous mass and flexural stiffness distributions that are isospectral to a uniform non-rotating beam (UNRB) of the same length under the same boundary condition. Barcilon has provided a transformation that converts the fourth order governing equation of a non-rotating beam to a canonical fourth order eigenvalue problem [17]. This transformation can also be used to convert the fourth-order governing equation of a NURB to another canonical fourth-order eigenvalue problem. If some of the coefficients vanish, then the canonical equation simplifies to the governing equation of the UNRB. This vanishing condition on the coefficients leads to a pair of coupled differential equations. We solve these equations for a particular case and thereby obtain a class of analytically solvable non-uniform rotating beams. The boundary conditions considered are clamped-free and hinged-free with an elastic hinge spring. This is because, generally a rotating cantilever beam serves as a useful mathematical model to analyze the vibrations of a hingeless rotor, while articulated rotors are modeled using a rotating hinged-free beam with a hinge spring. Most newer helicopters have hingeless blades while articulated blades constitute a more classical design. However, to obtain isospectral systems, the transformation must leave the boundary conditions invariant. We prove that the clamped boundary condition is always invariant. However, we need to impose certain specific conditions on the beam characteristics for the hinged and free boundary conditions to be invariant. We will also confirm the frequencies of the NURB obtained using the finite element method (FEM) with that of the exact frequencies of the UNRB.

### 7.2.1 MATHEMATICAL FORMULATION

The governing equation for the out-of-plane vibrations $W(X,T)$ of a rotating Euler-Bernoulli beam shown in (Fig. 7.12) is given by [104]

$$\frac{\partial^2}{\partial X^2}\left[EI(X)\frac{\partial^2 W}{\partial X^2}\right] - \frac{\partial}{\partial X}\left[S(X)\frac{\partial W}{\partial X}\right] + M(X)\ddot{W} = 0 \quad , \quad 0 \le X \le L \quad (7.90)$$

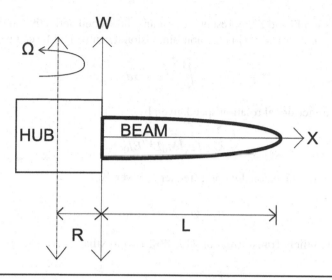

**Figure 7.12**
Schematic of a rotating beam

where $EI(X)$ is the flexural stiffness, $M(X)$ is the mass/length, centrifugal force $S(X) = \int_X^L M(X)\Omega^2(X+R)dX$, $R$ is the hub radius, $L$ is the length of the beam, $\Omega$ is the rotation speed and $T$ is the time variable. Here, $\ddot{W}$ is the second time derivative of $W$. In this section, we assume $R = 0$. Using separation of variables for the displacement $W$,

$$W(X,T) = Y(X)e^{i\omega T} \qquad (7.91)$$

Substituting Eq. (7.91) in Eq. (7.90), yields

$$\frac{d^2}{dX^2}\left[EI(X)\frac{d^2Y}{dX^2}\right] - \frac{d}{dX}\left[S(X)\frac{dY}{dX}\right] - \omega^2 M(X)Y = 0 \ , \ 0 \le X \le L \qquad (7.92)$$

For a given uniform non-rotating beam, the rotation speed $\Omega = 0$, flexural stiffness $EI(X) = EI_0$ and mass/length $M(X) = M_0$ and hence, the above equation leads to

$$EI_0\frac{d^4Y}{dX^4} - \omega^2 M_0 Y = 0 \qquad (7.93)$$

which is the governing differential equation for the vibrations of a non-rotating uniform beam with natural frequency $\omega$.

Now, non-dimensional variables $f$, $m$ and $x$ are introduced as

$$f(x) = \frac{EI(X)}{EI_0}, \quad m(x) = \frac{M(X)}{M_0}, \quad x = \frac{X}{L}, \qquad (7.94)$$

Eq. (7.92) is rewritten as

$$(f(x)Y'')'' - (t(x)Y')' = \eta^2 m(x)Y, \ 0 \le x \le 1 \qquad (7.95)$$

Here, the notation $Y'$ and $Y''$ represent the first and the second derivatives respectively of $Y$ with respect to $x$ and $t(x)$ is the non-dimensional centrifugal force given by

$$t(x) = \int_x^1 m(x)\lambda^2 x\, dx, \tag{7.96}$$

$\lambda$ is the non-dimensional rotation speed given by

$$\lambda = \Omega \sqrt{M_0 L^4 / E I_0} \tag{7.97}$$

and $\eta$ is the non-dimensional natural frequency given by

$$\eta = \omega \sqrt{M_0 L^4 / E I_0} \tag{7.98}$$

In Barcilon-Gottlieb Transformation [17, 86], two auxiliary variables $p$ and $q$ are defined as

$$p = \left(\frac{m}{f}\right)^{\frac{1}{4}} \; ; \; q = \left(\frac{1}{m^3 f}\right)^{\frac{1}{8}} \tag{7.99}$$

which also mandates that

$$f = p^{-3} q^{-2} \; ; \; m = pq^{-2} \tag{7.100}$$

The transformation of the independent variable $(x)$ is given implicitly as

$$z = \int_0^x p(z(x))\, dx \tag{7.101}$$

which implies that

$$x = \int_0^z \frac{dz}{p(z)} \tag{7.102}$$

$$\frac{d}{dx} = p(z)\left(\frac{d}{dz}\right) \; \Rightarrow \; \frac{1}{p(z)} = \frac{dx}{dz} \tag{7.103}$$

Now note that
$$x = 0 \Leftrightarrow z = 0 \tag{7.104}$$

Here, the symbol "$\Leftrightarrow$" is the bi-implication symbol which implies that if $x = 0$ if and only if $z = 0$. Define $z_0$ such that

$$x = 1 \Leftrightarrow z = z_0 \tag{7.105}$$

The transformation of the dependent variable $Y$ is

$$Y(x) = q(z)U(z) \tag{7.106}$$

We observe that Eq. (7.102) and Eq. (7.106) transform the variables $x$ and $Y$ into $z$ and $U$, respectively. This completes the Barcilon-Gottlieb transformation. Using this transformation, we get

$$\frac{d^2}{dx^2}\left[f\frac{d^2Y}{dx^2}\right] - \eta^2 mY = pq^{-1}\left[\frac{d^4U}{dz^4} + \frac{d}{dz}\left[A\frac{dU}{dz}\right] + BU - \eta^2 U\right] \qquad (7.107)$$

where

$$A = 4\frac{q_{zz}}{q} - 6\frac{q_z^2}{q^2} - 2\frac{q_z p_z}{qp} + \frac{p_{zz}}{p} - 2\frac{p_z^2}{p^2}, \qquad (7.108)$$

$$B = \frac{q_{zzzz}}{q} - 4\frac{q_z q_{zzz}}{q^2} - 2\frac{q_{zz}^2}{q^2} + 6\frac{q_z^2 q_{zz}}{q^3} + \frac{q_z p_{zzz}}{qp} + \frac{q_{zz} p_{zz}}{qp} - 4\frac{q_z^2 p_{zz}}{q^2 p}$$
$$- 5\frac{q_z p_z p_{zz}}{qp^2} - 4\frac{q_z q_{zz} p_z}{q^2 p} + 6\frac{q_z^3 p_z}{q^3 p} + 6\frac{q_z^2 p_z^2}{q^2 p^2} + 4\frac{q_z p_z^3}{qp^3} - 2\frac{q_{zz} p_z^2}{qp^2} \qquad (7.109)$$

These expressions for $A$ and $B$ are originally given by Barcilon [17] and later corrected by Gottlieb [86].

Here, the notation $q(z)$ indicates the value of $q$ at the coordinate $z$. The first, second, third and the fourth derivatives, respectively of $q$ w.r.t $z$ are denoted by $q_z$, $q_{zz}$, $q_{zzz}$ and $q_{zzzz}$.

For a non-rotating beam, $t(x) = 0$. We are interested in a rotating beam and therefore we most also include $t(x)$ in Eq. (7.107). The original Barcilon-Gottlieb transformation converts the governing differential equation of a non-uniform non-rotating beam into that of a uniform non-rotating beam. We now modify this transformation so that the governing differential equation of a non-uniform rotating beam is transformed into that of a uniform non-rotating beam. For this purpose, we need to consider the centrifugal force term $t(x)$. Hence, we have

$$\frac{d}{dx}\left[t\frac{dY}{dx}\right] = p\frac{d}{dz}\left[tp\frac{d}{dz}[qU]\right]$$

$$= p\left\{tqp\left(\frac{d^2U}{dz^2}\right) + \left(tpq_z + \frac{d}{dz}[tpq]\right)\left(\frac{dU}{dz}\right) + \left(\frac{d}{dz}[tq_z p]\right)U\right\}$$

$$= pq^{-1}\left\{\frac{d}{dz}\left[tq^2 p\frac{dU}{dz}\right] + \left(q\frac{d}{dz}[tq_z p]\right)U\right\} \qquad (7.110)$$

Now, we include Eq. (7.110) in Eq. (7.107), yielding

$$\frac{d^2}{dx^2}\left[f\frac{d^2Y}{dx^2}\right] - \frac{d}{dx}\left[t\frac{dY}{dx}\right] - \eta^2 mY = pq^{-1}\left(\frac{d^4U}{dz^4} + \frac{d}{dz}\left[\tilde{A}\frac{dU}{dz}\right] + \tilde{B}U - \eta^2 U\right) \qquad (7.111)$$

where

$$\tilde{A} = A - tq^2 p \qquad (7.112)$$

$$\tilde{B} = B - q\frac{d}{dz}[tq_z p] \qquad (7.113)$$

If $\tilde{A}(z) = 0$ and $\tilde{B}(z) = 0$, then

$$\frac{d^2}{dx^2}\left[f\frac{d^2Y}{dx^2}\right] - \frac{d}{dx}\left[t\frac{dY}{dx}\right] = \eta^2 mY \ , \ \ 0 \leq x \leq 1 \tag{7.114}$$

$$\Leftrightarrow \ \frac{d^4U}{dz^4} = \eta^2 U \ , \ \ 0 \leq z \leq z_0 \tag{7.115}$$

Eq. (7.114) is the governing differential equation of the non-uniform rotating beam. The governing differential equation of the uniform non-rotating beam is Eq. (7.115). Hence, both the equations (Eq. (7.114) and Eq. (7.115)) have the same non-dimensional natural frequencies($\eta$). Thus, a non-uniform rotating beam isospectral to a given non-rotating uniform beam is obtained by finding the auxiliary functions $p(z)$ and $q(z)$ satisfying the following equations

$$\tilde{A}(z) = 0 \tag{7.116}$$
$$\tilde{B}(z) = 0 \tag{7.117}$$

The above equations are nonlinear fourth order coupled differential equations which are onerous to solve. However, for $q(z) = q_0$ which is a constant, it is possible to obtain the results analytically. $q(z) = q_0$ implies that the displacement function requires a constant multiplicative factor [86]. Furthermore, from Eq. (7.99), we can observe that if $q =$ constant, then $m^3 f$ is also a constant. This indicates that at any location along the beam, the product of mass cubed and stiffness at that location, is a constant i.e. the stiffness varies inversely with mass cubed.

When $q(z) = q_0$, we can observe that $\tilde{B}(z) = 0$ as the derivatives $q_z, q_{zz}, q_{zzz}$ and $q_{zzzz}$ vanish. When $\tilde{A}(z) = 0$, then from Eq. (7.112), we obtain

$$A = q_0^2 t p \tag{7.118}$$

Since $q_z = q_{zz} = 0$, the value of $A$ from Eq. (7.108) is deciphered as

$$A = -\frac{2p_z^2}{p^2} + \frac{p_{zz}}{p} \tag{7.119}$$

Substituting this value of $A$ in Eq. (7.118), results in

$$-\frac{2p_z^2}{p^2} + \frac{p_{zz}}{p} = q_0^2 t p \tag{7.120}$$

The above equation is divided throughout by $p$ and rewritten as

$$-\frac{d^2\hat{p}}{dz^2} = q_0^2 t \tag{7.121}$$

where $\hat{p}$ is defined using Eq. (7.103) as

$$\hat{p} = p^{-1} = \frac{dx}{dz} \tag{7.122}$$

Eq. (7.121) is differentiated w.r.t $z$ and written as

$$-\frac{d^3\hat{p}}{dz^3} = q_0^2\frac{dt}{dz} \tag{7.123}$$

We also differentiate $t(x)$ given in Eq. (7.96) w.r.t $x$ and obtain

$$\frac{dt}{dx} = -m\lambda^2 x \tag{7.124}$$

Since $(m = pq_0^{-2})$ and $(d/dx) = p(d/dz)$, the above equation can be written as

$$\frac{dt}{dx} = p\frac{dt}{dz} = -pq_0^{-2}\lambda^2 x \tag{7.125}$$

which indicates that

$$q_0^2\frac{dt}{dz} = -\lambda^2 x \tag{7.126}$$

From Eq. (7.126) and Eq. (7.123), we observe that

$$q_0^2\frac{dt}{dz} = -\lambda^2 x = -\frac{d^3\hat{p}}{dz^3} \tag{7.127}$$

Substituting $\hat{p}$ from Eq. (7.122) in the above equation, we get

$$\frac{d^4x}{dz^4} = \lambda^2 x \tag{7.128}$$

The solution for the above differential equation is available from textbooks such as [120] as

$$x(z) = k_1\cos(nz) + k_2\cosh(nz) + k_3\sin(nz) + k_4\sinh(nz), \ n = \sqrt{\lambda} \tag{7.129}$$

where $k_1, k_2, k_3, k_4$ are the constants of integration. Now, the transformation $x(z)$ has to satisfy the initial conditions of the differential equation and also the physical boundary conditions (clamped-free and hinged free with a root spring) considered in this section. Another condition we must enforce is that the beams that are determined should have realistic values of mass and stiffness i.e.

$$m = pq_0^{-2} > 0 \quad \text{if and only if} \quad p > 0 \tag{7.130}$$

$$\text{if and only if} \quad f = p^{-3}q_0^{-2} > 0 \tag{7.131}$$

We know from Eq. (7.103), $p > 0$ if and only if $x_z > 0$. This mandates that $x_z$ is strictly non-decreasing.

The initial conditions are provided in Eq. (7.104) and Eq. (7.105). Note that the other initial conditions imposed on the problem are obtained as follows. Recall that we are using Eq. (7.123) in our analysis which is obtained by differentiating Eq.

(7.121). But if we integrate back Eq. (7.123), we obtain Eq. (7.121) plus an integration constant. To eliminate this integration constant, we should impose the following initial condition:

$$t|_{x=1} = \left[\frac{d^2\hat{p}}{dz^2}\right]_{z=z_0} = \left[\frac{d^3x}{dz^3}\right]_{z=z_0} = 0 \qquad (7.132)$$

Similarly, we differentiated Eq. (7.96) to obtain Eq. (7.124) which indicates the need to impose the following initial condition:

$$t|_{x=1} = 0 \qquad (7.133)$$

It can be observed that equations Eq. (7.132) and Eq. (7.133) are identical.

Also, using $q(z) = q_0$ in Eq. (7.106) and using Eq. (7.103), we get the following

$$Y = q_0 U \qquad (7.134)$$

$$Y_x = q_0 p U_z \qquad (7.135)$$

$$Y_{xx} = q_0 p(p_z U_z + p U_{zz}) \qquad (7.136)$$

$$Y_{xxx} = q_0 p p_z(p_z U_z + p U_{zz}) + q_0 p^2(p_{zz} U_z + 2 p_z U_{zz} + p U_{zzz}) \qquad (7.137)$$

### 7.2.1.1 Case(i): A cantilever beam

In this section, we search for a rotating cantilever beam which is isospectral to a given uniform non-rotating cantilever beam of the same length and hence, $z_0 = 1$. The rotating and the non-rotating beams are clamped at the root ($x = 0 = z$) and free at the tip ($x = 1 = z$). Hence, we have the following boundary conditions. At the root: $Y = 0 = Y_x; U = 0 = U_z$, and at the tip: $Y_{xx} = 0 = Y_{xxx}; U_{zz} = 0 = U_{zzz}$. The exact solution of the mode shapes of the uniform non-rotating cantilever beam is provided in Appendix-A.

The clamped boundary condition is satisfied in the Barcilon-Gottlieb transformation. So we need to satisfy only the free boundary condition for the cantilever beam. This mandates that the coefficients of the term $U_z$ in Eq. (7.136) and Eq. (7.137) should be zero at $z = 1$. Therefore, in Eq. (7.137), $p_z$ and $p_{zz}$ at $z = 1$ should become zero. We know from Eq. (7.103) that $p = 1/x_z$ and therefore, we obtain

$$p_z = -x_{zz} x_z^{-2} \qquad (7.138)$$

$$p_{zz} = 2x_{zz}^2 x_z^{-3} - x_{zzz} x_z^{-2} \qquad (7.139)$$

We understand that $p > 0 \Rightarrow x_z > 0$. Moreover, $x_{zzz} = 0$ at $z = 1$ (Eq. (7.132)). Hence, if $x_{zz} = 0$ at $z = 1$, then $p_z = 0$ and $p_{zz} = 0$ at $z = 1$. Therefore, another initial condition $x(z)$ should satisfy along with the initial conditions Eq. (7.104), Eq. (7.105) and Eq. (7.132) is

$$x_{zz}(z = 1) = 0 \tag{7.140}$$

The solution given in Eq. (7.129) for the governing differential equation of the transformation (Eq. (7.128)) has four unknowns $(k_1, k_2, k_3, k_4)$, and has to satisfy the all the above initial conditions. These above mentioned conditions are four linear equations in $(k_1, k_2, k_3, k_4)$ and can be expressed as

$$\bar{P}\bar{K} = \bar{Q} \tag{7.141}$$

where $\bar{P}$, $\bar{K}$ and $\bar{Q}$ are matrices written as

$$\bar{P} = \begin{bmatrix} \cos(n) & \cosh(n) & \sin(n) & \sinh(n) \\ \sin(n) & \sinh(n) & -\cos(n) & \cosh(n) \\ -\cos(n) & \cosh(n) & -\sin(n) & \sinh(n) \\ \cos(0) & \cosh(0) & \sin(0) & \sinh(0) \end{bmatrix}, \quad \bar{K} = \begin{bmatrix} k_1 \\ k_2 \\ k_3 \\ k_4 \end{bmatrix}, \quad \bar{Q} = \begin{bmatrix} 1 \\ 0 \\ 0 \\ 0 \end{bmatrix} \tag{7.142}$$

Solving the matrix equation Eq. (7.142), for $\lambda = \sqrt{n} = 1$, we decipher that

$$\bar{K} = [k_1, k_2, k_3, k_4]^T = [-0.1645, 0.1645, 0.6998, 0.2095]^T \tag{7.143}$$

Substituting the above values of $k_1, k_2, k_3$ and $k_4$ in Eq. (7.129), yields

$$x(z) = -0.1645\cos(z) + 0.1645\cosh(z) + 0.6998\sin(z) + 0.2095\sinh(z) \tag{7.144}$$

Differentiating the above expression with respect to $z$, yields

$$x_z(z) = 0.1645\sin(z) + 0.1645\sinh(z) + 0.6998\cos(z) + 0.2095\cosh(z) \tag{7.145}$$

We understand from Eq. (7.103) that $p = 1/x_z$. Hence,

$$p(z) = (0.1645\sin(z) + 0.1645\sinh(z) + 0.6998\cos(z) + 0.2095\cosh(z))^{-1} \tag{7.146}$$

We also know that $m = pq_0^{-2}$, and $f = p^{-3}q_0^{-2}$, thus

$$m(z) = \left(q_0^2(0.1645\sin(z) + 0.1645\sinh(z) + 0.6998\cos(z) + 0.2095\cosh(z))\right)^{-1} \tag{7.147}$$

$$f(z) = q_0^{-2}\left((0.1645\sin(z) + 0.1645\sinh(z) + 0.6998\cos(z) + 0.2095\cosh(z))\right)^3 \tag{7.148}$$

Similarly, using Eq. (7.121), $t(z)$ is expressed as

$$t(z) = -(-0.1645\sin(z) + 0.1645\sinh(z) - 0.6998\cos(z) + 0.2095\cosh(z)) \tag{7.149}$$

We seek to find the distribution of mass $(m)$, stiffness $(f)$, in terms of $x$. To accomplish this, we must transform $m(z)$ and $f(z)$ to $m(x)$ and $f(x)$. It is not possible to find $m(x)$ and $f(x)$ from $m(z)$ and $f(z)$ analytically. Hence, we have to resort to a numerical scheme such as the Newton-Raphson iteration to calculate $z$ if given any $x$. The general iterative formula using Newton-Raphson method indexNewton-Raphson method to calculate $z^*$ for any given $x^*$ is

$$z_{i+1} = z_i - \left( \frac{x(z_i) - x^*}{x_z(z_i)} \right) \tag{7.150}$$

where $x(z_i)$ and $x_z(z_i)$ can be calculated using Eq. (7.144) and Eq. (7.145). The initial estimate needed for this iterative formula to start is taken as $x^*$ i.e. $z_1 = x^*$. Therefore, for any given $x$, we can calculate $z$ and from the above mentioned equations, $m(x), f(x), t(x)$ and $p(x)$ can be calculated. The corresponding graphs are provided in Fig. 7.13. Recall that $m(x)$ and $f(x)$ are the non-dimensional mass and stiffness distributions, respectively. We observe from Fig. 7.13 that the isospectral rotating beam has a higher mass near the root and a lower mass near the tip, relative to the non-rotating beam. Conversely, the rotating beam is less stiff near the root and more stiff near the tip, compared to the non-rotating beam. To negate the effect of the centrifugal stiffening, this functional adjustment of the mass and stiffness distributions for the rotating beam is sufficient.

In the above analysis we have used $\lambda = 1$. However, there is an upper limit for $\lambda$ which we discuss below. For any value of $\lambda$ within this limit, we can apply the above method and obtain $m(x)$ and $f(x)$.

**Limiting value of rotation speed $(\lambda)$**

We prove that for high values of the rotation speed $(\lambda)$, there cannot exist rotating beams which are isospectral to a given uniform non-rotating beam of the same length. In order to prove this, we make use of the fact that the lowest non-dimensional natural frequency $(\eta_1)$ of a rotating beam is always greater than or equal to the rotation speed $(\eta_1 \geq \lambda)$ (see Appendix-B).

For a given non-rotating uniform cantilever beam, the first frequency is $\bar{\eta}_1 = 3.516$. If the non-uniform rotating beam is rotating with a speed greater than this first frequency of the uniform non-rotating beam, then $\lambda > \bar{\eta}_1$. We have mentioned earlier that the first frequency of the rotating beam $(\eta_1)$ is always greater than or equal to the rotation speed. Hence, $\eta_1 \geq \lambda$. Therefore, combining the above mentioned conditions, we have $\eta_1 \geq \lambda > \bar{\eta}_1 \implies \eta_1 > \bar{\eta}_1$.

Hence, we can say that there cannot exist rotating cantilever beams isospectral to a given uniform non-rotating cantilever beam of the same length for $\lambda > \bar{\eta}_1$. We now prove that for $\lambda < \bar{\eta}$, there does exist a non-uniform rotating beam isospectral to a given uniform non-rotating beam if we analyze the problem in a different paradigm.

For a given $\lambda$, we can solve Eq. (7.141) and obtain $x(z)$. We appreciate that $x(z)$ should be non-decreasing. If $x(z)$ is non-decreasing that implies $x_z(z) > 0 \,\forall\, z \in [0, 1]$. If $x(z)$ is not non-decreasing (i.e $x_z(z) < 0$ for some $z \in [0, 1]$), then the resulting values of $f(x)$ and $m(x)$ would be negative which reveals that the obtained beam is non-realistic.

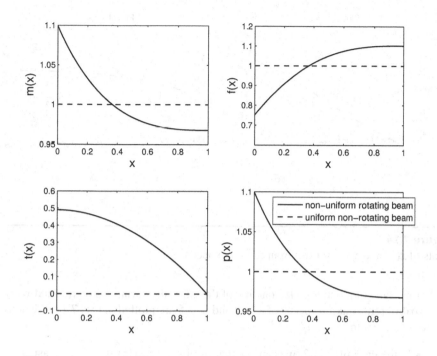

**Figure 7.13**
Variations of $m(x), f(x), t(x)$ and $p(x)$ with $x$ of isospectral cantilever beams

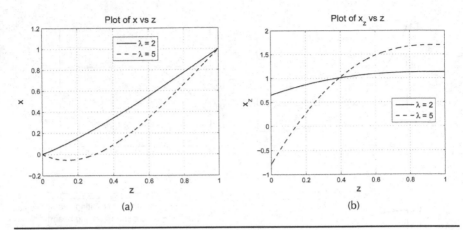

**Figure 7.14**
Plots of $x(z)$ and $x'(z)$ for different rotation speeds

For the sake of a better understanding of this idea, the variations of $x(z)$ and $x_z(z)$ for two different rotation speeds of $\lambda = 2$ and $\lambda = 5$ are plotted in Fig. 7.14. We note the following from Fig. 7.14.

- In the case of $\lambda = 2$, we can see that in Fig. 7.12a $x(z)$ is non decreasing and hence, the minimum value of $x_z$ (Fig. 7.14b) is positive (occurs at $z = 0$).
- In the case of $\lambda = 5$, we can see that in Fig. 7.12a $x(z)$ is decreasing initially and later increases and hence, the minimum value of $x_z$ (Fig. 7.14b) is negative (occurs at $z = 0$)

Therefore, $x(z)$ derived for the case of $\lambda = 2$ yields realistic values of $m(x)$ and $f(x)$, but, $x(z)$ derived for the case of $\lambda = 5$ leads to unrealistic values of $m(x)$ and $f(x)$. Therefore, the condition to be imposed on $\lambda$ for the existence of a rotating beam isospectral to a given uniform non-rotating beam of the same length is that the minimum value of $x_z$ for $z \in [0, 1]$ should be positive

We have found that the minimum value of $x_z$ is negative for $\lambda = 5$ and positive for $\lambda = 2$. Therefore, the minimum of $x(z)$ changes as $\lambda$ changes, i.e it is positive for some values of $\lambda$ and negative for other values. Therefore, we plot the minimum of $x_z$ for different $\lambda$ in Fig. 7.15. From Fig. 7.15, we decipher the following:

- for $\lambda < \lambda_0 = 3.516$, the minimum of $x_z$ is positive ($x(z)$ is non-decreasing)
- for $\lambda > \lambda_0 = 3.516$, the minimum of $x_z$ is negative ($x(z)$ is not non-decreasing)
- for $\lambda = \lambda_0 = 3.516$, the minimum of $x_z$ is zero .

Hence, we can appreciate that for $\lambda < \lambda_0$, there always exists a rotating beam isospectral to the uniform non-rotating beam of the same length while for $\lambda > \lambda_0$, this is impossible.

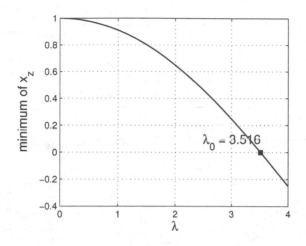

**Figure 7.15**
Plot of minimum of $x_z$ vs the rotation speed

For rotation speeds higher than $\lambda_0$, we can obtain rotating beams isospectral to a given uniform non-rotating beam of smaller length ($z_0 < 1$). We know that the first frequency of a unit coefficient($m(z) = f(z) = 1$) beam of length $z_0$ is given by $\bar{\eta}_1 = 3.516/z_0^2$. Hence, the rotation speed should be less than $\bar{\eta}_1$. Therefore,

$$\lambda < \left( \frac{3.516}{z_0^2} \right) \Leftrightarrow z_0 < \sqrt{\left( \frac{3.516}{\lambda} \right)} \qquad (7.151)$$

For example, if $\lambda = 4$, we can select any $z_0 < 0.9375$ and by solving Eq. (7.141), we can determine the beam characteristics $m(x), f(x), t(x)$ and $p(x)$. For $\lambda = 4$, we select three values of $z_0 = \{0.6, 0.7, 0.8\}$ with $q_0 = 1$, and the variations of $m(x), f(x), t(x)$ and $p(x)$ are shown in Fig. 7.16. From Fig. 7.16 we can decipher that as $z_0$ increases the stiffness of the beam decreases at the root and becomes zero for the threshold value of the reduced length ($z_0 = 0.9375$). Similarly, the mass/length becomes progressively steeper at the root as $z_0$ increases and becomes infinity for $z_0 = 0.9375$.

### 7.2.1.2 Case(ii): A hinged-free beam having an elastic hinge spring

In this section, we will find rotating hinged-free beams with a root spring isospectral to a given non rotating uniform hinged-free beam with a root spring of a non-dimensional spring constant $K_s$. The rotating and the non-rotating beams are hinged at the root ($x = 0 = z$) with a root spring and free at the tip ($x = 1 = z$). Hence, we obtain the following boundary conditions: hinged at the root with an elastic hinge spring: $Y = 0$ ; $K_s Y_x = f Y_{xx}; U = 0$ ; $K_s U_z = U_{zz}$, and free at the tip $Y_{xx} = 0 = Y_{xxx}; U_{zz} = 0 = U_{zzz}$. The exact solution of the mode shapes of the uniform hinged-free beam with a root spring is provided in Appendix-A.

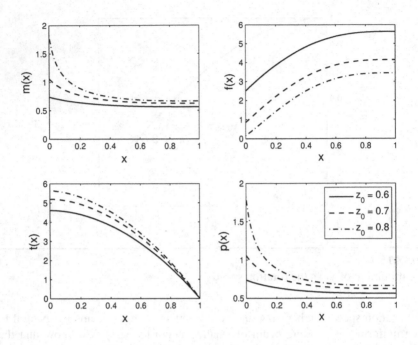

**Figure 7.16**
Variations of $m(x), f(x), t(x)$ and $p(x)$ of isospectral cantilever beams for $z_0 = 0.6, 0.7, 0.8$

Following the same methodology as in the case of a cantilever beam for satisfying the free boundary condition, the transformation, $x(z)$ should satisfy Eq. (7.140). For the boundary condition at the root to be satisfied, we must substitute $Y_x$ from Eq. (7.135) and $Y_{xx}$ from Eq. (7.136) in $K_s Y_x = f Y_{xx}$ which yields

$$(p_z - \frac{K_s}{f})U_z + pU_{zz} = 0 \quad at \; z = 0 \tag{7.152}$$

Combining $K_s U_z = U_{zz}$ and Eq. (7.152), results in

$$\frac{U_{zz}}{U_z} = K_s = \frac{K_s - f p_z}{f p} \quad at \; z = 0 \tag{7.153}$$

Substituting $f = p^{-3} q_0^{-2}$ in the above equation, yields

$$q_0^2 = \frac{1}{p^3}\left(p + \frac{p_z}{K_s}\right) \tag{7.154}$$

In this case, we have one more additional condition on $x(z)$ given in Eq. (7.154) along with the conditions mentioned in the case of a cantilever beam. We observe that a unique solution exists for the matrix equation Eq. (7.141), for $n = \sqrt{\lambda} = 1.25$, given by

$$[k_1, k_2, k_3, k_4] = [-0.2523, 0.2523, 0.6107, 0.0148] \tag{7.155}$$

Using these values of $k_1, k_2, k_3$ and $k_4$, we can decipher $x(z)$ using Eq. (7.129). Once $x(z)$ is known, $p$, $p_z$ can be determined using Eq. (7.138) and Eq. (7.139). We can observe that from Eq. (7.154), the value of $q_0$ can be determined. We obtain $q_0 = 0.5$ when $K_s = 1.7061$.

Since $x(z)$ is now known, $m(x), f(x), t(x)$ and $p(x)$ are determined in way similar to the case of a cantilever beam. The corresponding graphs are shown in Fig. 7.17. We can understand that the variations of the mass and stiffness are similar to the mass and stiffness functions derived in the case of a cantilever beam. Specifically, the mass distribution is more at the root and less at the tip while the stiffness distribution is more at the tip and less at the root, to negate the centrifugal stiffness.

Note that $K_s = \infty$ is the case of a cantilever beam and we have already discussed this case. Furthermore, $K_s = 0$ is the case of a hinged-free beam with no spring at the hinge. For this case, isospectral beams do not exist as the first frequency of the uniform non-rotating beam is zero while that of the rotating beam is the rotation speed itself.

We observe that there are many values of $\lambda$ resulting in real values of $q_0$ and hence, there exists a family of rotating beams with varying $(m(x), f(x), \lambda)$ isospectral to the non-rotating uniform beam under the hinged-free with a hinge spring boundary condition. Further discussion on the bounds which need to be imposed on $\lambda$ are presented in the next section.

**Limiting value of rotation speed** We can calculate the limiting speeds for the hinge-free beam with a root hinge in a manner that is similar to the case of a cantilever beam. For example, assume that the spring constant $K_s = 1.7061$. Now the exact

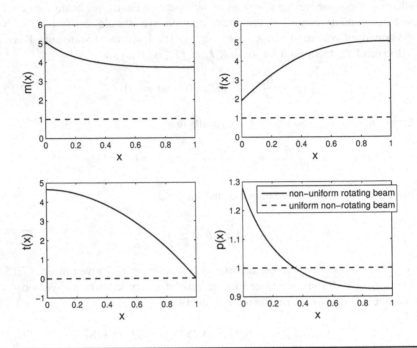

**Figure 7.17**
Variations of $m(x), f(x), t(x)$ and $p(x)$ of isospectral hinged-free beams with a root spring

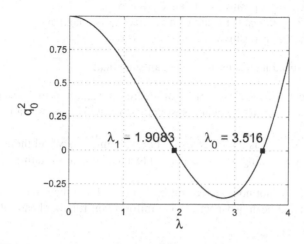

**Figure 7.18**
Plot of $q_0^2$ vs $\lambda$

solution of the first frequency of the uniform non-rotating beam of unit length is revealed as $\bar{\eta}_1 = 1.9083$ [118]. Moreover, the lowest frequency of a rotating beam is always greater than the rotation speed (see Appendix-B). Thus, the rotating speed $\lambda$ should be less than $\bar{\eta}_1$ for the existence of rotating beams isospectral to a uniform non-rotating beam of the same length.

We can prove the above mentioned statement in an alternate way. Note that for a given rotation speed, the transformation $x(z)$ for determining isospectral beams is the same for the rotating cantilever and rotating hinged-free beams with a root spring. Therefore, if we plot the minimum of $x_z$ against the rotation speed, for the hinged-free beam with a root spring, we obtain the same plot as shown in Fig. 7.15. From Fig. 7.15, we observe that:

- for $\lambda < \lambda_0 = 3.516$, the minimum of $x_z$ is positive ($x(z)$ is non-decreasing)
- for $\lambda > \lambda_0 = 3.516$, the minimum of $x_z$ is negative ($x(z)$ is not non-decreasing)

Therefore, for $\lambda > \lambda_0 = 3.516$, there exists no rotating beam isospectral to a non-rotating uniform beam *of the same length* under the hinged-free with a root spring boundary condition.

We have now established that for a given value of spring constant and rotation speed, there exists *only one rotating beam* isospectral to the non-rotating uniform beam for a specific value of $q_0^2$ given by Eq. (7.154). Note that the right hand side of Eq. (7.154) must be positive for real values of $q_0$. Now we endeavor to calculate the limit on the rotation speed by plotting the value of $q_0^2$ deciphered from Eq. (7.154) for different rotation speeds. This graph is shown in Fig. 7.18. From Fig. 7.18, we observe the following:

- For $\lambda < \lambda_1$, $q_0^2$ is positive and hence, $q_0$ is real.
- For $\lambda_1 \leq \lambda \leq \lambda_0$, $q_0^2$ is negative and hence, $q_0$ is imaginary.
- For $\lambda > \lambda_0$, $q_0^2$ is positive and hence, $q_0$ is real.

Based on the preceding discussions, we can state that:

i. For $\lambda < \lambda_1$, the transformation $x(z)$ is non-decreasing and $q_0$ is real, and therefore, there always exists a rotating beam isospectral to a uniform non-rotating beam of the same length.

ii. For $\lambda_1 \leq \lambda \leq \lambda_0$, $x(z)$ is non-decreasing but $q_0$ is imaginary and therefore, there cannot exist any a rotating beam isospectral to a uniform non-rotating beam of the same length.

iii. For $\lambda > \lambda_0$, $x(z)$ is not non-decreasing but $q_0$ is real and therefore, there cannot exist any a rotating beam isospectral to a uniform non-rotating beam of the same length.

Therefore, we can understand that only for $\lambda < \lambda_1$, there always exists a rotating beam isospectral to a uniform non-rotating beam of the same length.

For rotation speeds higher than $\lambda_1$, we can discover rotating beams isospectral to a uniform non-rotating beam of *smaller length* ($z_0 < 1$). For instance, let the length of the reduced uniform-non-rotation beam be $z_0 = 0.75$ and let $K_s = 1.7061$. The first frequency of this non-rotating uniform beam can be calculated as $\bar{\eta} = 3.051$. Thus, for any rotation speed $\lambda < \bar{\eta}$, we can determine $m(x), f(x), t(x)$ and $p(x)$ for $z \leq 0.75$. In Fig. 7.19, for $\lambda = 2.5$ we select three values of $z_0 = \{0.65, 0.7, 0.75\}$ and the variation of $m(x), f(x), t(x)$ and $p(x)$ is shown in Fig. 7.19. From Fig. 7.19, we can appreciate that as $z_0$ increases, the stiffness of the beam increases at the tip to almost six times the stiffness of the uniform beam for $z_0 = 0.75$ and eventually reaches infinity as the reduced length ($z_0$) approaches the threshold limit. Furthermore, $m(x)$ becomes progressively steeper at the root as $z_0$ increases and eventually encounters a singularity of $m(x)$ at the root as $z_0$ approaches the threshold limit. Note that these graphs are drawn for a fixed value of the spring constant and therefore, the value of $q_0$ changes as $z_0$ changes.

## 7.2.2 NUMERICAL SIMULATIONS

In the earlier section we established a method for obtaining non-uniform rotating beams that are isospectral to a given uniform non-rotating beam. In order to verify that these two beams have the same spectral properties, the finite element formulation is used for verification.

### 7.2.2.1 Finite element formulation

In the finite element formulation, the beam is partitioned into a number of finite elements. The displacement, $Y(x)$ is assumed to have a cubic distribution along each finite element and therefore, has 4-dof. Two of these 4-dof are displacement DOF and two are slope DOF at the ends of the finite element. Specifically, along the $i^{th}$

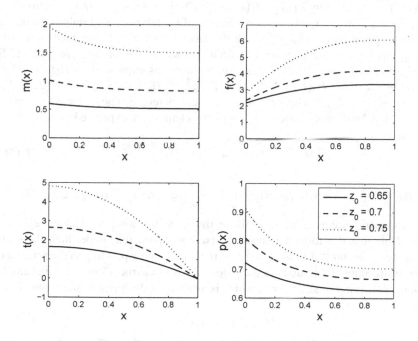

**Figure 7.19**
Variations of $m(x), f(x), t(x)$ and $p(x)$ of isospectral hinged-free beams with a root spring for $z_0 = 0.65, 0.7, 0.75$

element, $Y(x)$ is given by $Y(x) = H_1(x - x_i)q_{1,i} + H_2(x - x_i)q_{2,i} + H_3(x - x_i)q_{3,i} + H_4(x - x_i)q_{4,i}$ where $x_i$ and $x_{i+1}$ coordinates of the left node and right node of the element, respectively $q_{1,i}$ and $q_{2,i}$ are the displacement DOF and slope DOF at the left node, $q_{3,i}$ and $q_{4,i}$ are the displacement DOF and slope DOF at the right node [122] and $H_1, H_2, H_3, H_4$, are the Hermitian shape functions expressed as $H_1(\xi) = 1 - (3\xi^2)/l^2 + (2\xi^3)/l^3$; $H_2(\xi) = \xi - (2\xi^2)/l + \xi^3/l^2$; $H_3(\xi) = (3\xi^2)/l^2 - (2\xi^3)/l^3$; $H_4(\xi) = -\xi^2/l + \xi^3/l^2$ where $l = x_{i+1} - x_i$ element length. The element level mass matrix($\tilde{M}_i$) and stiffness matrices ($\tilde{K}_i$) of the $i^{th}$ element are expressed as

$$(\tilde{M}_{jk})_i = \int_0^l m(\xi + x_i)H_j(\xi)H_k(\xi)d\xi \tag{7.156}$$

$$(\tilde{K}_{jk})_i = \int_0^l f(\xi + x_i)H_j''(\xi)H_k''(\xi)d\xi + \int_0^l t(\xi + x_i)H_j'(\xi)H_k'(\xi)d\xi \tag{7.157}$$

Gauss quadrature is used to evaluate the integrals using 10 Gauss-Legendre points. Note that if a torsional spring is fixed at a particular node, then the stiffness of the spring must be added to the stiffness term corresponding to the rotational degree of freedom at that node in the global stiffness matrix. Consider the example of a torsional spring of stiffness $K_s$ which is attached at the hinge of the beam. Then

$$(\tilde{K'}_{jk})_i = (\tilde{K}_{jk})_i + K_s \tag{7.158}$$

where $(\tilde{K'}_{jk})_i$ is the stiffness term calculated including the stiffness of the spring, $(\tilde{K}_{jk})_i$ is the stiffness term calculated without including the spring, and $j = k = 2$ i.e. indices corresponding to the rotational degrees of freedom at the left node. These element level mass and stiffness matrices are assembled appropriately to obtain global mass ($\hat{M}$) and stiffness ($\hat{K}$) matrices. The natural frequencies ($\eta$) and mode shapes ($V$) are then obtained by solving the general eigenvalue problem.

$$\hat{K}V = \eta^2 \hat{M}V \tag{7.159}$$

### 7.2.2.2   Results of FEM

The number of finite elements to be used in FEM is an important factor in convergence and optimizing computational time. We have used $k = 150$ elements in the FEM analysis for Case(i) and Case(ii), a number which is sufficiently large to gaurantee convergence. The convergence graphs of the frequencies versus the number of finite elements used in the finite element formulation is provided in Fig. 7.20 for the cantilever beam and in Fig. 7.21 for the hinged-free beam with a root spring. The value of $(\eta_k/\eta_{exact})$ is plotted against the total number of finite elements for the first eight modes. Here $\eta_k$ is the frequency calculated using $k$ finite elements and $\eta_{exact}$ is the exact frequency. We can observe that all the modes start converging from around $k = 70$ elements onwards. Ergo, we have used $k = 150$ elements in the FEM analysis.

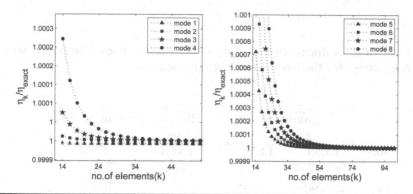

**Figure 7.20**
Convergence plots of $\eta_k/\eta_{exact}$ vs number of finite elements for the cantilever beam

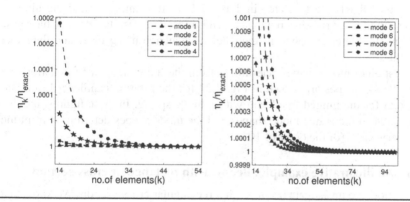

**Figure 7.21**
Convergence plots $\eta_k/\eta_{exact}$ vs number of finite elements for the hinged-free beam
with a root spring

**Table 7.1**

**Exact Analytical Uniform Non-rotating Beam Frequencies Compared with FEM Frequencies for the Isospectral Rotating Beam**

| Mode | Cantilever | | Hinged-free with a Root Spring | |
|:---:|:---:|:---:|:---:|:---:|
| | Exact | FEM | Exact | FEM |
| 1 | 3.5160 | 3.5160 | 1.9083 | 1.9083 |
| 2 | 22.0345 | 22.0345 | 16.7253 | 16.7253 |
| 3 | 61.6972 | 61.6972 | 51.4760 | 51.4760 |
| 4 | 120.9019 | 120.9019 | 105.8141 | 105.8141 |
| 5 | 199.8595 | 199.8595 | 179.8683 | 179.8684 |
| 6 | 298.5555 | 298.5556 | 273.6497 | 273.6497 |
| 7 | 416.9908 | 416.9908 | 387.1640 | 387.1640 |
| 8 | 555.1652 | 555.1654 | 520.4137 | 520.4138 |

The exact values of the frequencies of the UNRB [118] and the frequencies (using FEM) of the NURB are tabulated in Table 7.1 for the cantilever and the hinged-free beam with a root spring boundary conditions. The results in this table show a nearly negligible error between frequencies calculated using FEM and the exact frequencies.

The first eight mode shapes determined from the finite element formulation and the exact mode shapes are shown in Fig. 7.22 for the rotating cantilever beam and in Fig. 7.23 for the hinged-free beam with a hinge-spring. In these figures, the difference between the actual mode shapes and the mode shapes determined from the FEM is very small (of the order of $10^{-7}$).

### 7.2.2.3   An illustrative example: Beams with rectangular cross-section

In this section, we investigate beams with a rectangular cross-section. We know that the mass/length $(M_0)$ and flexural stiffness $EI_0$ of the given UNRB is given by:

$$M_0 = \rho_0 B_0 H_0 \tag{7.160}$$

$$EI_0 = E_0 \frac{B_0 H_0^3}{12} \tag{7.161}$$

where $\rho_0$ is the density of the beam, $E_0$ is the tensile modulus of the beam, $B_0$ and $H_0$ are the breadth and height of the cross-section, respectively. Similarly, assuming that

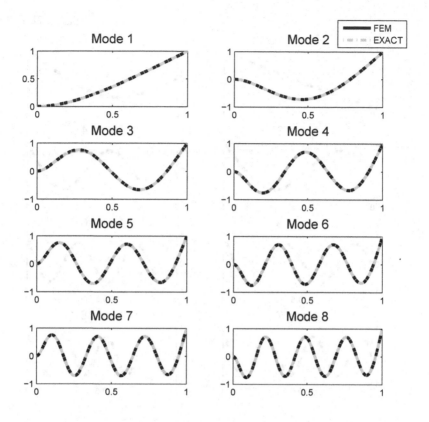

**Figure 7.22**
First eight mode shapes of the rotating cantilever beam (Exact and FEM)

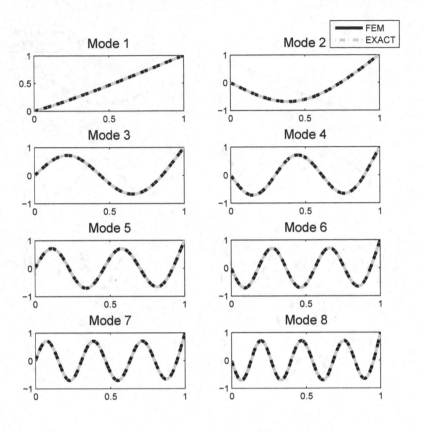

**Figure 7.23**
First eight mode shapes of the rotating spring hinge-free beam (Exact and FEM)

the isospectral NURB are of the same material, the mass/length $M(X)$ and flexural stiffness $EI(X)$ of a non-uniform rotating beam is expressed by:

$$M(X) = \rho_0 BH \tag{7.162}$$

$$EI(X) = E_0 \frac{BH^3}{12} \tag{7.163}$$

where $B$ and $H$ are the breadth and height of the NURB, respectively. Introducing non-dimensional variables $b$ and $h$,

$$b = \frac{B}{B_0}; \quad h = \frac{H}{H_0} \tag{7.164}$$

Therefore, using the non-dimensional variables defined above, Eq. (8.6) can be rewritten as

$$f(x) = \frac{EI(X)}{EI_0} = bh^3; \quad m(x) = \frac{M(X)}{M_0} = bh \tag{7.165}$$

which implies that

$$b = \sqrt{\frac{m^3}{f}}; \quad h = \sqrt{\frac{f}{m}} \tag{7.166}$$

Therefore, using $m(x)$ and $f(x)$ derived before, one can manufacture isospectral beams which have a rectangular cross-section. The breadth and height distribution of these beams can be determined from the above equation.

### 7.2.2.4 Case(i): A cantilever beam ($q_0 = 1$)

The breadth and height distribution of a UNRB is shown in Fig. 7.24a. The breadth and height distribution of isospectral NURB of the same length as that of the given UNRB, are shown in Fig. 7.24b, Fig. 7.23c and Fig. 7.24d for $\lambda = 1$, $\lambda = 1.5$ and $\lambda = 2$, respectively.

When $\lambda > \lambda_0 = 3.516$, we already know that there exist cantilever NURB isospectral to a UNRB of smaller length. Therefore, choosing the length of the UNRB as $z_0 = 0.6$, from Eq. (7.151), we understand that $\lambda < 6.25$. The breadth and height distribution of such isospectral beams are shown in Fig. 7.25a and Fig. 7.25b for $\lambda = 4$ and $\lambda = 5$, respectively.

**Case(ii): A hinged free beam with a root spring** ($K_s = 1.7061$) The breadth and height distribution of isospectral NURB of the same length as that of the given UNRB under the hinged-free with a root spring boundary condition are shown in Fig. 7.26a and Fig. 7.26b for $\lambda = 1$ and $\lambda = 1.25$, respectively.

When $\lambda > \lambda_1 = 1.9083$, we appreciate that there exist NURB isospectral to a UNRB of smaller length under the hinged-free with a root spring boundary condition. We have also proved that for $z_0 = 0.75$ and $\lambda < 3.051$, isospectral beams always

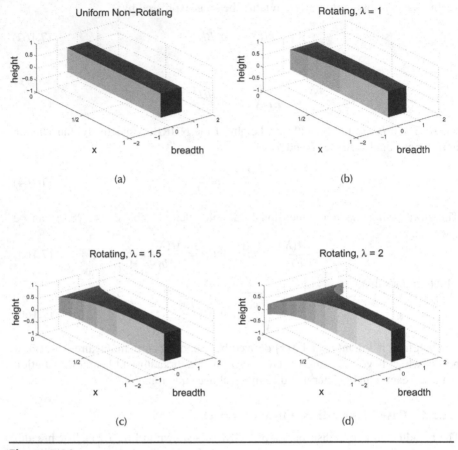

**Figure 7.24**
Breadth and height distributions for isospectral cantilever beams of equal length

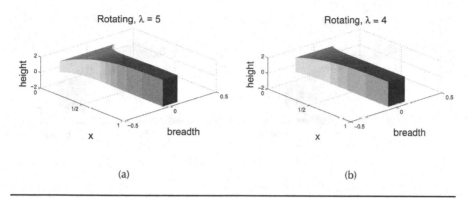

**Figure 7.25**
Breadth and height distributions for rotating cantilever beams isospectral to a uniform non-rotating beam of smaller length

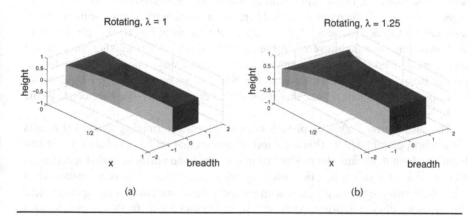

**Figure 7.26**
Breadth and height distributions of the isospectral rotating hinged-free beams with a root spring

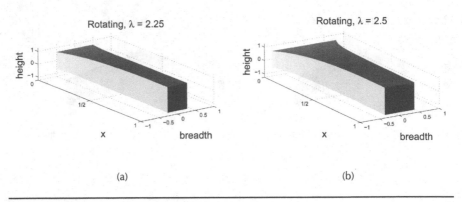

(a)                                                                                   (b)

**Figure 7.27**
Breadth and height distributions of the rotating hinged-free beams with a root spring isospectral to a uniform non-rotating beam of smaller length

exist. The breadth and height distribution of such isospectral beams are shown in Fig. 7.27a and Fig. 7.27b for $\lambda = 2.25$ and $\lambda = 2.5$, respectively.

The proposed procedure to determine a NURB isospectral to a given UNRB, is particularly attractive when the shape of the NURB is mandated by the modal characteristics. However, in some practical applications, the shape of the NURB is mandated by the aeroelastic requirements. Therefore a solution to the inverse of our problem, namely: to determine a UNRB isospectral to a given NURB, is very useful in such situations. For such inverse problems, if the given NURB satisfies the required equations derived for the isospectrality, then we can always apply this procedure to find an isospectral UNRB. However, if the NURB does not satisfy the isospectrality equations, then we will realize that there does not exist a UNRB isospectral to the given NURB.

The assumption $q(z) = q_0$ plays a critical role in determining analytical results for the isospectral NURB. This assumption implies that $m^3 f$ is a constant i.e. at any location along the beam, the product of mass cubed and stiffness at that location, is a constant i.e. the stiffness varies inversely with mass cubed. If $q_0$ is a constant, then we cannot analyze rotating beams with uniform tapers (or compound tapers) where there exists a direct relationship between the mass and the stiffness, i.e. $q(z)$ is not a constant for such tapered beams.

## 7.3    SUMMARY

Two different beam structures for which closed form solutions for natural frequencies can be obtained are presented in this chapter.

In the first case, an analytical procedure for determining non-uniform beams under no axial loads, which are isospectral to an axially loaded uniform beam, is presented.

The non-uniform beam equation is converted to the axially loaded beam equation, by a suitable adjustment of the Barcilon-Gottlieb transformation. Analytical expressions for the mass and stiffness variations of such non-uniform beams are developed for two specific cases ($pq_z = k_0$ and $q = q_0$). This analysis is then applied to beams with rectangular cross-sections, showing a practical application of this procedure. The results in this case present some non-uniform beams whose frequencies are known exactly, since uniform axially loaded beams have closed form solutions. It is also found that non-uniform stiff-strings exist, which are isospectral to uniform piano strings with hinged-hinged boundary conditions. Also, tensionless non-uniform stiff strings, whose frequencies match with that of a given piano string, are presented in this section.

In the second case, an analytical method of obtaining non-uniform rotating beams (NURB) that are isospectral to a given uniform non-rotating beam (UNRB) is presented. For this purpose, the governing equation of an Euler-Bernoulli NURB is converted to that of a UNRB Euler-Bernoulli beam using the auxiliary variables defined in Barcilon-Gottlieb transformation. The special conditions under which such beams exist are shownb. The additional condition on the auxiliary variable $p(z)$ so that it satisfies the free boundary condition is given. Two cases are considered here (clamped-free and hinged-free with a root spring). An upper bound on the rotating speed for discovering NURB isospectral to a UNRB of the same length is presented for the cantilever and hinged free with a root spring boundary boundary conditions. For rotating speeds greater than this bound, a method for finding NURB isospectral to a UNRB of a smaller length is proposed. The finite element formulation is presented for the verification of the spectral properties. A numerical example of a beam with a rectangular cross section is provided to illustrate the application of this analysis.

## 7.4   APPENDIX A

The analytical solution for the mode shapes of a UNRB under different boundary conditions is provided in [118, 139] as the following: For the cantilever boundary condition, the $i^{th}$ mode shape $U_i^c(z)$ is expressed as

$$U_i^c(z) = \cosh(\sqrt{\eta_i}z) - \cos(\sqrt{\eta_i}z) - \sigma_i(\sinh(\sqrt{\eta_i}z) - \sin(\sqrt{\eta_i}z)) \ , \ 0 \leq z \leq 1$$
(7.167)

where

$$\sigma_i = \left( \frac{\sinh(\sqrt{\eta_i}) - \sin(\sqrt{\eta_i})}{\cosh(\sqrt{\eta_i}) + \cos(\sqrt{\eta_i})} \right)$$
(7.168)

and $\eta_i$ is the non-dimensional frequency of the $i^{th}$ mode.

For the hinged-free boundary condition with a root spring boundary condition, the $i^{th}$ mode shape $U_i^h(z)$ is expressed as

$$U_i^h(z) = -(\sin(\sqrt{\eta_i}(1-z)) + \sinh(\sqrt{\eta_i}(1-z))) + \gamma(\cosh(\sqrt{\eta_i}(1-z)) \quad (7.169)$$
$$+ \cos(\sqrt{\eta_i}(1-z)))$$

where

$$\gamma = \left( \frac{\sin(\sqrt{\eta_i}) + \sinh(\sqrt{\eta_i})}{\cos(\sqrt{\eta_i}) + \cosh(\sqrt{\eta_i})} \right) \qquad (7.170)$$

and $\eta_i$ is the non-dimensional frequency of the $i^{th}$ mode.

## 7.5   APPENDIX B

We prove that the frequencies of a rotating beam are always greater than or equal to the rotation speed under the cantilever and hinged-free with a root spring boundary conditions. Recall that the governing equation of the out-of-plane vibrations of a rotating beam is given by Eq. (8.7). The physical boundary conditions for the hinged-free (with no hinge spring) beam are

Hinged at the root ($x = 0$)

$$Y = 0 = Y_{xx} \qquad (7.171)$$

Free at the tip ($x = 1$)

$$Y_{xx} = 0 = Y_{xxx} \qquad (7.172)$$

So, for the hinged-free beam, we can appreciate that $Y = x$ is a solution of Eq. (8.7) as it satisfies these boundary conditions. Substituting $Y = x$ in Eq. (8.7), we get

$$-t'(x) = \eta^2 m(x)x \qquad (7.173)$$

Substituting $t'(x)$ from Eq. (7.124) in the above equation, we obtain

$$\lambda^2 m(x)x = \eta^2 m(x)x \quad \Rightarrow \quad \eta = \lambda \qquad (7.174)$$

Ergo, $Y = x$ is a mode shape (rigid body mode) of the hinged-free rotating beam, and the corresponding frequency is the rotating speed ($\eta = \lambda$). This mode shape exhibits only one node at $x = 0$, it is the first mode shape and its frequency ($\lambda$) is the smallest frequency. Therefore, we can state that the frequencies of a rotating beam are always greater than or equal to the rotation speed under the hinged-free boundary condition.

The frequencies of a rotating beam under the cantilever and hinged-free with a root spring boundary condition would be greater than that of the hinged-free boundary condition. Therefore, the frequencies of a rotating beam are always greater than or equal to the rotation speed under the cantilever and hinged-free with a root spring boundary conditions.

# 8 Isospectral Systems for Testing

Typically, experiments need to be performed to determine the natural rotating frequencies of beam structures . Such experiments need expensive rotor rigs to rotate the beam and assorted complicated equipment. Despite the enormous progress in finite element methods, experiments need to be performed to ensure that the any of the rotating natural frequencies do not come close to the multiples of the rotation speeds, as such a condition results in high vibration levels. A novel approach to reduce the cost and complexity of rotating beam tests is to find isospectral non-rotating beams which have the same natural frequencies. Experiments on non-rotating beams can be easily performed and there is established knowledge in this area available among structural dynamicists. In this chapter, we introduce non-rotating beams which are isospectral to rotating beams .

An undamped vibrating system has a set of natural frequencies, forming a spectrum, at which it vibrates freely when external forces are absent. An important class of problems for such vibrating systems can be broadly classified into inverse problems and isospectral problems. In inverse problems [73], one determines the material properties of a system for a given frequency spectrum. Typically, more than one frequency spectrum is required for a reconstruction procedure, to determine the material properties of the system [138]. Systems, that have the same vibrating frequencies, but have different material properties are called isospectral systems. Isospectral systems are of enormous interest in mechanics as they offer alternative usable designs. The existence of isospectral systems also proves that a system cannot be uniquely identified from its spectrum. The contents of this chapter are adapted from Ref. [114] and [116].

## 8.1 NON-ROTATING BEAMS ISOSPECTRAL TO A GIVEN ROTATING UNIFORM BEAM

### 8.1.1 ISOSPECTRAL SPRING-MASS SYSTEMS

Examples of discrete isospectral systems are in-line spring mass systems. A schematic of a 2-dof spring mass system is shown in Fig. 8.1, where $m_1, m_2$ are the masses and $k_1, k_2$ are the spring constants of the springs. Let us consider one such 2-dof system (System-A) [57] where the values of masses $m_1$ and $m_2$, and spring constants $k_1$ and $k_2$, and the natural frequencies of System-A ($\omega_1$ and $\omega_2$), are tabulated in Table 8.1. Similarly, let us consider one more system (System-A' ) where $m_1'$ and $m_2'$ are the masses, $k_1'$ and $k_2'$, are the spring constants and $\omega_1'$ and $\omega_2'$ are the natural frequencies, whose values are tabulated in Table 8.1. From Table 8.1, we can

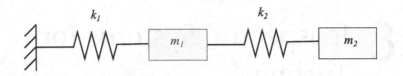

**Figure 8.1**
Schematic of a 2-dof spring-mass system.

**Table 8.1**
**Material and Spectral Properties of System-$A$ and System-$A'$**

| | System-$A$ | | System-$A'$ |
|---|---|---|---|
| $m_1$ | 6.25 kg | $m_1'$ | 1.286 kg |
| $m_2$ | 4 kg | $m_2'$ | 1 kg |
| $k_1$ | 92.5 kN/m | $k_1'$ | 15.857 kN/m |
| $k_2$ | 20 kN/m | $k_2'$ | 6kN/m |
| $\omega_1$ | 62.2 rad/s | $\omega_1'$ | 62.2 rad/s |
| $\omega_2$ | 138.3rad/s | $\omega_2'$ | 138.3 rad/s |

see that both System-$A$ and System-$A'$ have the same frequency spectrum as $\omega_1 = \omega_1'$ and $\omega_2 = \omega_2'$ . Therefore, System-$A$ and System-$A'$ are isospectral.

Experimental determination of rotating beam frequencies is difficult. For example, Senatore [177] experimentally determined the frequencies of a rotating beam, using lumped parameter axial loads on a uniform non-rotating beam, in order to simulate an approximation of the centrifugal force field, acting on the rotating beam. However, this method could not exactly predict the first natural frequency and mode shape due to the lumped axial loads. Therefore, it is interesting to investigate, if we can find any non-rotating beam which is isospectral to the uniform rotating beam. It is important to take into account the exact centrifugal force acting on the rotating beam. The reasons for obtaining isospectral beams are as follows: (a) It is onerous to conduct experiments on rotating beams to obtain its natural frequencies. If we find an isospectral non-rotating beam, one can easily conduct experiments on the non-uniform beam, to obtain the natural frequencies; (b) Such isospectral beams can provide insight by predicting the stiffening effect of the centrifugal force; and (c) provide benchmark problems for finite element analysis.

In this section of this chapter, we seek non-rotating beams with continuous mass and flexural stiffness distributions, that are isospectral to a given uniform rotating beam. The mass and stiffness functions of non-rotating beams, isospectral to a uniform beam rotating at different speeds are derived. We note that for high rotating speeds, the derived mass and stiffness functions of the non-rotating beam are not physically realizable, owing to stiffening effect of the centrifugal force. In such situations, if we attach a torsional spring, of a spring constant $K_R$, at the free end

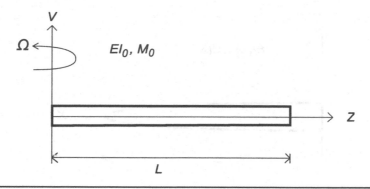

**Figure 8.2**
Schematic of a uniform rotating beam

of the non-rotating beam, the obtained mass and stiffness functions become physically realizable. We also numerically verify that the frequencies of the non-rotating beam obtained using the finite element method(FEM) are the exact frequencies of the uniform rotating beam. This confirms numerically that the non-uniform beam obtained in this method is indeed isospectral to the given uniform rotating beam. We also present an example of realistic beams having a rectangular cross section, to show a physically realizable application of our analysis.

### 8.1.2 MATHEMATICAL ANALYSIS

The governing differential equation for the transverse free vibrations $V(Z)$ of a uniform rotating beam (Fig. 8.2) of length $L$, stiffness $EI_0$ and mass $M_0$ rotating with an angular speed $\Omega$ is given in [104] as:

$$EI_0 \left[ \frac{d^4V}{dZ^4} \right] - \frac{d}{dZ} \left[ M_0\Omega^2 \left( \frac{L^2 - Z^2}{2} \right) \frac{dV}{dZ} \right] - \omega^2 M_0 V = 0 \quad , \quad 0 \le Z \le L \quad (8.1)$$

A non-dimensional variable $z = Z/L$ is introduced so that the above equation can be rewritten as

$$\frac{d^4V}{dz^4} - \frac{d}{dz} \left[ \lambda^2 \left( \frac{1 - z^2}{2} \right) \frac{dV}{dz} \right] - \eta^2 V = 0 \quad , \quad 0 \le z \le 1 \quad (8.2)$$

where $\eta$ is the non-dimensional frequency expressed as

$$\eta = \omega \sqrt{M_0 L^4 / EI_0} \quad (8.3)$$

and $\lambda$ is the non-dimensional rotation speed expressed as

$$\lambda = \Omega \sqrt{M_0 L^4 / EI_0} \quad (8.4)$$

Similarly, the governing equation for the out-of-plane free vibrations $Y(X)$ of a non-uniform non-rotating Euler-Bernoulli beam (Fig. 8.3), which is isospectral to

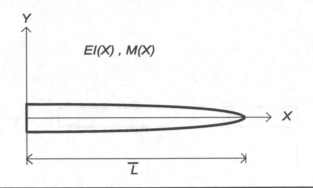

**Figure 8.3**
Schematic of a non-uniform non-rotating beam

the rotating beam, is provided in [139] as

$$\frac{d^2}{dX^2}\left[EI(X)\frac{d^2Y}{dX^2}\right] - \omega^2 M(X)Y = 0 \quad , \quad 0 \le X \le \bar{L} \tag{8.5}$$

where $EI(X)$ is the flexural stiffness , $M(X)$ is the mass/length, $\bar{L}$ is the length of the beam, which is identical to that of the uniform rotating beam ($\bar{L} = L$).

Now, non-dimensional variables $f$, $m$ and $x$ are introduced as

$$f(x) = \frac{EI(X)}{EI_0} \quad , \quad m(x) = \frac{M(X)}{M_0} \quad , \quad x = \frac{X}{L} \tag{8.6}$$

Eq. (8.5) can be rewritten as

$$(f(x)Y'')'' = \eta^2 m(x)Y \quad , \quad 0 \le x \le 1 \tag{8.7}$$

The notation $Y'$ and $Y''$ represent the first and the second derivatives respectively of $Y$ with respect to $x$. The transformation, which converts the above non-rotating non-uniform beam equation (8.5) to the canonical fourth order equation for the function $U(z)$

$$\frac{d^4U}{dz^4} + \frac{d}{dz}\left[A\frac{dU}{dz}\right] + BU = \eta^2 U \quad , \quad 0 \le z \le 1 \tag{8.8}$$

is postulated by Barcilon [17]. When $A = -\lambda^2(1-z^2)/2$ and $B = 0$, $U$ corresponds to $V$ in Eq. (8.2). In the Barcilon-Gottlieb transformation [17, 86], two auxiliary variables $p$ and $q$ are defined as

$$p = \left(\frac{m}{f}\right)^{\frac{1}{4}} \; ; \; q = \left(\frac{1}{m^3 f}\right)^{\frac{1}{8}} \tag{8.9}$$

which also implies that

$$f = p^{-3}q^{-2} \; ; \; m = pq^{-2} \tag{8.10}$$

The transformation of the independent variable $(x)$ is given implicitly as

$$z = \int_0^x p(z(x))\,dx \tag{8.11}$$

which implies that

$$x = \int_0^z \frac{dz}{p(z)} \tag{8.12}$$

$$\frac{d}{dx} = p(z)\left(\frac{d}{dz}\right) \Rightarrow \frac{1}{p(z)} = \frac{dx}{dz} \tag{8.13}$$

We note that

$$x = 0 \Leftrightarrow z = 0 \tag{8.14}$$

Also, we have

$$x = 1 \Leftrightarrow z = 1 \tag{8.15}$$

The transformation of the dependent variable $Y$ is expressed as

$$Y(x) = q(z)U(z) \tag{8.16}$$

Now Eq. (8.12) and Eq. (8.16) transform the variables $x$ and $Y$ into $z$ and $U$, respectively. This completes the Barcilon-Gottlieb transformation. Using this transformation, we obtain

$$\frac{d^2}{dx^2}\left[f\frac{d^2Y}{dx^2}\right] - \eta^2 mY = pq^{-1}\left[\frac{d^4U}{dz^4} + \frac{d}{dz}\left[A\frac{dU}{dz}\right] + BU - \eta^2 U\right] \tag{8.17}$$

where

$$A = 4\frac{q_{zz}}{q} - 6\frac{q_z^2}{q^2} - 2\frac{q_z p_z}{qp} + \frac{p_{zz}}{p} - 2\frac{p_z^2}{p^2} \tag{8.18}$$

$$\begin{aligned}B = \frac{q_{zzzz}}{q} &- 4\frac{q_z q_{zzz}}{q^2} - 2\frac{q_{zz}^2}{q^2} + 6\frac{q_z^2 q_{zz}}{q^3} + \frac{q_z p_{zzz}}{qp} + \frac{q_{zz} p_{zz}}{qp} - 4\frac{q_z^2 p_{zz}}{q^2 p} \\ &- 5\frac{q_z p_z p_{zz}}{qp^2} - 4\frac{q_z q_{zz} p_z}{q^2 p} + 6\frac{q_z^3 p_z}{q^3 p} + 6\frac{q_z^2 p_z^2}{q^2 p^2} + 4\frac{q_z p_z^3}{qp^3} - 2\frac{q_{zz} p_z^2}{qp^2}\end{aligned} \tag{8.19}$$

These expressions for $A$ and $B$ are originally expounded by Barcilon [17] and later corrected by Gottlieb [86]. Here, the notation $q(z)$ refers to the value of $q$ at the coordinate $z$. The first, second, third and the fourth derivatives, respectively of $q$ with respect to $z$ are denoted by $q_z$, $q_{zz}$, $q_{zzz}$ and $q_{zzzz}$. Therefore, we can close the above analysis by establishing the following.

The governing differential equation of the non-uniform non-rotating beam is

$$\frac{d^2}{dx^2}\left[f\frac{d^2Y}{dx^2}\right] - \eta^2 mY = 0 \ , \ 0 \leq x \leq 1 \tag{8.20}$$

The above equation can be transformed into a canonical form using Barcilon-Gottlieb transformation as

$$\frac{d^4U}{dz^4} + \frac{d}{dz}\left[A(z)\frac{dU}{dz}\right] + B(z)U - \eta^2 U = 0 \ , \ \ 0 \le z \le 1 \tag{8.21}$$

The expressions $A(z)$ and $B(z)$, for this non-rotating beam, are provided in Eq. (8.90) and Eq. (8.91). The governing differential equation of the uniform rotating beam which is rotating with a non-dimensional speed $\lambda$, is from Eq. (8.1),

$$\frac{d^4V}{dz^4} + \frac{d}{dz}\left[-\frac{\lambda^2}{2}(1-z^2)\frac{dV}{dz}\right] - \eta^2 V = 0 \ , \ \ 0 \le z \le 1 \tag{8.22}$$

If the coefficients $A(z)$ and $B(z)$ of the non-rotating beam in Eq. (8.21) coincide with the coefficients of the uniform rotating beam Eq. (8.22), then the non-rotating beam is isospectral to the given rotating beam; i.e. $A(z) = -\lambda^2(1-z^2)/2$ and $B(z) = 0$. Thus, a non-uniform non-rotating beam isospectral to a given rotating uniform beam is generated when $U$ corresponds to $V$. Specifically, Eq. (8.20) and Eq. (8.22) have identical natural frequencies if and only if the auxiliary functions $p(z)$ and $q(z)$ satisfy the following conditions:

$$A(z) = -\lambda^2(1-z^2)/2 \tag{8.23}$$
$$B(z) = 0 \tag{8.24}$$

We must solve the above equations to get non-rotating beams isospectral to a given uniform rotating beam. These equations are nonlinear fourth order coupled differential equations which are onerous to solve analytically for a general $q(z)$. However, for $q(z) = q_0$ which is a constant, it is feasible to obtain the results analytically. Furthermore, note that if $q = q_0$, then $m$ and $f$ are simply related throughout by $f = q_0^{-8}m^{-3}$. When $q(z) = q_0$, we can observe that $B(z) = 0$ as the derivatives $q_z, q_{zz}, q_{zzz}$ and $q_{zzzz}$ vanish. When $A(z) = -\lambda^2(1-z^2)/2$, then from Eq. (8.90) we get

$$-\frac{\lambda^2}{2}(1-z^2) = \frac{p_{zz}}{p} - \frac{2p_z^2}{p^2} \tag{8.25}$$

The above equation is divided throughout by $p$ and rewritten as

$$\frac{d^2\hat{p}}{dz^2} - \frac{\lambda^2}{2}(1-z^2)\hat{p} = 0 \tag{8.26}$$

where $\hat{p}$ is defined using Eq. (8.13) as

$$\hat{p} = p^{-1} = \frac{dx}{dz} \tag{8.27}$$

Note that at $z = 1$, from Eq. (8.26), we get

$$\left[\frac{d^2\hat{p}}{dz^2}\right]_{z=1} = \left[\frac{\lambda^2}{2}(1-z^2)\hat{p}\right]_{z=1} = 0 \tag{8.28}$$

Moreover, using $q(z) = q_0$ in Eq. (8.16) and using Eq. (8.13), yields the following

$$Y = q_0 U \tag{8.29}$$

$$Y_x = q_0 p U_z \tag{8.30}$$

$$Y_{xx} = q_0 p (p_z U_z + p U_{zz}) \tag{8.31}$$

$$Y_{xxx} = q_0 p p_z (p_z U_z + p U_{zz}) + q_0 p^2 (p_{zz} U_z + 2 p_z U_{zz} + p U_{zzz}) \tag{8.32}$$

The initial conditions that should be imposed on Eq. (8.26) depend on the physical boundary conditions (clamped, free, hinged etc.). In the next section, we solve this equation for cantilever rotating and non-rotating beams.

### 8.1.2.1  Cantilever Rotating and Non-rotating Beams

In this section, we seek a non-uniform non-rotating cantilever beam which is isospectral to a given uniform rotating cantilever beam.

The rotating and the non-rotating beams are clamped at the root ($x = 0 = z$) and free at the tip ($x = 1$, $z = 1$). Therefore, we have the following boundary conditions:
At the root:

$$Y = 0 = Y_x \tag{8.33}$$

$$U = 0 = U_z \tag{8.34}$$

At the tip:

$$(f Y_{xx})_x = 0 = Y_{xx} \tag{8.35}$$

In the above equation, since $Y_{xx} = 0$, we have $(f Y_{xx})_x = 0 \Rightarrow f_x Y_{xx} + f Y_{xxx} = 0 \Rightarrow Y_{xxx} = 0$, Therefore, the Eq. (8.35) can be modified as

$$Y_{xxx} = 0 = Y_{xx} \tag{8.36}$$

$$U_{zz} = 0 = U_{zzz} \tag{8.37}$$

We can observe from Eq. (8.107), Eq. (8.108), Eq. (8.109) and Eq. (8.32), that the clamped boundary condition is satisfied in the Barcilon-Gottlieb transformation. Thus we need to satisfy only the free boundary condition for the cantilever beam. This implies that the coefficients of the term $U_z$ in Eq. (8.109) should be zero at $z = 1$ (since at $z = 1$, $U_{zz} = 0 = U_{zzz}$). Therefore, in Eq. (8.109), $p_z$ at $z = 1$ should become zero. Similarly in Eq. (8.32), the coefficients of the term $U_z$ should become zero at $z = 1$. Thus $p_{zz}$ should also become zero at $z = 1$. We understand from Eq. (8.27) that $p = 1/\hat{p}$ and therefore, we obtain

$$p_z = -\hat{p}_z \hat{p}^{-2} \tag{8.38}$$

$$p_{zz} = 2 \hat{p}_z^2 \hat{p}^{-3} - \hat{p}_{zz} \hat{p}^{-2} \tag{8.39}$$

Therefore, from the above equations and Eq. (8.28), we obtain at $z = 1$

$$\hat{p}_z = 0 \Leftrightarrow p_z = 0 \Leftrightarrow p_{zz} = 0 \tag{8.40}$$

Another initial condition which $\hat{p}$ should satisfy is obtained from Eq. (8.12), which is

$$1 = \int_0^1 \frac{dz}{p} \quad \Leftrightarrow \quad 1 = \int_0^1 \hat{p}\, dz \tag{8.41}$$

Therefore, Eq. (8.26) can be solved for any given rotation speed ($\lambda$), using the two initial conditions Eq. (8.40) and Eq. (8.41). For example, for $\lambda = 1$, we can find a corresponding isospectral non-rotating beam. The solution of Eq. (8.26), for $\lambda = 1$, is given as

$$\hat{p}(z) = k_1 \sigma_1 \, {}_1F_1(\alpha_1; \beta_1; \sigma_2) + z\, k_2 \sigma_1 \, {}_1F_1(\alpha_2; \beta_2; \sigma_2) \tag{8.42}$$

where

$$\alpha_1 = 2\sqrt{2}\left(\frac{\sqrt{2}}{16} - \frac{1}{16}i\right) \;;\; \alpha_2 = 2\sqrt{2}\left(\frac{3\sqrt{2}}{16} - \frac{1}{16}i\right) \tag{8.43}$$

$$\beta_1 = \frac{1}{2} \;;\; \beta_2 = \frac{3}{2} \tag{8.44}$$

$$\sigma_1 = e^{-\frac{\sqrt{2}x^2 i}{4}} \;;\; \sigma_2 = \frac{\sqrt{2}x^2 i}{2} \tag{8.45}$$

$$k_1 = 1.0911 \;;\; k_2 = -0.3376 \tag{8.46}$$

and ${}_1F_1(\alpha_1; \beta_1; \sigma_2)$ and ${}_1F_1(\alpha_2; \beta_2; \sigma_2)$ are confluent hypergeometric functions [149, 181] of the first kind given by

$$ {}_1F_1(\alpha; \beta; \sigma) = \sum_{k=0}^{\infty} \frac{(\alpha)_k}{(\beta)_k} \frac{\sigma^k}{k!} \tag{8.47}$$

where $(a)_k$ and $(b)_k$ are Pochhammer symbols [52].

Since $\hat{p}(z)$ is now known, $m(z)$ and $f(z)$ can be determined using Eq. (8.10), by choosing $q_0 = 1$, and are expressed as

$$m(z) = (k_1 \sigma_1 \, {}_1F_1(\alpha_1; \beta_1; \sigma_2) + z\, k_2 \sigma_1 \, {}_1F_1(\alpha_2; \beta_2; \sigma_2))^{-1} \tag{8.48}$$

$$f(z) = (k_1 \sigma_1 \, {}_1F_1(\alpha_1; \beta_1; \sigma_2) + z\, k_2 \sigma_1 \, {}_1F_1(\alpha_2; \beta_2; \sigma_2))^3 \tag{8.49}$$

Note that Eq. (8.49) can also be derived from Eq. (8.48) by using the relation $f = (q_0^{-8})m^{-3}$, which simplifies to $f = m^{-3}$, as $q_0 = 1$. Similarly, $x(z)$ can be determined using Eq. (8.12), and given by

$$x(z) = \int_0^z \hat{p}\, dz \tag{8.50}$$

The above equation (8.50) possesses no exact solution, and therefore we must integrate $\hat{p}$ (Eq. (8.42)) using Guass-Legendere quadrature and get $x(z)$. The derived $m(z)$, $f(z)$ and $x(z)$ functions for $\lambda = 1$ are plotted in Fig. 8.4.

Our interest is to determine the mass and stiffness functions in terms of $x$ (i.e. $m(x)$ and $f(x)$). It is not feasible to obtain $m(x)$ and $f(x)$ from $m(z)$ and $f(z)$ analytically, as there exists no closed form expression for $z(x)$. Here, for a given value of $x$, we cannot derive the value of $z$ analytically. If we were given a value of $x$, say $\bar{x}$, we need to determine the value of $z$, say $\bar{z}$. Thus we use the well known Newton-Raphson method to calculate $\bar{z}$ given any $\bar{x}$, by using the iterative formula [120]

$$z_{i+1} = z_i - \left( \frac{x(z_i) - \bar{x}}{x_z(z_i)} \right) \tag{8.51}$$

where

$$x_z(z_i) = \left[ \frac{dx}{dz} \right]_{z=z_i} = \hat{p}(z_i) \tag{8.52}$$

and $x(z_i)$ is obtained from Eq. (8.50). The initial estimate for this iterative formula can be considered as $z_1 = \bar{x}$. For example, for $\bar{x} = 0.6$, we can apply the above analysis to calculate $\bar{z}$, which is $\bar{z} = 0.588$. Thus, in Fig. 8.4, if we draw a horizontal line $x = \bar{x} = 0.6$, it cuts the graph $x(z)$ at the point B(0.588, 0.6). Using Eq. (8.48) and Eq. (8.49), $m(\bar{z})$ and $f(\bar{z})$ can be calculated, and are given by $m(\bar{z}) = 1.023$ and $f(\bar{z}) = 0.934$. Thus, in Fig. 8.4, if we draw a vertical line from point B($z = \bar{z} = 0.588$), it cuts the graph $m(z)$ at the point C(0.588, 1.023) and cuts the graph $f(z)$ at the point D(0.588, 0.934). Therefore, at the non-dimensional length $\bar{x} = 0.6$, the mass and the stiffness of the non-rotating beam isospectral to the given rotating beam, are given by $m(\bar{x}) = 1.023$ and $f(\bar{x}) = 0.934$, respectively. Moreover, note also that for $q_0 = 1$, we have $f = m^{-3}$. Therefore, the graphs $f(z)$ and $m(z)$ crossover at point E ($z = 0.385$), where $f = m = 1$, thus confirming $f = m^{-3}$.

Thus, for any given $x$, $z(x)$, and hence mass ($m(x)$) and stiffness ($f(x)$) can be determined. These $m(x)$ and $f(x)$ distributions are plotted in Fig. 8.5a. From Fig. 8.5a, we can surmise that the non-rotating isospectral beam should be more stiff at the root and less stiff at the tip compared to the uniform rotating beam. Similarly, the mass distribution should be less at the root and more at the tip compared to the uniform rotating beam.

In the above analysis, we have determined the non-rotating beam isospectral to the rotating beam for $\lambda = 1$. The mass and stiffness variations of beams isospectral to a uniform beam rotating at other speeds can be obtained similarly. For $\lambda = 2$ and $\lambda = 3$, $m(x)$ and $f(x)$ variations of isospectral beams are calculated and plotted in Fig. 8.5b and Fig. 8.5c, respectively.

From the above figures, we can observe that as $\lambda$ increases, the stiffness of the isospectral non-rotating beam increases at the root and decreases at the tip and asymptotically goes to zero for higher values of $\lambda$. These mass and stiffness distributions have been plotted in Fig. 8.6 for $\lambda = 4, 5, 6$ and $q_0 = 1$. From Fig. 8.6, we can observe that the isospectral non-rotating beam, on an average, has a high value of stiffness and a low value of mass, as compared to the uniform rotating beam. Thus we can postulate that, as $\lambda$ increases, it becomes more and more difficult for the variation in mass and the stiffness functions, to counter the stiffening effect of the centrifugal force, acting on the uniform rotating beam. Therefore, for higher values of $\lambda$, we can attach an elastic support, such as a torsional spring, at the free end of the non-rotating beam, in order to increase the stiffness of the beam. In the next section,

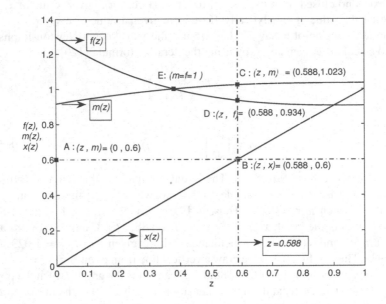

**Figure 8.4**
Variations of $m(z)$, $f(z)$ and $x(z)$ of the isospectral non-rotating beam ($\lambda = 1$)

we seek to find non-rotating beams with a torsional spring attached at the free end, that are isospectral to a uniform beam rotating at high speeds.

### 8.1.2.2  Isospectral Non-rotating Beams with a Torsional Spring at the Free End

Consider a non-rotating beam clamped at one end and free at the other end with a torsional spring, of a spring constant, $K_R$, attached at the free end, as shown in Fig. 8.7. The method for finding these type of beams isospectral to the rotating beam (no spring), is analogous to that discussed in the previous section, except that the physical boundary conditions should be managed in a different way. The boundary conditions for this beam are given in [118] as follows:

Clamped at the root ($x = 0$)

$$Y = 0 = Y_x \tag{8.53}$$

Free at the tip with a torsional spring ($x = 1$)

$$(fY_{xx})_x = 0 \tag{8.54}$$

$$fY_{xx} + k_rY_x = 0 \tag{8.55}$$

where $k_r$ is the non-dimensional spring constant, given by $k_r = K_RL/EI_0$. Expanding

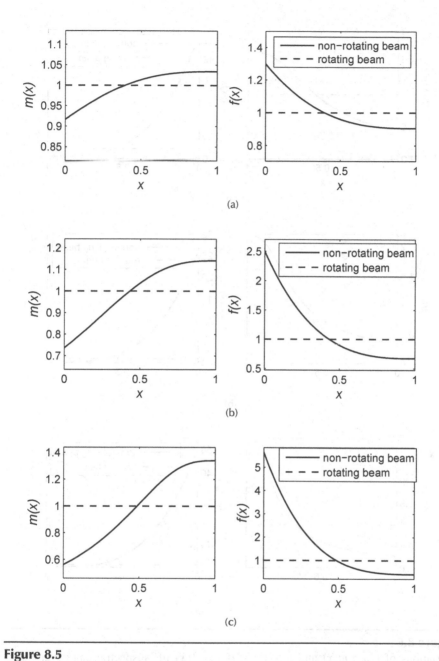

**Figure 8.5**
Variations of mass $m(x)$ and flexural stiffness $f(x)$ of isospectral cantilever non-rotating ($q_0 = 1$) and uniform rotating beams at different rotation speeds $\lambda$. (a) $\lambda = 1$; (b) $\lambda = 2$; (c) $\lambda = 3$

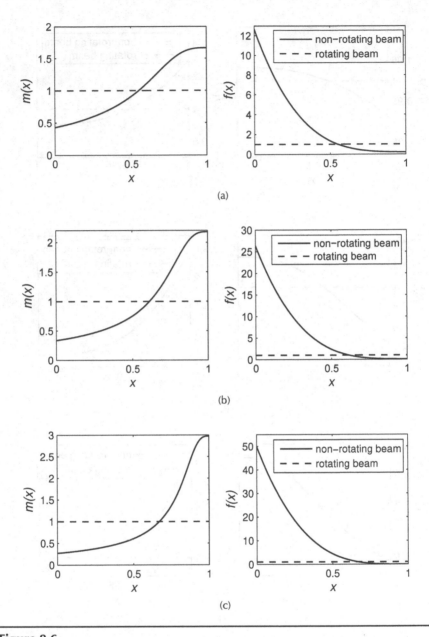

**Figure 8.6**
Variations of mass $m(x)$ and flexural stiffness $f(x)$ of isospectral cantilever non-rotating ($q_0 = 1$) and uniform rotating beams at different rotation speeds $\lambda$ . (a) $\lambda = 4$; (b) $\lambda = 5$; (c) $\lambda = 6$

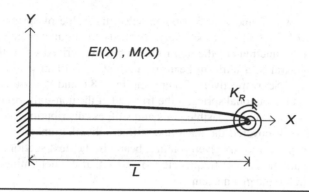

**Figure 8.7**
Schematic of a non-uniform non-rotating beam with a torsional spring at the free end

Eq. (8.54), we get

$$(fY_{xx})_x = f_x Y_{xx} + fY_{xxx} = pf_z Y_{xx} + fY_{xxx}$$

$$= p(q_0^{-2} p^{-3})_z Y_{xx} + (q_0^{-2} p^{-3}) Y_{xxx} = q_0^{-2} p^{-3} (Y_{xxx} - 3Y_{xx} p_z) \tag{8.56}$$

The boundary conditions given by Eq. (8.54) and Eq. (8.55), can be transformed from the variable $Y$ into the variable $U$, which is the variable corresponding to the uniform rotating beam, using Eq. (8.108), Eq. (8.109), Eq. (8.32) and Eq. (8.56). Specifically, Eq. (8.55) and Eq. (8.54) can be rewritten in variable $U$ as (at $x = 1; z = 1$)

$$q_0^{-2} p^{-3} (q_0 p(p_z U_z + pU_{zz})) + k_r q_0 pU_z = 0 \tag{8.57}$$

$$(q_0^{-2} p^{-3})[q_0 pp_z(p_z U_z + pU_{zz}) + q_0 p^2 (p_{zz} U_z + 2p_z U_{zz} + pU_{zzz}) \tag{8.58}$$
$$- 3q_0 p(p_z U_z + pU_{zz})p_z] = 0$$

The above equations can be simplified using $U_{zz} = U_{zzz} = 0$ (Eq. (8.37)), to yield (at $z = 1$)

$$p_z = -k_r q_0^2 p^3 \tag{8.59}$$

$$pp_{zz} = 2p_z^2 \tag{8.60}$$

We can observe that from Eq. (8.25), we get at $z = 1$, $pp_{zz} - 2p_z^2 = 0$. Therefore, that Eq. (8.60) is already satisfied. Thus, we must choose a spring constant $k_r$, such that Eq. (8.59) is also satisfied. Therefore, Eq. (8.26), subject to the boundary conditions, Eq. (8.41) and Eq. (8.59), can be solved for any given $\lambda$. For example, for $\lambda = 4$, we can select $q_0 = 1$ and $k_r = 2.013$, so that the boundary conditions are satisfied.

The mass and stiffness functions of the isospectral non-rotating beam, derived using this approach, are plotted in Fig. 8.8a. We can observe from Fig. 8.8a that adding a torsional spring, has considerably brought down the average value of stiffness, as compared to that derived without a torsional spring (Fig. 8.6a).

Similarly, for $\lambda = 5$ and $\lambda = 6$, we can select the value of spring constant as $k_r = 2.6318$ and $k_r = 3.3843$, respectively, to satisfy the boundary conditions. The mass and stiffness functions of the non-rotating beams, derived using this method, which are isospectral to a uniform beam rotating at $\lambda = 5$ and $\lambda = 6$ are plotted in Figs. 8.8b and 8.8c, respectively. Comparing Figs. 8.6 and 8.8, we can conclude that the addition of a torsional spring at the free end of the non-uniform non-rotating beam, has yielded physically feasible mass and stiffness distributions.

Thus, we have discovered the mass and stiffness distributions of a non-uniform beam that is isospectral to the given rotating beam. In the next section, we will numerically compute the natural frequencies to verify that the determined beams are indeed isospectral to the given beam.

### 8.1.2.3  Finite element formulation

For the given rotating uniform beam, the natural frequencies are determined using the method of Frobenius, given in [207]. In order to compute the natural frequencies of the non-uniform non-rotating beams, we use the finite element method. In the finite element formulation, we divide the beam into a number of finite elements. The displacement, $Y(x)$ is assumed to have a cubic distribution along each finite element and hence, has four degrees of freedom (4-dof). Two of these 4-dof are displacement DOF and the other two are slope DOF, at the ends of each finite element. Specifically, $Y(x)$, along the $i^{th}$ element, is given by

$$Y(x) = H_1(x - x_i)q_{1,i} + H_2(x - x_i)q_{2,i} + H_3(x - x_i)q_{3,i} + H_4(x - x_i)q_{4,i} \quad (8.61)$$

where $x_i$ is the coordinate of the left node, $x_{i+1}$ is the coordinate of the right node, $q_{1,i}$ and $q_{2,i}$ are the displacement DOF and slope DOF at the left node, $q_{3,i}$ and $q_{4,i}$ are the displacement DOF and slope DOF at the right node [122]. The Hermite shape functions ($H_1, H_2, H_3$ and $H_4$) are given by $H_1(\xi) = 1 - (3\xi^2)/l^2 + (2\xi^3)/l^3$ ; $H_2(\xi) = \xi - (2\xi^2)/l + \xi^3/l^2$; $H_3(\xi) = (3\xi^2)/l^2 - (2\xi^3)/l^3$; $H_4(\xi) = -\xi^2/l + \xi^3/l^2$, where $l = x_{i+1} - x_i$ element length.

The expressions for the kinetic energy $T_k$ and potential energy $U_p$ of the non-rotating non-uniform beam, are given by

$$T_k = \int_0^1 m(x)\dot{Y}^2 dx \quad (8.62)$$

$$U_p = \int_0^1 f(x)(Y''(x))^2 dx \quad (8.63)$$

The natural frequencies ($\eta$) and mode shapes ($\hat{V}$) are then obtained as follows. First, we substitute the expression for displacement $Y(x)$, from Eq. (8.129), into equations (8.130) and (8.131). Next we apply Lagrange's equations of motion, which yields the well known eigenvalue problem:

$$\hat{K}\hat{V} = \eta^2 \hat{M}\hat{V} \quad (8.64)$$

where, $\hat{M}$ and $\hat{K}$ are the global mass and stiffness matrices. These global mass and stiffness matrices are obtained by assembling the element level mass ($\tilde{M}$) and

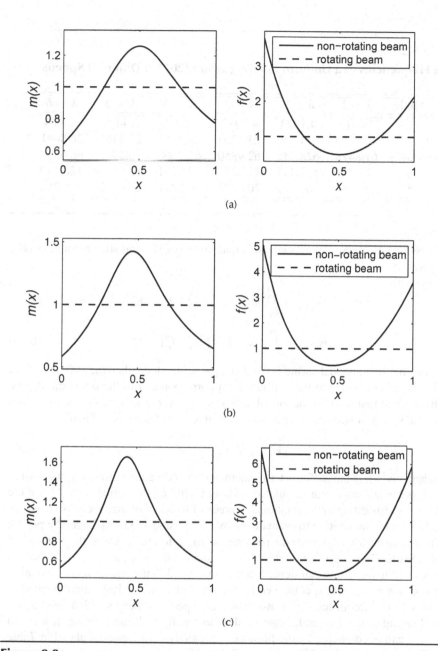

**Figure 8.8**

Variations of mass $m(x)$ and stiffness $f(x)$ of clamped-free non-rotating beams ($q_0 = 1$) with a torsional spring, isospectral to a cantilever rotating beam for different rotation speeds $\lambda$. (a) $\lambda = 4$; (b) $\lambda = 5$; (c) $\lambda = 6$

## Table 8.2
## The Frequencies of a Uniform Rotating Beam [207] at Different Speeds

| Mode | $\lambda = 1$ | $\lambda = 2$ | $\lambda = 3$ | $\lambda = 4$ | $\lambda = 5$ | $\lambda = 6$ |
|------|-------|-------|-------|-------|-------|-------|
| 1 | 3.6817 | 4.1373 | 4.7973 | 5.5850 | 6.4495 | 7.3604 |
| 2 | 22.1810 | 22.6149 | 23.3203 | 24.2734 | 25.4461 | 26.8091 |
| 3 | 61.8418 | 62.2732 | 62.9850 | 63.9668 | 65.2050 | 66.6840 |
| 4 | 121.051 | 121.497 | 122.236 | 123.261 | 124.566 | 126.140 |
| 5 | 200.012 | 200.467 | 201.223 | 202.277 | 203.622 | 205.253 |

stiffness ($\tilde{\mathbf{K}}$) matrices. The $i^{th}$ element's mass matrix ($\tilde{\mathbf{M}}_i$) and stiffness matrix ($\tilde{\mathbf{K}}_i$) are given by

$$(\tilde{\mathbf{M}}_{jk})_i = \int_0^l m(\xi + x_i) H_j(\xi) H_k(\xi) d\xi \tag{8.65}$$

$$(\tilde{\mathbf{K}}_{jk})_i = \int_0^l f(\xi + x_i) H_j''(\xi) H_k''(\xi) d\xi \tag{8.66}$$

Note that if a torsional spring is fixed at a particular node, then the stiffness of the spring should be added to the stiffness term corresponding to the rotational degree of freedom at that node in the global stiffness matrix. Let us take the example of a torsional spring of stiffness $k_r$ attached at the free end of the beam. Then

$$(\tilde{\mathbf{K}}'_{jk})_i = (\tilde{\mathbf{K}}_{jk})_i + k_r \tag{8.67}$$

where $(\tilde{\mathbf{K}}'_{jk})_i$ is the stiffness term calculated including the stiffness of the spring, $(\tilde{\mathbf{K}}_{jk})_i$ is the stiffness term calculated without including the spring, and $j, k$ are the indices corresponding to the rotational degrees of freedom of that node. We have applied the Gauss quadrature to evaluate the integrals using 10 Gauss-Legendre points.

The frequencies of a uniform rotating beam, determined using the method of Frobenius [207], are given in Table 8.2, for rotational speeds $\lambda = 1, ..., 6$. The frequencies of non-rotating cantilever beams, isospectral to this uniform beam rotating at a speed $\lambda = 1, 2$ and 3, computed by the finite element analysis, are tabulated in Table 8.3. The frequencies of non-rotating clamped free beams, which have a torsional spring at the free end, isospectral to this uniform beam rotating at a speed $\lambda = 4, 5$ and 6, computed by the finite element analysis, are also tabulated in Table 8.3. By comparing tables Table 8.2 and Table 8.3, we can conclude that there is an almost negligible error between frequencies of the rotating beam and that of the non-rotating non-uniform beam. This provides numerical verification, of the isospectral properties of analytical mass and stiffness functions, derived in the earlier section.

**Table 8.3**

**Frequencies of Isospectral Non-rotating Non-uniform Beams, Calculated Using FEM**

| Mode | $\lambda = 1$ | $\lambda = 2$ | $\lambda = 3$ | $\lambda = 4$ | $\lambda = 5$ | $\lambda = 6$ |
|------|---------|---------|---------|---------|---------|---------|
| 1 | 3.6816 | 4.1373 | 4.7973 | 5.5850 | 6.4495 | 7.3604 |
| 2 | 22.1810 | 22.6149 | 23.3203 | 24.2733 | 25.4461 | 26.8091 |
| 3 | 61.8418 | 62.2732 | 62.9850 | 63.9668 | 65.2050 | 66.6839 |
| 4 | 121.051 | 121.497 | 122.236 | 123.262 | 124.566 | 126.141 |
| 5 | 200.012 | 200.467 | 201.223 | 202.277 | 203.622 | 205.253 |

### 8.1.3  REALISTIC BEAMS

We consider beams having a rectangular cross-section. The mass/length $M_0$ and stiffness $EI_0$ of the given uniform rotating beam is expressed as

$$M_0 = \rho_0 \hat{B}_0 \hat{H}_0 \tag{8.68}$$

and

$$EI_0 = E_0 \frac{\hat{B}_0 \hat{H}_0^3}{12} \tag{8.69}$$

where $\rho_0$ is the density of the beam, $E_0$ is the tensile modulus, $\hat{B}_0$ and $\hat{H}_0$ are the breadth and height of the rectangular cross section, respectively. Similarly, the mass $M(X)$ and stiffness $EI(X)$ of the non-rotating beams, which are made of the same material as that of the rotating beam, is given by

$$M(X) = \rho_0 \hat{B}(X) \hat{H}(X) \tag{8.70}$$

and

$$EI(X) = E_0 \frac{\hat{B}(X) \hat{H}(X)^3}{12} \tag{8.71}$$

where $\hat{B}(X)$ and $\hat{H}(X)$ are the breadth and height of the rectangular cross section, respectively, of the non-rotating beams. We propose non-dimensional variables $\hat{b}$ (for breadth) and $\hat{h}$ (for height) as

$$\hat{b} = \hat{B}(X)/\hat{B}_0 \;\; ; \;\; \hat{h} = \hat{H}(X)/\hat{H}_0 \tag{8.72}$$

Invoking these non-dimensional breadth and height variables, the mass and stiffness functions are given by

$$m = \frac{M(X)}{M_0} = \hat{b}\hat{h} \;\; ; \;\; f = \frac{EI(X)}{EI_0} = \hat{b}\hat{h}^3 \tag{8.73}$$

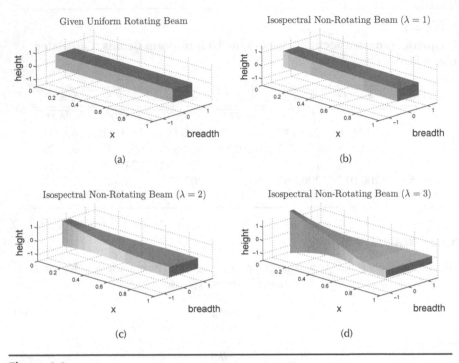

**Figure 8.9**
Variations of $\hat{b},\hat{h}$ of non-rotating beams ($q_0 = 1$) isospectral to a uniform beam rotating at different rotation speeds $\lambda$

which mandates that

$$\hat{b} = \sqrt{\frac{m^3}{f}} \;\; ; \;\; \hat{h} = \sqrt{\frac{f}{m}} \tag{8.74}$$

Therefore, based on the $m(x)$ and $f(x)$ functions derived in earlier sections, we can find beams that have a rectangular cross-section. The breadth ($\hat{b}$) and height ($\hat{h}$) distribution of these beams can be determined from Eq. (8.137). These $\hat{b}$ and $\hat{h}$ distributions, for a non-uniform non-rotating beam, which are isospectral to a uniform beam rotating at a speed $\lambda = 1, 2$ and 3, are plotted in Figs. 8.9b, 8.9c and 8.9d, respectively. Similarly, the $\hat{b}$ and $\hat{h}$ distributions, for non-uniform non-rotating clamped-free beams, having a torsional spring at the free end, which are isospectral to a uniform cantilever beam rotating at a speed $\lambda = 4$ and 5, are plotted in Figs. 8.10a and 8.10b, respectively. It can be observed that these are physically realizable beams.

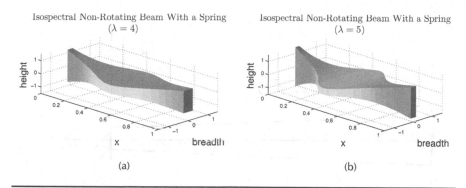

**Figure 8.10**
Variations of $\hat{b}, \hat{h}$ of clamped-free non-rotating beams ($q_0 = 1$), which have a torsional spring at the free end, isospectral to a uniform cantilever beam rotating at rotation speeds $\lambda$

## 8.2  NON-ROTATING BEAMS ISOSPECTRAL TO TAPERED ROTATING BEAMS

In this section, we seek to find non-rotating beams which are isospectral to a given *tapered rotating beam*. Such beams, if they exist, have important applications in the dynamics of rotating systems since helicopter blades and wind turbine blades are well modeled as tapered beams. Unlike uniform rotating beams, which were discussed in the previous section, tapered beams have enormous practical application.

Attarnejad and Shabha [5] investigated the free vibration analysis of rotating tapered beams by developing the concept of basic displacement functions. Ozdemir and Kaya [152] established the vibration characteristics of a rotating tapered cantilever beam with linearly varying rectangular cross-section of area proportional to $x^n$ where $n = 1, 2$. Vinod et al. [194] formulated an approximate spectral element for uniform and tapered rotating beams. Deepak et al. [51] studied the vibration of nanocomposite rotating beams. Thus, considerable amount of research has been done on the models of a rotating beam.

Experimental determination of rotating beam frequencies is arduous. Direct experiments on the rotating beam are onerous and expensive [34]. Therefore, we shall seek to find any non-rotating beam which is isospectral to the tapered rotating beams. Typically, rotating beams used in applications have polynomial taper.

In this section, we discover non-rotating beams with continuous mass and stiffness distributions, that are isospectral to a given tapered rotating beam. The given tapered rotating beam mass and stiffness functions are polynomial functions of the span. Such polynomial variations of mass and stiffness are widespread in the design of helicopter and wind turbine blades [199]. For high rotating speeds, the derived mass and stiffness functions of the non-rotating beam are not physically realizable, due to stiffening effect of the centrifugal force. In such special situations, if we attach torsional and translational springs at the free end of the non-rotating beam, the

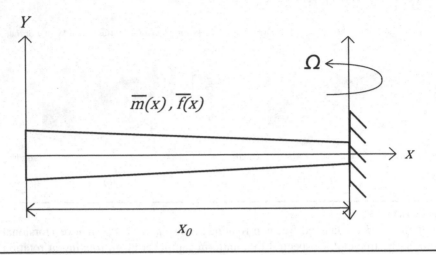

**Figure 8.11**
Schematic of a tapered rotating beam

resulting mass and stiffness functions become physically realizable. It is also shown numerically that the frequencies of the non-rotating beam obtained using the finite element method (FEM) are the frequencies of the tapered rotating beam. We again provide an example of realistic beams having a rectangular cross section, to show a physically realizable application of the analysis.

## 8.2.1 MATHEMATICAL FORMULATION

In this section, we present the mathematical analysis of the given rotating beam and isospectral non-rotating beam. We also present the Barcilon-Gottlieb transformation for the beams, and provide the details of the fourth-order Runge-Kutta scheme.

### 8.2.1.1 The given rotating beam

The governing differential equation for the transverse free vibrations of the given tapered rotating beam as shown in Fig. 8.11 given as [104]:

$$\frac{d^2}{dx^2}\left[\bar{f}(x)\frac{d^2Y}{dx^2}\right] - \frac{d}{dx}\left[t(x)\frac{d^2Y}{dx^2}\right] = \eta^2\bar{m}(x)Y \tag{8.75}$$

where $\bar{f}(x)$ and $\bar{m}(x)$ are the non-dimensional stiffness and mass given by polynomials:

$$\bar{m}(x) \quad = \quad a_0 + a_1x + a_2x^2 + \cdots + a_nx^n \tag{8.76}$$
$$\bar{f}(x) \quad = \quad b_0 + b_1x + b_2x^2 + \cdots + b_kx^k \tag{8.77}$$

$t(x)$ is the axial force due to centrifugal stiffening, given by

$$t(x) = \lambda^2 \int_0^x \bar{m}(x)(1-x)dx \tag{8.78}$$

$\lambda$ and $\eta$ are the non-dimensional rotating speed and natural frequency, respectively.

Using the Barcilon-Gottlieb transformation [17, 86], Eq. (8.75) can be transformed into the canonical form:

$$\frac{d^4 U}{dz^4} + \frac{d}{dz}\left[\bar{A}\frac{dU}{dz}\right] + \bar{B}U = \eta^2 U \tag{8.79}$$

The coefficients in the $(x, Y)$ frame are transformed into the $(z, U)$ frame, using the following relations.

$$Y = \bar{q}U \tag{8.80}$$

$$z = \int_0^x \bar{p}\,dx \tag{8.81}$$

where $\bar{p}$ and $\bar{q}$ are the auxiliary variables involved in the transformation, given by

$$\bar{p} = \left(\frac{\bar{m}(x)}{\bar{f}(x)}\right)^{0.25} \quad ; \quad \bar{q} = \left(\bar{m}(x)^3\,\bar{f}(x)\right)^{-0.125} \tag{8.82}$$

$$\bar{A} = 4\frac{\bar{q}_{zz}}{\bar{q}} - 6\frac{\bar{q}_z^2}{\bar{q}^2} - 2\frac{\bar{q}_z\bar{p}_z}{\bar{q}\bar{p}} + \frac{\bar{p}_{zz}}{\bar{p}} - 2\frac{\bar{p}_z^2}{\bar{p}^2} - t\bar{q}^2\bar{p} \tag{8.83}$$

$$\bar{B} = \frac{\bar{q}_{zzzz}}{\bar{q}} - 4\frac{\bar{q}_z\bar{q}_{zzz}}{\bar{q}^2} - 2\frac{\bar{q}_{zz}^2}{\bar{q}^2} + 6\frac{\bar{q}_z^2\bar{q}_{zz}}{\bar{q}^3} + \frac{\bar{q}_z\bar{p}_{zzz}}{\bar{q}\bar{p}} + \frac{\bar{q}_{zz}\bar{p}_{zz}}{\bar{q}\bar{p}} - 4\frac{\bar{q}_z^2\bar{p}_{zz}}{\bar{q}^2\bar{p}}$$
$$- 5\frac{\bar{q}_z\bar{p}_z\bar{p}_{zz}}{\bar{q}\bar{p}^2} - 4\frac{\bar{q}_z\bar{q}_{zz}\bar{p}_z}{\bar{q}^2\bar{p}} + 6\frac{\bar{q}_z^3\bar{p}_z}{\bar{q}^3\bar{p}} + 6\frac{\bar{q}_z^2\bar{p}_z^2}{\bar{q}^2\bar{p}^2} + 4\frac{\bar{q}_z\bar{p}_z^3}{\bar{q}\bar{p}^3} - 2\frac{\bar{q}_{zz}\bar{p}_z^2}{\bar{q}\bar{p}^2} - \bar{q}\frac{d}{dz}[t\bar{p}\bar{q}_z] \tag{8.84}$$

### 8.2.1.2   The isospectral non-rotating beam

Let us assume that the isospectral non-rotating beam, shown in Fig. 8.12 has a non-dimensional stiffness and mass functions given by $f(x)$ and $m(x)$. The governing equation of free vibration for a non-rotating Euler-Bernoulli beam, in its non-dimensional form is

$$\frac{d^2}{dx^2}\left[f(x)\frac{d^2 Y}{dx^2}\right] = \eta^2 m(x)Y \tag{8.85}$$

where $Y$ is the transverse displacement and $\eta$ is the non-dimensional natural frequency. The Barcilon-Gottlieb transformation is used to transform Eq. (8.85) into a canonical form:

$$\frac{d^4 U}{dz^4} + \frac{d}{dz}\left[A\frac{dU}{dz}\right] + BU = \eta^2 U \tag{8.86}$$

where the variables $Y$ and $x$ are replaced by $U$ and $z$, respectively, using the transformation:

$$Y = qU \tag{8.87}$$

$$z = \int_0^x p\,dx \tag{8.88}$$

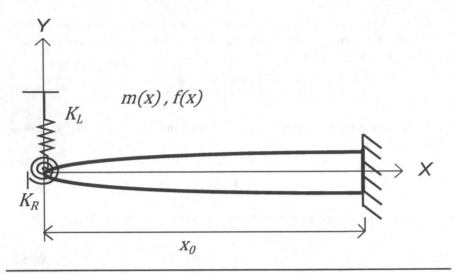

**Figure 8.12**
Schematic of a non-rotating beam

Here, $p$ and $q$ are the auxiliary variables, defined as:

$$p = \left(\frac{m}{f}\right)^{0.25} \; ; \; q = (m^3 f)^{-0.125} \tag{8.89}$$

$A$ and $B$ are the coefficients of the canonical equation (8.86), given by

$$A = 4\frac{q_{zz}}{q} - 6\frac{q_z^2}{q^2} - 2\frac{q_z p_z}{qp} + \frac{p_{zz}}{p} - 2\frac{p_z^2}{p^2} \tag{8.90}$$

$$B = \frac{q_{zzzz}}{q} - 4\frac{q_z q_{zzz}}{q^2} - 2\frac{q_{zz}^2}{q^2} + 6\frac{q_z^2 q_{zz}}{q^3} + \frac{q_z p_{zzz}}{qp} + \frac{q_{zz} p_{zz}}{qp} - 4\frac{q_z^2 p_{zz}}{q^2 p}$$

$$- 5\frac{q_z p_z p_{zz}}{qp^2} - 4\frac{q_z q_{zz} p_z}{q^2 p} + 6\frac{q_z^3 p_z}{q^3 p} + 6\frac{q_z^2 p_z^2}{q^2 p^2} + 4\frac{q_z p_z^3}{qp^3} - 2\frac{q_{zz} p_z^2}{qp^2} \tag{8.91}$$

Thus, the rotating beam equation given by Eq. (8.75) and the non-rotating beam equation given by Eq. (8.85),which are in the $(x, Y)$ frame, can be transformed into Eqs. (8.79) and (8.86), respectively, which are in the $(z, U)$ frame. Therefore, Eq. (8.75) and Eq. (8.85) are isospectral to each other if and only if the coefficients in Eqs. (8.79) and (8.86) match with each other. Therefore , if

$$A = \bar{A} \; ; B = \bar{B} \tag{8.92}$$

then the non-rotating beam is isospectral to the given tapered rotating beam. Thus, the unknown entities $A$ and $B$ are to be determined using the known entities $\bar{A}$ and $\bar{B}$. The equations $A = \bar{A}; B = \bar{B}$ are coupled ordinary differential equations, and can be solved numerically using the fourth-order Runge-Kutta scheme for solving ordinary differential equations.

### 8.2.1.3 The fourth order Runge-Kutta scheme

The equations $A = \bar{A}; B = \bar{B}$ are coupled differential equations for which an analytical solution may not exist. Therefore, we solve these equations using the fourth-order Runge-Kutta scheme. Let us introduce the variables $q_1, q_2, q_3, p_1, p_2$ defined as:

$$\frac{dq}{dz} = q_1 \; ; \; \frac{dq_1}{dz} = q_2 \; ; \; \frac{dq_2}{dz} = q_3 \; ; \; \frac{dp}{dz} = p_1 \; ; \; \frac{dp_1}{dz} = p_2 \tag{8.93}$$

Substituting the above equations into Eqs. (8.90) and (8.91), we have:

$$\begin{aligned}
\frac{dp_2}{dz} &= -p \left[ \frac{4p_1^3}{p^3} + \frac{2p_1^2 q_1}{p^2 q} + \frac{2p_1 q_1^2}{pq^2} + \frac{12q_1^3}{q^3} - \frac{5p_1 p_2}{p^2} \right. \\
&\quad \left. - \frac{2q_1 p_2}{pq} - \frac{2p_1 q_2}{pq} - \frac{16q_1 q_2}{q^2} + \frac{4q_3}{q} - \bar{A}_z \right] = p_3
\end{aligned} \tag{8.94}$$

$$\begin{aligned}
\frac{dq_3}{dz} &= -q \left[ \frac{q_4}{q} - 4\frac{q_2 q_3}{q^2} - 2\frac{q_2^2}{q^2} + 6\frac{q_1^2 q_2}{q^3} + \frac{q_1 p_3}{qp} + \frac{q_2 p_2}{qp} - 4\frac{q_1^2 p_2}{q^2 p} \right. \\
&\quad \left. - 5\frac{q_1 p_1 p_2}{qp^2} - 4\frac{q_1 q_2 p_1}{q^2 p} + 6\frac{q_z^3 p_z}{q^3 p} + 6\frac{q_z^2 p_z^2}{q^2 p^2} + 4\frac{q_z p_z^3}{qp^3} - 2\frac{q_2 p_1^2}{qp^2} - \bar{B} \right]
\end{aligned} \tag{8.95}$$

Note that Eqs. (8.93) to (8.95) are in the form of

$$\frac{d\mathbf{X}}{dz} = \mathbf{g}(z, \mathbf{X}) \tag{8.96}$$

where $\mathbf{X} = [p_2, p_1, p, q_3, q_2, q_1, q]^T$. In summary, the equations $A = \bar{A}; B = \bar{B}$ are modified into Eqs. (8.93) to (8.95), which are in the form of $\mathbf{X}_z = g(\mathbf{X}, z)$. The fourth-order Runge-Kutta method can be used to numerically solve such differential equations. The initial conditions for the fourth-order Runge-Kutta method are at $z = 0$ can be given as:

$$\mathbf{X}|_{(z=0)} = [\bar{p}_{zz}, \bar{p}_z, \bar{p}, \bar{q}_{zzz}, \bar{q}_{zz}, \bar{q}_z, \bar{q}]^T|_{(z=0)} \tag{8.97}$$

The step size is taken to be $h = 1/10000$.

### 8.2.1.4 Boundary conditions

While transforming the equations from the $(x, Y)$ frame to $(z, U)$ frame, the boundary conditions also get transformed. The clamped boundary condition remains invariant in the Barcilon-Gottlieb transformation. For all the other boundary conditions, the initial conditions on the stiffness and mass must be chosen in such a way that the boundary conditions remain invariant. The Barcilon-Gottlieb transformation transforms the derivatives $Y', Y'', Y'''$ in terms of $U_z, U_{zz}, U_{zzz}$ as follows:

$$Y = qU \tag{8.98}$$

$$\frac{dY}{dx} = p \left[ qU_z + q_z U \right] \tag{8.99}$$

$$\frac{d^2Y}{dx^2} = p \left[ (p_z q_z + p q_{zz})U + (q p_z + 2 p q_z)U_z + (pq)U_{zz} \right] \tag{8.100}$$

$$\frac{d^3Y}{dx^3} = p \left[ (p_z^2 q_z + p q_z p_{zz} + 3 p p_z q_{zz} + p^2 q_{zzz})U \right. \tag{8.101}$$
$$+ (q p_z^2 + 6 p p_z q_z + p q p_{zz} + 3 p q_{zz})U_z$$
$$\left. + (3 p q p_z + 3 p^2 q_z)U_{zz} + (p^2 q)U_{zzz} \right]$$

### Free Boundary Condition (Rotating Beam)

The rotating beam is clamped at the root and free at the tip. The corresponding boundary conditions are: At $x = 1$:

$$Y(x) = 0 = Y'(x) \tag{8.102}$$

At $x = 0$:

$$Y''(x) = 0 = Y'''(x) \tag{8.103}$$

The Barcilon-Gottlieb transformation transforms the above equations into the following

$$U = U_z = 0 \tag{8.104}$$

$$(\bar{p}_z \bar{q}_z + \bar{p} \bar{q}_{zz})U + (\bar{q} \bar{p}_z + 2 \bar{p} \bar{q}_z)U_z + (\bar{p} \bar{q})U_{zz} = 0 \tag{8.105}$$

$$(\bar{p}_z^2 \bar{q}_z + \bar{p} \bar{q}_z p_{zz} + 3 \bar{p} \bar{p}_z \bar{q}_{zz} + \bar{p}^2 \bar{q}_{zzz})U \tag{8.106}$$
$$+ (\bar{q} \bar{p}_z^2 + 6 \bar{p} \bar{p}_z \bar{q}_z + \bar{p} \bar{q} p_{zz} + 3 \bar{p} \bar{q}_{zz})U_z$$
$$+ (3 \bar{p} \bar{q} \bar{p}_z + 3 \bar{p}^2 \bar{q}_z)U_{zz} + (\bar{p}^2 \bar{q})U_{zzz} = 0$$

**Boundary Conditions on the Non-Rotating Beam** The non-rotating beam is clamped at the root and has translational and torsional springs at the tip. The corresponding boundary conditions are at $x = 1$:

$$Y(x) = 0 = Y'(x) \tag{8.107}$$

Free boundary condition at $x = 0$:

$$Y''(x) = k_R Y'(x) \tag{8.108}$$

$$Y'''(x) = k_L Y(x) \tag{8.109}$$

where $k_L$ and $k_R$ are the spring constants of the translational and torsional springs, respectively. The Barcilon-Gottlieb transformation transforms the above four equations into the $(z, U)$ frame as follows:

$$U = U_z = 0 \tag{8.110}$$

$$(p_z q_z + p q_{zz})U + (q p_z + 2 p q_z)U_z + (pq)U_{zz} = k_R(q U_z + q_z U) \tag{8.111}$$

$$(p_z^2 q_z + pq_z p_{zz} + 3pp_z q_{zz} + p^2 q_{zzz})U \tag{8.112}$$
$$+(qp_z^2 + 6pp_z q_z + pqp_{zz} + 3pq_{zz})U_z$$
$$+(3pqp_z + 3p^2 q_z)U_{zz} + (p^2 q)U_{zzz} = k_L(qU)$$

The spring constants $k_L$ and $k_R$ must be chosen in such a way that the boundary conditions in $(z, U)$ frame of both the rotating and the non-rotating beams are identical. In other words, the coefficients of Equations (8.104)-(8.106) and (8.110)-(8.112) must match with each other.

### 8.2.1.5 Illustrative Example 1: Linear Mass and cubic stiffness

Let us consider a tapered rotating beam with $\bar{m}(x)$ and $\bar{f}(x)$, given as

$$\bar{m}(x) = (0.5x + 0.5) \tag{8.113}$$
$$\bar{f}(x) = (0.5x + 0.5)^3 \tag{8.114}$$

This type of tapered rotating beam has a solid rectangular cross section with constant width and linearly varying height [104]. The auxiliary variables $\bar{p}(x)$ and $\bar{q}(x)$ for this beam are given as:

$$\bar{p}(x) = \left(\frac{\bar{m}}{\bar{f}}\right)^{0.25} = \sqrt{\frac{2}{(x+1)}} \tag{8.115}$$

$$\bar{q}(x) = \left(\frac{2}{(x+1)}\right)^{0.75} \tag{8.116}$$

Also, the relation between $z$ and $x$ is given by

$$z = \int_0^x \bar{p}(x)dx = 2\sqrt{2}(\sqrt{x+1} - 1) \tag{8.117}$$

$$\Rightarrow x = z^2/8 + z/\sqrt{2} \tag{8.118}$$

Using the above relations, the mass and stiffness functions in the $z, U$ frame can be computed as:

$$\bar{m}(z) = (z^2/16 + z/(2\sqrt{2}) + 1/2) \tag{8.119}$$
$$\bar{f}(z) = (z^2/16 + z/(2\sqrt{2}) + 1/2)^3 \tag{8.120}$$
$$\bar{p}(z) = \left(\frac{\bar{m}}{\bar{f}}\right)^{0.25} = (z^2/16 + z/(2\sqrt{2}) + 1/2)^{-0.5} \tag{8.121}$$
$$\bar{q}(z) = (z^2/16 + z/(2\sqrt{2}) + 1/2)^{0.75} \tag{8.122}$$

The expressions for $\bar{A}(z)$ and $\bar{B}(z)$ can be computed by substituting the above expressions of $\bar{q}(z)$ and $p(z)$ in Eqs. (8.83) and (8.84).

$$\bar{A}(z) = \frac{-288 + \lambda^2 z\left(-768\sqrt{2} - 192z + 128\sqrt{2}z^2 + 96z^3 + 12\sqrt{2}z^4 + z^5\right)}{192\left(8 + 4\sqrt{2}z + z^2\right)} \tag{8.123}$$

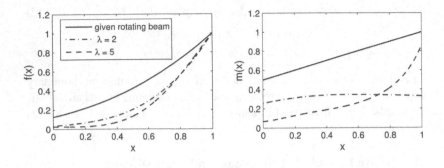

**Figure 8.13**
Variations of mass $m(x)$ and stiffness $f(x)$ of non-rotating beams isospectral to the given rotating tapered beam, rotating at $\lambda = 2, 5$, for illustrative example 1

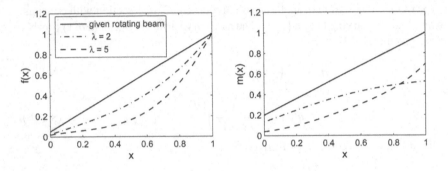

**Figure 8.14**
Variations of mass $m(x)$ and stiffness $f(x)$ of non-rotating beams isospectral to the given rotating tapered beam, rotating at $\lambda = 2, 5$, for illustrative example 2

$$\bar{B}(z) = -\frac{1008}{256\left(8 + 4\sqrt{2}z + z^2\right)^4}\left[\left(64 + 64\sqrt{2}z + 48z^2 + 8\sqrt{2}z^3 + z^4\right) + \quad (8.124)\right.$$

$$\lambda^2\left(8 + 4\sqrt{2}z + z^2\right)\left(-49152 - 18432\sqrt{2}z + 26112z^2 + 27136\sqrt{2}z^3 + \right.$$

$$(8.125)$$

$$\left.\left. 22720z^4 + 5408\sqrt{2}z^5 + 1496z^6 + 112\sqrt{2}z^7 + 7z^8\right)\right] \quad (8.126)$$

The above expressions are used as inputs for the fourth-order Runge-Kutta method, and the mass $m(x)$ and stiffness $f(x)$ distributions of a non-rotating beam isospectral to the tapered beam rotating at $\lambda$, are obtained as outputs. These $m(x)$ and $f(x)$ distributions are plotted in Fig. 8.13 for $\lambda = 2$ with no elastic springs.

As $\lambda$ increases, it becomes arduous for the variation in mass and the stiffness functions, to counter the stiffening effect of the centrifugal force, acting on the ta-pered rotating beam. For higher values of $\lambda$, we can attach an elastic support, such as a torsional and a translational spring, at the free end of the non-rotating beam, in order to increase the stiffness of the beam.

These $m(x)$ and $f(x)$ distributions are plotted for $\lambda = 5$ are in Fig. 8.13. The values of the spring constants for this case are $k_L = 0.5, k_R = 0.75$.

### 8.2.1.6    Illustrative Example 2: Linear mass and linear stiffness

In this example, the rotating tapered beam has a linearly varying mass and stiffness distributions such that

$$\bar{m}(x) = 0.2 + 0.8x \tag{8.127}$$
$$\bar{f}(x) = 0.05 + 0.95x \tag{8.128}$$

These type of tapered rotating beams are used as models for wind turbine blades [207].

Similar analysis as shown in the previous example can be used to calculate $\bar{A}(z)$ and $\bar{B}(z)$. These $\bar{A}(z)$ and $\bar{B}(z)$ functions are used as inputs for the fourth-order Runge-Kutta method to calculate the mass $m(x)$ and stiffness $f(x)$ distributions of a non-rotating beam isospectral to the tapered beam rotating at $\lambda$. These $m(x)$ and $f(x)$ distributions are plotted for $\lambda = 2$ and $\lambda = 5$ in Fig. 8.14. The values of the spring constants for the $\lambda = 5$ case are $k_L = 0.75, k_R = 0.5$, while no springs are used for the $\lambda = 2$ case.

### 8.2.1.7    Finite element formulation

In order to compute the natural frequencies of the non-uniform non-rotating beams, we use the finite element method [36]. In the finite element formulation, we divide the beam into a number of finite elements. The displacement, $Y(x)$ is assumed to have a cubic distribution along each finite element and hence, has four degrees of freedom (4-dof). Two of these 4-dof are displacement DOF and the other two are slope DOF, at the ends of each finite element. Specifically, $Y(x)$, along the $i^{th}$ element, is given by

$$Y(x) = H_1(x - x_i)q_{1,i} + H_2(x - x_i)q_{2,i} + H_3(x - x_i)q_{3,i} + H_4(x - x_i)q_{4,i} \tag{8.129}$$

where $x_i$ is the coordinate of the left node, $x_{i+1}$ is the coordinate of the right node, $q_{1,i}$ and $q_{2,i}$ are the displacement DOF and slope DOF at the left node, $q_{3,i}$ and $q_{4,i}$ are the displacement DOF and slope DOF at the right node [122]. The Hermite shape functions $(H_1, H_2, H_3$ and $H_4)$ are given by $H_1(\xi) = 1 - (3\xi^2)/l^2 + (2\xi^3)/l^3$ ; $H_2(\xi) = \xi - (2\xi^2)/l + \xi^3/l^2$ ; $H_3(\xi) = (3\xi^2)/l^2 - (2\xi^3)/l^3$ ; $H_4(\xi) = -\xi^2/l + \xi^3/l^2$, where $l = x_{i+1} - x_i$ element length.

The expressions for the kinetic energy $T_k$ and potential energy $U_p$ of the non-rotating beam, are given by

$$T_k = \int_0^1 m(x)\dot{Y}^2 dx \tag{8.130}$$

$$U_p = \int_0^1 f(x)(Y''(x))^2 dx + 0.5 * k_L Y(0)^2 + 0.5 * k_R Y'(0)^2 \tag{8.131}$$

The natural frequencies($\eta$) and mode shapes($\hat{V}$) are then obtained as follows. First, we substitute the expression for displacement $Y(x)$, from Eq. (8.129), into equations (8.130) and (8.131). Next we apply Lagrange's equations of motion, which yields the well known eigenvalue problem:

$$\hat{K}\hat{V} = \eta^2 \hat{M}\hat{V} \tag{8.132}$$

where, $\hat{M}$ and $\hat{K}$ are the global mass and stiffness matrices. These global mass and stiffness matrices are obtained by assembling the element level mass ($\tilde{M}$) and stiffness ($\tilde{K}$) matrices. The $i^{th}$ element's mass matrix ($\tilde{M}_i$) and stiffness matrix ($\tilde{K}_i$) are given by

$$(\tilde{M}_{jk})_i = \int_0^l m(\xi + x_i)H_j(\xi)H_k(\xi)d\xi \tag{8.133}$$

$$(\tilde{K}_{jk})_i = \int_0^l f(\xi + x_i)H_j''(\xi)H_k''(\xi)d\xi \tag{8.134}$$

The first three frequencies of the tapered rotating beam given in illustrative example 1, are given in [207], and are tabulated in Table 8.4, for rotational speed $\lambda = 2, 5$. The first three frequencies of non-rotating beam, isospectral to this tapered beam rotating at speed $\lambda = 2, 5$, calculated by the finite element analysis, are tabulated in Table 8.4. Since results for higher modes are not available in literature, a comparison of only the first three modes is presented here.

The frequencies of the tapered rotating beam given in illustrative example 2, calculated using the method of Frobenius are given in [207], and are tabulated in Table 8.5, for rotational speed $\lambda = 2, 5$. The frequencies of non-rotating beam, isospectral to this tapered beam rotating at speed $\lambda = 2, 5$, calculated by the finite element analysis, are tabulated in Table 8.5. These results provide numerical verification of the isospectral properties of the mass and stiffness functions, derived in the earlier section.

## 8.2.2  RECTANGULAR CROSS-SECTION

In this section, we analyze beams having a rectangular cross-section. The mass $m(x)$ and stiffness $f(x)$ of such a beam is related to the breadth and height of the rectangular cross section, as follows:

$$m = k_1 bh \tag{8.135}$$

and

$$f = k_2 bh^3 \tag{8.136}$$

**Table 8.4**

**The Frequencies of the Tapered Rotating Beam ($\eta$) [104], and the Frequencies of the Isospectral Non-rotating Beam Calculated Using FEM($\hat{\eta}$) (illustrative example 1)**

| Mode | Rotating Beam ($\eta$) | Non-Rotating Beam ($\hat{\eta}$) | Rotating Beam ($\eta$) | Non-Rotating Beam ($\hat{\eta}$) |
|------|------|------|------|------|
| | | $\lambda = 2$ | | $\lambda = 5$ |
| 1 | 4.437 | 4.437 | 6.743 | 6.743 |
| 2 | 18.937 | 18.937 | 21.905 | 21.905 |
| 3 | 47.871 | 47.872 | 50.934 | 50.933 |

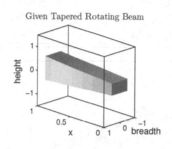

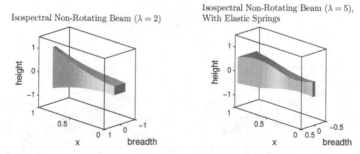

**Figure 8.15**

Variations of breadth and height of non-rotating beams isospectral to a given tapered rotating beam rotating at $\lambda = 2, 5$ (in illustrative example 1)

$$b = \sqrt{\frac{m^3 k_2}{k_1^3 f}} \; ; \; h = \sqrt{\frac{f k_1}{m k_2}} \tag{8.137}$$

where $k_1$ and $k_2$ are the non-dimensional constants. Also $b$ and $h$ are the breadth

**Table 8.5**

**The Frequencies of the Tapered Rotating Beam ($\eta$) [207], and the Frequencies of the Isospectral Non-rotating Beam Calculated Using FEM($\hat{\eta}$) (Illustrative Example 2)**

| Mode | Rotating Beam ($\eta$) | Non-Rotating Beam ($\hat{\eta}$) | Rotating Beam ($\eta$) | Non-Rotating Beam ($\hat{\eta}$) |
|------|------------------------|----------------------------------|------------------------|----------------------------------|
|      | $\lambda = 2$          |                                  | $\lambda = 5$          |                                  |
| 1    | 5.725                  | 5.725                            | 7.644                  | 7.644                            |
| 2    | 24.413                 | 24.413                           | 26.458                 | 26.458                           |
| 3    | 60.366                 | 60.367                           | 62.406                 | 62.407                           |
| 4    | 113.306                | 113.307                          | 115.370                | 115.372                          |
| 5    | 183.422                | 183.424                          | 185.504                | 185.509                          |

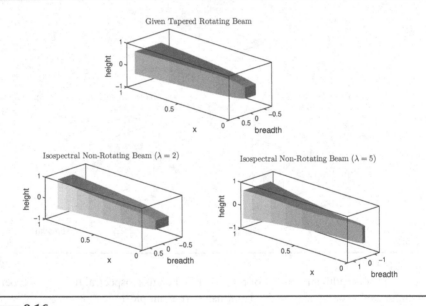

**Figure 8.16**
Variations of breadth and height of non-rotating beams isospectral to a given tapered rotating beams rotating at $\lambda = 2, 5$ (illustrative example 2)

and height of the rectangular cross section, respectively. Therefore, using $m(x)$ and $f(x)$ functions derived in earlier sections, one can discover beams that have a rectangular cross-section. The breadth ($b$) and height ($h$) distribution of these beams can be determined from Eq. (8.137). The $b$ and $h$ distributions for non-rotating beams, which are isospectral to a tapered beam rotating at a speed $\lambda = 2, 5$, for the illustrative examples 1 and 2, are plotted in Figs. 8.15 and 8.16, respectively. It can be seen that these are physically realizable beams, for example through numerical machining [170].

## 8.3 SUMMARY

Two rotating beam cases are considered in this chapter. The uniform rotating beam is found to have an isospectral non-rotating counterpart. The tapered rotating beams are also found to have isospectral non-rotating counterparts.

A method of obtaining non-uniform non-rotating beams that are isospectral to a given *uniform rotating beam* is first proposed in this chapter. The governing equation of the non-uniform non-rotating beam, is recast to that of a uniform rotating beam, using the auxiliary variables defined in Barcilon-Gottlieb transformation, by matching the coefficients in both the equations. We have provided the necessary conditions, on the auxiliary variables, so that the boundary conditions remain invariant. For high rotation speeds, we generate physically realizable mass and stiffness functions of isospectral non-rotating beams, that have a torsional spring at the free end. The natural frequencies of the derived non-rotating beams, are also calculated using FEM, and are found to be identical to the frequencies of the given uniform rotating beam. A numerical example of a beam with a rectangular cross section is invoked to show the application of this analysis. The breadth and height variations, of the rectangular cross-section, with span, of isospectral rotating and non-rotating beams, are obtained in this chapter.

A semi-analytical method of obtaining non-rotating beams that are isospectral to a given *tapered rotating beam* is presented in this chapter as the second case. The governing equation of the non-rotating and the rotating beams are modified using the Barcilon-Gottlieb transformation, and the coefficients in both the equations are matched with each other. Isospectral non-rotating beams are obtained for rotating beams with (*a*) Linear mass and cubic stiffness variation and (*b*) Linear mass and stiffness variations. The natural frequencies of the derived non-rotating beams, are also calculated using FEM, and are found to be same as the frequencies of the given rotating beam. A practical example of a beam having a rectangular cross section is presented to show the application of this analysis. The breadth and height variations, of the rectangular cross-section, with span, of isospectral rotating and non-rotating beams, are obtained in our study.

Since experimental determination of rotating beam frequencies is an onerous task, the isospectral non-rotating beams could be tested instead. Modern numerical machining and 3D printing approaches allow such non-rotating beams to be manufactured. Recent advancements in computerized machining techniques such as additive manufacturing or rapid-prototyping [106, 121], facilitate the manufacturing of any complex 3D shape. Therefore, it is possible to manufacture isospectral rotating and non-rotating beams, using the breadth and height variations of the cross-section, with the help of computerized machining.

# 9 Beams with Shared Eigenpairs

Rotating beams are useful mathematical models of helicopter rotor blades, wind turbine blades, propeller blades, gas turbine blades and other important mechanical structures. Such beams can undergo vibration problems if their rotating natural frequencies coincide with multiples of the rotation speed. Therefore, the accurate computation of natural frequencies and mode-shapes of a rotating beam is an important problem [104]. The dynamics of a rotating beam is dictated by a fourth-order partial-differential equation which cannot be solved exactly for natural frequencies and mode-shapes, even for a uniform rotating beam [207]. Therefore, the frequencies of a rotating beam are typically estimated by approximate methods like the Rayleigh-Ritz method [180, 219], Galerkin method [8, 126] etc, or by the finite element method [11, 97, 98, 99, 146, 192, 194, 199]. However, if an exact solution to the governing equation is not available, the approximate methods are verified by comparing with other numerical methods or with series solutions obtained using the Frobenius method of solving differential equations [130]. The series solution approach is also termed as a semi-analytical approach and is the closest available result to the exact solution of the rotating beam equation. Another semi-analytical approach is the dynamic stiffness method [13, 14]. The dynamic stiffness method utilizes frequency dependent shape functions obtained from the solution of the governing equations of the structure. One advantage of the dynamic stiffness method is that only one element can yield all the natural frequencies and mode-shapes to any desired level of accuracy [16]. Banerjee et al. [16] exhort the dynamic stiffness method for checking the results of finite element analysis of rotating beams. Bambill et al. [10] used the differential quadrature method for vibration analysis of rotating Timoshenko beams. In this method, the derivatives are approximated using the weighted linear summation of functional values at all sample points in overall domain. Using this approximation, the differential equation is transformed into set of algebraic equations. The number of equations depends on number of sample points taken in the domain.

Attarnejad & Shahba [4] studied the free vibration of rotating non-prismatic Euler-Bernoulli beams by using differential transform method to evaluate the dynamic stiffness matrix. They also studied the free vibration analysis of rotating tapered beams by introducing the novel concept of dynamic basic displacement functions [6]. Ozgumus & Kaya [153, 154] studied the bending vibration characteristics of rotating Euler-Bernoulli beams by using the differential transform method (DTM). The differential transform is a method to determine the coefficients of the Taylor series of a function by solving the induced recursive equation obtained from the given differential equation. So although there is no simple closed form solution of the governing differential equation of the rotating beams, the approximate methods are checked by comparing them with the series solutions or numerical methods discussed above.

The finite element method, in its various avatars, has emerged as a popular approach to the vibration analysis of rotating beams. Most of the early works on finite element methods were based on the h-version, which uses Hermite cubic polynomials as shapes functions [192]. In this version, the stiffness and mass distribution are considered uniform within an element, hence to cater for non-uniform structures, a large number of elements are required for accurate determination of frequencies and mode shapes. But in recent years, new types of finite element methods emerged which make uses of only one element like the p-version [104], Fourier-p super element [99] and spectral methods [194]. In p-version FEM, the order of the basis polynomial is increased for convergence of results and this approach is highly efficient for structures with non-uniformity. Some recent works recommend the use of a single element using p-version, Fourier-p, dynamic finite element and spectral approaches [102].

In the h-version FEM, the number of elements is increased until convergence is achieved. Typically, the stiffness and mass distribution are assumed to be uniform within the element. Therefore, a large number of elements are needed for non-uniform beams. In other perspectives based on a single element, the order of the basis functions is increased for convergence. Practical structures posses highly non-uniform variations in mass and stiffness properties. Use of a single element allows easy inclusion of the non-uniformity as functions $EI(x)$ and $m(x)$, for flexural stiffness and mass per unit length, respectively. Furthermore, the model order using spectral and Fourier-p FEM approaches is typically less than that with h-version, which is useful for control applications.

As new single element methods for rotating beam analysis are created, they should be validated without requiring a different analysis. Similarly, the h-version codes need to be validated. The lacuna here lies in the absence of an exact solution for vibration analysis of rotating beams. However, a uniform non-rotating beam has exact solutions to its frequencies and mode-shapes, determined by solving a transcendental equation [139].

In this chapter, we seek to find out an equivalent rotating beam with the same frequency and mode-shape of a uniform non-rotating beam for a particular mode. Such beams, if they exist, would be of interest from a fundamental science perspective and also yield test functions, which could be used to check approximate methods for rotating beams and for the design of such beams to match the frequencies and mode-shapes of non-rotating beams. The first section solves this problem for a cantilever beam and the second section solves the problem for a pinned-free beam. The material in this chapter is adapted from Ref. [122] and Ref. [172].

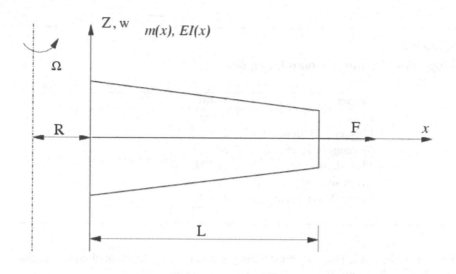

**Figure 9.1**
Schematic of a rotating cantilever beam

## 9.1 ROTATING BEAMS AND NON-ROTATING BEAMS WITH SHARED EIGENPAIR: WITH FIXED-FREE BOUNDARY CONDITIONS

### 9.1.1 FORMULATION

Contemplate a beam rotating about the vertical axis with uniform angular velocity $\Omega$ and with dimensions being given in Fig. 9.1. The governing equation for the out-of-plane vibrations of the rotating beam is expressed as,

$$(EI_1(x)w'')'' + m(x)\,\ddot{w} - (T(x)w')' = 0 \tag{9.1}$$

where $T(x) = \int_x^L m(x)\,\Omega^2\,(R+x)\,dx + F$, $EI_1(x)$ is the flexural stiffness, $m(x)$ is the mass per unit length of the beam, $w(x,t)$ is the out-of-plane bending displacement, $R$ is the hub radius, $L$ is the beam length and $F$ is the axial force applied at the free end of the beam.

Now, ponder upon a non-rotating beam of the same length, $L$, mass-per-unit-length, $m(x)$, and boundary conditions, as the rotating beam in Eq. (9.1). This beam is regulated by the equation [139],

$$(EI_2(x)w'')'' + m(x)\ddot{w} = 0 \tag{9.2}$$

where $EI_2(x)$ is the flexural stiffness of this beam.

Let the $j^{th}$ mode-shape and the frequency of these beams be the same and equal to $\phi_j$ and $\omega_j$, respectively. The bending displacement $w_j$ for the $j^{th}$ mode is given by

$$w_j = \phi_j e^{i\omega_j t} \tag{9.3}$$

**Table 9.1**

**Properties of Uniform Non-rotating Beam**

| Property | Value |
|----------|-------|
| Length ($L$) | $0.6\,m$ |
| Elasticity Modulus ($E$) | $2 \times 10^{11}\,Pa$ |
| Moment of Inertia ($I$) | $2 \times 10^{-9}\,m^4$ |
| Mass density ($\rho$) | $7840\,kg/m^3$ |
| Cross-section Area ($A$) | $2.4 \times 10^{-6}\,m^2$ |
| Linear Mass density ($m$) | $1.8816\,kg/m$ |

Evidently, the mode-shape $\phi_j$ must satisfy the governing differential equation. Substituting Eq. (9.3) into Eqs. (9.1) and (9.2), we obtain

$$(EI_1(x)\phi_j'')'' - m(x)\omega_j^2 \phi_j - (T(x)\phi_j')' = 0 \tag{9.4}$$

$$(EI_2(x)\phi_j'')'' - m(x)\omega_j^2 \phi_j = 0 \tag{9.5}$$

Subtracting Eq. (9.5) from Eq. (9.4) yields

$$(EI_1(x)\phi_j'')'' - (EI_2(x)\phi_j'')'' - (T(x)\phi_j')' = 0 \tag{9.6}$$

Eq. (9.6) provides the relationship between the flexural stiffness of a rotating beam and that of a non-rotating beam with same eigen-pair. Notice that the expression $\phi_j$ is different for different modes and thus different expressions for $EI_1(x)$ are derived for different modes. Explicitly, $EI_1(x)$ in terms of $EI_2(x)$ is

$$EI_1(x) = \frac{\int (\int ((EI_2(x)\phi_j'')'' + (T(x)\phi_j')') \, dx) \, dx}{\phi_j''} \tag{9.7}$$

If we consider a non-rotating beam of uniform flexural stiffness, then, $m(x) = m$, $EI_2(x) = EI_2$ and Eq. (9.7) becomes

$$EI_1(x) = \frac{\int (\int (EI_2\phi_j^{(4)} + (T(x)\phi_j')') \, dx) \, dx}{\phi_j''} \tag{9.8}$$

The mode-shapes of a uniform cantilever beam can be solved exactly and are expressed as the beam function [55],

$$\phi_j(x) = C_j(\cos(\alpha_j x) - \cosh(\alpha_j x)) +$$
$$C_j \left( \frac{\sin(\alpha_j L) - \sinh(\alpha_j L)}{\cos(\alpha_j L) + \cosh(\alpha_j L)} (\sin(\alpha_j x) - \sinh(\alpha_j x)) \right) \tag{9.9}$$

where $\alpha'_j s$ are solutions to the characteristic equation $\cos(\alpha_j L)\cosh(\alpha_j L) = -1$ and $C'_j s$ are arbitrary constants. The frequencies $\omega_j$ are expressed as [139],

$$\omega_j = (\alpha_j L)^2 \sqrt{\frac{EI_2}{mL^4}} \tag{9.10}$$

where $(\alpha_1 L)^2 = 3.5160$, $(\alpha_2 L)^2 = 22.0345$, $(\alpha_3 L)^2 = 61.6972$, $(\alpha_4 L)^2 = 120.902$, are the exact values up to four decimal places for the first four modes.

### 9.1.2 FLEXURAL STIFFNESS FUNCTIONS

We want to find flexural stiffness variations $EI_1(x)$ for each mode such that the frequency and mode-shape for the corresponding mode of the rotating beam are identical to that of a non-rotating beam. Eqs. (9.9) and (9.10) can be substituted in Eq. (9.8) to derive the corresponding $EI_1(x)$ of a rotating beam. We consider a non-rotating beam with uniform properties and dimensions as shown in Table 9.1 [157]. The first four natural frequencies of this beam, up to four decimal places are given as 142.4015, 892.4150, 2498.7879, and 4896.6271 radians per second, respectively. The corresponding non-dimensional frequencies ($\mu = \sqrt{\frac{m\omega^2 L^4}{EI}}$) are 3.5160, 22.0345, 61.6972 and 120.902, respectively. The functions $EI_1(x)$ derived from Eq. (9.8) are expressed in Eq. (9.11) for the first four mode-shapes. Here $(EI_1)_i$ corresponds to the $i^{th}$ mode.

$$(EI_1)_i = \frac{N_i}{D_i} \tag{9.11}$$

where

$$
\begin{aligned}
N_1 = & \left( \frac{-3.5160EI_2}{L^2} + 0.78441m\Omega^2 L^2 \right) \cos(z_1) - \frac{3.5160EI_2}{L^2} \cosh(z_1) \\
& + \frac{2.5811EI_2}{L^2} \sinh(z_1) + \left( \frac{2.5811EI_2}{L^2} - 0.57583m\Omega^2 L^2 \right) \sin(z_1) \\
& + 0.53330m\Omega^2 Lx\sin(z_1) + 0.3915m\Omega^2 Lx\cos(z_1) - 0.02866m\Omega^2 L^2 e^{z_1} \\
& + 0.36705m\Omega^2 x^2 \sin(z_1) - 0.5m\Omega^2 x^2 \cos(z_1) - 0.18692m\Omega^2 L^2 e^{-z_1} \\
& - 0.07094m\Omega^2 Lxe^{z_1} + 0.46240m\Omega^2 Lxe^{-z_1} + 0.06647m\Omega^2 x^2 e^{z_1} \\
& + 0.43352m\Omega^2 x^2 e^{-z_1} + c_1 x + c_2
\end{aligned} \tag{9.12}
$$

$$D_1 = \frac{-3.5160}{L^2} \cos(z_1) + \frac{-3.5160}{L^2} \cosh(z_1) + \frac{2.5811}{L^2} \sin(z_1) + \frac{2.5811}{L^2} \sinh(z_1) \tag{9.13}$$

$$N_2 = \left(\frac{-22.034EI_2}{L^2} + 0.54538m\Omega^2L^2\right)\cos(z_2) - \frac{-22.034EI_2}{L^2}\cosh(z_2)$$

$$+ \frac{22.441EI_2}{L^2}\sinh(z_2) + \left(\frac{22.441EI_2}{L^2} - 0.55546m\Omega^2L^2\right)\sin(z_2)$$

$$+ 0.21303m\Omega^2Lx\sin(z_2) + 0.21697m\Omega^2Lx\cos(z_2) + 0.00420m\Omega^2L^2e^{z_2}$$

$$+ 0.50923m\Omega^2x^2\sin(z_2) - 0.5m\Omega^2x^2\cos(z_2) - 0.45881m\Omega^2L^2e^{-z_2}$$

$$+ 0.00197m\Omega^2Lxe^{z_2} + 0.21500m\Omega^2Lxe^{-z_2} + 0.00462m\Omega^2x^2e^{z_2}$$

$$+ 0.50462m\Omega^2x^2e^{-z_2} + c_1x + c_2$$

$$(9.14)$$

$$D_2 = \frac{-22.034}{L^2}\cos(z_2) + \frac{-22.034}{L^2}\cosh(z_2) + \frac{22.441}{L^2}\sin(z_2) + \frac{22.441}{L^2}\sinh(z_2)$$

$$(9.15)$$

$$N_3 = \left(\frac{-61.697EI_2}{L^2} + 0.51621m\Omega^2L^2\right)\cos(z_3) - \frac{61.697EI_2}{L^2}\cosh(z_3)$$

$$+ \frac{61.649EI_2}{L^2}\sinh(z_3) + \left(\frac{61.649EI_2}{L^2} - 0.51581m\Omega^2L^2\right)\sin(z_3)$$

$$+ 0.12731m\Omega^2Lx\sin(z_3) + 0.12721m\Omega^2Lx\cos(z_3) - 0.00018m\Omega^2L^2e^{z_3}$$

$$+ 0.49961m\Omega^2x^2\sin(z_3) - 0.5m\Omega^2x^2\cos(z_3) - 0.48360m\Omega^2L^2e^{-z_3}$$

$$- 0.49365 \times 10^{-4}m\Omega^2Lxe^{z_3} + 0.12726m\Omega^2Lxe^{-z_3} + 0.49981m\Omega^2x^2e^{-z_3}$$

$$+ 0.19388 \times 10^{-3}m\Omega^2x^2e^{z_3} + c_1x + c_2$$

$$(9.16)$$

$$D_3 = \frac{-61.697}{L^2}\cos(z_3) + \frac{-61.697}{L^2}\cosh(z_3) + \frac{61.649}{L^2}\sin(z_3) + \frac{61.649}{L^2}\sinh(z_3)$$

$$(9.17)$$

$$N_4 = \left( \frac{-120.90 E I_2}{L^2} + 0.50827 m \Omega^2 L^2 \right) \cos(z_4) - \frac{120.90 E I_2}{L^2} \cosh(z_4)$$

$$+ \frac{120.91 E I_2}{L^2} \sinh(z_4) + \left( \frac{120.91 E I_2}{L^2} - 0.50829 m \Omega^2 L^2 \right) \sin(z_4)$$

$$+ 0.09095 m \Omega^2 L x \sin(z_4) + 0.09095 m \Omega^2 L x \cos(z_4) - 0.49174 m \Omega^2 L^2 e^{z_4}$$

$$+ 0.50001 m \Omega^2 x^2 \sin(z_4) - 0.5 m \Omega^2 x^2 \cos(z_4) - 0.49174 m \Omega^2 L^2 e^{-z_4}$$

$$- 0.15258 \times 10^{-5} m \Omega^2 L x e^{z_4} + 0.09095 m \Omega^2 L x e^{-z_4} + 0.50002 m \Omega^2 x^2 e^{-z_4}$$

$$+ 0.83883 \times 10^{-5} m \Omega^2 x^2 e^{z_4} + c_1 x + c_2$$

$$(9.18)$$

$$D_4 = \frac{-120.90}{L^2} \cos(z_4) + \frac{-120.90}{L^2} \cosh(z_4) + \frac{120.91}{L^2} \sin(z_4) + \frac{120.91}{L^2} \sinh(z_4)$$

$$(9.19)$$

Here $z_1 = \frac{1.875x}{L}$, $z_2 = \frac{4.694x}{L}$, $z_3 = \frac{7.855x}{L}$, $z_4 = \frac{10.996x}{L}$ and $c_1$ and $c_2$ are arbitrary constants produced by integration. We notice that these functions contain trigonometric, hyperbolic, polynomial and exponential functions. It can also be seen that as $\Omega \to 0$, $E I_1 \to E I_2$.

The variation of $E I_1(x)$ versus $x$, with constants $c_1$ and $c_2$ equal to 0 is shown in Fig. 9.2, for a beam with properties listed in Table 9.1 and $\Omega = 360 rpm$ ($R = F = 0$), for the first four modes. The corresponding non-rotating beam has a uniform flexural stiffness, $E I_2 = 400 N m^2$. Notice that the $E I_1(x)$ variation for the $j^{th}$ mode has $j - 1$ internal singularities. The singularities are explained as follows. If the nodal coordinates of the $j^{th}$ mode of a uniform non-rotating beam are $x_1, x_2, \dots, x_{j-1}$, then the second derivative of the corresponding mode-shape, $\phi_j''$, vanishes at $x = L - x_1$, $L - x_2, \dots, L - x_{j-1}$. The proof is given in Appendix A. Also, at $x = L$, the second derivative must become zero because of the boundary condition $M = E I w'' = 0$, at the free end of the beam. Since Eqs. (9.7) and (9.8) have $\phi_j''$ in the denominator, the singularities occur at each of these points. The tip singularities are not shown in Fig. 9.2. We show later in the chapter that the tip singularities can be disregarded for numerical results.

### 9.1.3   NUMERICAL SIMULATIONS: P-VERSION FEM

The derived variation $E I_1(x)$ in a rotating beam must yield the same frequencies and mode-shapes as that of a non-rotating beam. For a uniform non-rotating beam, these frequencies and mode-shapes can be solved exactly, to any desired level of accuracy. If the functions $E I_1(x)$ obtained above are inset into a rotating beam finite element code, then we can determine the frequencies and mode-shapes of a non-rotating beam. This procedure can be used to test the veracity of the solutions obtained from the rotating beam code.

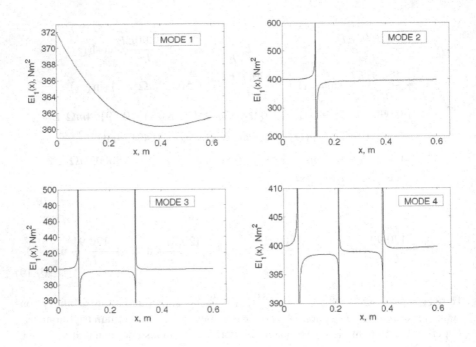

**Figure 9.2**
$EI_1(x)$ variation for first four modes for a beam with properties given in Table 9.1

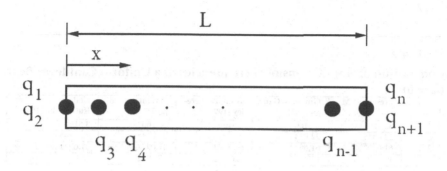

**Figure 9.3**
Beam element used for p-version FEM

For the FEM modeling, we use the p-version formulation, where only one element is used for the entire beam. Such approaches are popular in rotating beam analysis for handling non-uniformity and obtaining reduced order. The displacement $w$ is expanded in terms of an $n$th degree polynomial function as,

$$w = \sum_{i=0}^{n} p_i x^i \qquad (9.20)$$

Let the vector of generalized coordinates be $\mathbf{P} = \{p_i\}$. A set of $n-3$ nodes $-\alpha_1$, $\alpha_2, \ldots, \alpha_{n-3}''$ are identified in the interior of the beam at equidistant points as in Fig. 9.3. Define variables $q_i$ as $q_1 = w(0)$, $q_2 = w'(0)$, $q_3 = w(\alpha_1)$, $q_4 = w(\alpha_2), \ldots,$ $q_{n-1} = w(\alpha_{n-3})$, $q_n = w(L)$ and $q_{n+1} = w'(L)$.

Let $\mathbf{Q} = \{q_i\}$. Substituting Eq. (20) in these gives a relation between the $\mathbf{P}$ and $\mathbf{Q}$. The interpolation functions, $\psi_i$ are obtained for the $(n-1)$-node element by satisfying the conditions $Q_j = \delta_{ij}$, where $\delta$ is the Kronecker delta function and obtaining the corresponding constants, $p_{ij}$. The interpolation polynomials are then expressed as

$$\psi_i = \sum_{j=0}^{n} p_{ij} x^j \qquad (9.21)$$

The expressions for kinetic and potential/strain energy of the rotating beam are given as,

$$T = \frac{1}{2} \int_0^L m(x) \dot{w}^2 dx \qquad (9.22)$$

$$U = \frac{1}{2} \int_0^L EI_1(x)(w'')^2 dx + \frac{1}{2} \int_0^L T(x)(w')^2 dx \qquad (9.23)$$

After putting the displacement expressions into Eqs. (9.22) and (9.23), and applying Lagrange's equations of motion, the natural frequencies of the rotating beam are

**Table 9.2**

**Comparison of Non-dimensional Frequencies of a Uniform Cantilever Beam ($\lambda = 0$)**

| Mode | Present FEM | Gunda & Ganguli [97] | Wang & Wereley [199] | Wright et al. [207] | Hodges & Rutkowsky [104] |
|------|-------------|----------------------|----------------------|---------------------|--------------------------|
| 1 | 3.5160 | 3.5160 | 3.5160 | 3.5160 | 3.5160 |
| 2 | 22.0345 | 22.0345 | 22.0345 | 22.0345 | 22.0345 |
| 3 | 61.6972 | 61.6972 | 61.6972 | 61.6972 | 61.6972 |
| 4 | 120.902 | 120.902 | 120.902 | 120.902 | N/A |

computed from the general eigenvalue problem given by,

$$[\mathbf{K}]\{x\} = \omega^2 [\mathbf{M}]\{x\} \tag{9.24}$$

The (global) mass and stiffness matrices are obtained using,

$$(M_{ij})_n = \int_0^L m(x)\psi_i\psi_j\,dx \tag{9.25}$$

$$(K_{ij})_n = \int_0^L EI_1(x)\psi_i''\psi_j''\,dx + \int_0^L T(x)\psi_i'\psi_j'\,dx \tag{9.26}$$

Here $(M_{ij})_n$ and $(K_{ij})_n$ denote the global mass and stiffness matrices obtained by using an $n^{th}$ order expansion for $w$. The integration limits in Eqs. (9.25) and (9.26) are 0 and $L$, since only one element is used for the entire beam.

We first check the veracity of the p-version FEM code, for different non-dimensional rotating speeds ($\lambda = \sqrt{\frac{m\Omega^2 L^4}{EI}} = 0$ and 12), by comparing the computed results with published literature [97, 104, 199, 207]. The comparison is shown in Table 9.2 and 9.3. For the beam with properties in Table 9.1 and with $\Omega = 360rpm$, the first four non-dimensional rotating frequencies are 3.6600, 22.1615, 61.8225 and 121.0311, respectively.

The functions $EI_1(x)$ derived from Eqs. (9.11)–(9.19) may now be substituted into Eqs. (9.25) and (9.26) to test the veracity of the numerical code for a rotating beam. Since $EI_1(x)$ obtained from Eqs. (9.11)–(9.19) is a complicated function of $x$, the first term in Eq. (9.3) cannot be exactly calculated. Therefore, this integral needs to be evaluated using numerical methods. Gauss quadrature is used to evaluate the integral. Since the functions $EI_1(x)$ are not simple polynomial functions, a large number of Gauss-Legendre points are used for the quadrature. We have used 30 points in our analysis. It is ensured that the Gauss-quadrature nodes do not fall on the singularity points for accurate estimate of the integral. Such an approach is typically used to address singularities using numerical methods [24, 25, 112, 143]. With 30 Gauss-Legendre nodes in the domain, $(0\,m, 0.6\,m)$, it is observed that none of the nodes

### Table 9.3

**Comparison of Non-dimensional Frequencies of a Uniform Cantilever Beam ($\lambda = 12$)**

| Mode | Present FEM | Gunda & Ganguli [97] | Wang & Wereley [199] | Wright et al. [207] | Hodges & Rutkowsky [104] |
|---|---|---|---|---|---|
| 1 | 13.1702 | 13.1702 | 13.1702 | 13.1702 | 13.1702 |
| 2 | 37.6031 | 37.6032 | 37.6031 | 37.6031 | 37.6031 |
| 3 | 79.6145 | 79.6146 | 79.6145 | 79.6145 | 79.6145 |
| 4 | 140.534 | 140.535 | 140.534 | 140.534 | N/A |

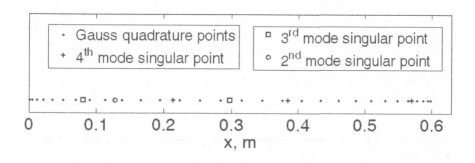

### Figure 9.4

Gauss quadrature points of order 30 for the domain $(0, 0.6)$ and the singularity points of the flexural stiffness functions

fall within a range of $\pm 0.01\,m$ of any singular points of the first four modes. This is illustrated in Fig. 9.4. However, this number of nodes provides sufficient sampling for an accurate evaluation of the integral as will be confirmed from numerical results. The same number of points can be used for any length of the beam as the property is maintained for all beam lengths. The algorithm for calculating the points and weights of Gauss quadrature for a general domain are provided in Appendix **B**.

Convergence of the code to the exact value of frequency, in rad/sec, within 4 decimal places of accuracy was obtained using polynomial order $n = 6, 8, 11$ and 14, for the first four modes, respectively. The utility of the derived functions is that they provide the exact value of the non-dimensional frequencies ($\mu = \sqrt{\frac{m\omega^2 L^4}{EI}}$) of a non-rotating beam, which are 3.5160, 22.0345, 61.6972 and 120.902 respectively for the first four modes. The code converges with alacrity, and is illustrated in Fig. 9.5. The nodal displacements are computed as eigenvectors, $\{x\}$ from Eq. (9.26). The mode-shapes are computed by using the interpolation polynomials to interpolate the displacements between nodal values. The normalized mode-shapes (with a value of 1 at $x = L$) calculated at convergence compare precisely with the exact mode-shapes

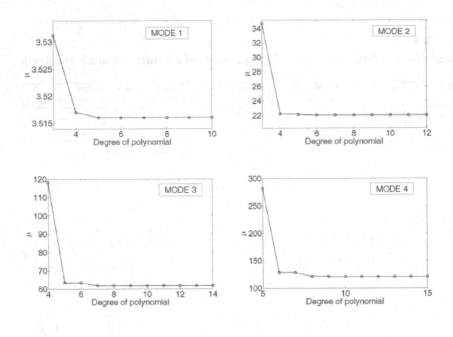

**Figure 9.5**
Convergence of the p-version finite element code for first four modes using derived flexural stiffness functions

from Eq. (9.9). The residue of mode-shapes obtained is of order of $10^{-7}$ or less and is shown in Fig. 9.16. Thus, it is numerically shown that the derived flexural stiffness functions give rotating beams with the same eigen-pair of the corresponding uniform non-rotating beam for a given mode. For the first mode, there exist no internal singularities and a physically feasible equivalent rotating beam exists. Moreover, the functions for modes higher than the first are integrable using Gauss quadrature despite the presence of singularities.

### 9.1.4  NUMERICAL SIMULATIONS: H-VERSION FEM

Though several rotating beam codes use the p-version type single element formulations, there exist legacy h-version codes, where an averaged value for the stiffness and mass per unit length is used for each element.

In the h-version FEM, the beam is discretized into several finite elements and relevant interpolation functions are used to set up equations in each element. We use Hermite cubic interpolation functions and 4-dof finite elements with displacement and slope dof at each node [157]. The natural frequencies of the rotating beam are obtained from the eigenvalue problem of Eq. (9.24).

**Table 9.4**

**Hermite Interpolation Polynomials**

| | |
|---|---|
| $\psi_1$ | $2(\frac{\bar{x}}{e})^3 - 3(\frac{\bar{x}}{e})^2 + 1$ |
| $\psi_2$ | $((\frac{\bar{x}}{e})^3 - 2(\frac{\bar{x}}{e})^2 + (\frac{\bar{x}}{e}))\, e$ |
| $\psi_3$ | $-2(\frac{\bar{x}}{e})^3 + 3(\frac{\bar{x}}{e})^2$ |
| $\psi_4$ | $((\frac{\bar{x}}{e})^3 - (\frac{\bar{x}}{e})^2)\, e$ |

Here, the global mass matrix $[\mathbf{M}]$ and the global stiffness matrix $[\mathbf{K}]$ of the beam are computed by assembling the elemental mass and stiffness matrices. These matrices for the $n^{th}$ finite element are expressed as,

$$(M_{ij})_n = \int_{x_n}^{x_{n+1}} m(\widetilde{x}_n)\psi_i\psi_j\, dx \tag{9.27}$$

$$(K_{ij})_n = \int_{x_n}^{x_{n+1}} EI_1(\widetilde{x}_n)\psi_i''\psi_j''\, dx + \int_{x_n}^{x_{n+1}} T(\widetilde{x}_n)\psi_i'\psi_j'\, dx \tag{9.28}$$

where $x_n$, $x_{n+1}$ and $\widetilde{x}_n$ are the coordinates of the left node, right node and the midpoint of the $n^{th}$ finite element and $\psi_{1,2,3,4}$ are the Hermite cubic interpolation polynomials defined in the interval $(x_n, x_{n+1})$. These are given in Table 9.4, where $\bar{x} = x - x_n$ refers to the local coordinate and $e = x_{n+1} - x_n$ refers to the element length.

The code is first validated by comparing the predictions with different rotation speeds with published literature [97, 104, 199, 207]. Next, we substitute the functions $EI_1(x)$ from Eqs. (9.11)-(9.19) in Eqs. (9.27) and (9.28). Since $EI_1(\widetilde{x}_n)$, $m(\widetilde{x}_n)$ and $T(\widetilde{x}_n)$ are constants and $\psi_{1,2,3,4}$ are simple polynomial functions, the integrals in Eqs. (9.25) and (9.26) can be determined in a closed form. The convergence characteristics for different modes are presented next.

### 9.1.4.1 First mode

For the first mode, we discretize the domain into a uniform mesh and then gradually refine the mesh by increasing the number of elements. Since the function $EI_1(x)$ for mode 1 is a benign function with a gradual variation and no internal singularities (see Fig. 9.2), the h-version code converges rapidly. With 16 uniform elements, the first mode non-dimensional frequency converges to the exact value up to 4 decimal places, which equals 3.5160. The convergence is illustrated in Fig. 9.6. The mode-shapes also converge to that of the corresponding uniform non-rotating beam. The residue of the nodal displacements obtained from the eigenvalue problem, with 16 uniform elements in the domain, is shown in Fig. 9.6. We notice that for the fundamental mode, a physically feasible equivalent rotating beam exists as internal singularities are absent.

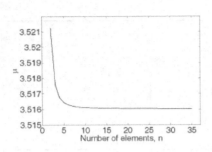 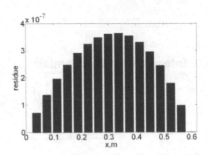

---

**Figure 9.6**
Convergence characteristics and residue of nodal displacements for first mode with
h-version finite element code

### 9.1.4.2  Second mode

The flexural stiffness function corresponding to the second mode has one internal
singularity. Unfortunately, the regular h-version FEM does not converge uniformly.
There is a flare up in the value of frequency when a certain number of elements are
used in the domain. This happens when the midpoint of an element coincides with
or falls adjacent to the point of singularity. In particular, the code blows up for 30
uniform elements in the domain where the mid-point of the $7^{th}$ element ($x = 0.13m$)
coincides with the singularity point. To alleviate this problem, we place a node at
the singularity point, which ensures that the midpoint of no element coincides with
the singularity point. This procedure is typical of finite element analysis in fracture
mechanics, where a node is placed at the crack for the analysis [133]. Here, we
create a fundamental mesh with a node at the singularity point and a fixed number of
elements in the domain. Then we keep refining the mesh by dividing each of these
elements into $m$ equal parts ($m = 1, 2, 3, \ldots$). As the number of elements increases,
the h-version code approaches the exact value of the frequency.

For mode 2, the nodes are identified by setting $\phi(x) = 0$. The nodes are found to
exist at a non-dimensional coordinate of $\eta = \frac{x}{L} = 0.783$, which for a beam of length,
$0.6\,m$, occurs at $x = 0.470\,m$. Hence, the singularity occurs at $x = 0.6\,m - 0.470m = 0.130\,m$. We develop an initial discretization for this mode, with a mesh as shown in
Fig. 9.7. The refined mesh for $m = 2$ and $m = 3$ is also presented in Fig. 9.7. Note
that the fundamental mesh has elements of more or less equal lengths. Elements of
equal lengths are created for faster and uniform convergence. With gradual refining
of the mesh, the total number of elements in the domain, $n$, increases as multiples of
5, i.e, $n = 5m$. The number of elements needed for the non-dimensional frequency
to converge to an accuracy of four decimal places ($\mu = 22.0345$) is $n = 55$. The
convergence rate and the residue of the nodal displacements obtained with $n = 55$
($m = 11$) are shown in Fig. 9.8.

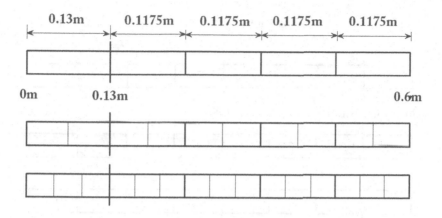

**Figure 9.7**
Fundamental mesh for the second mode h-version FEM, and the first two refinements

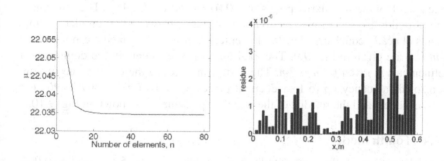

**Figure 9.8**
Convergence characteristics and residue of nodal displacements for second mode with h-version finite element code

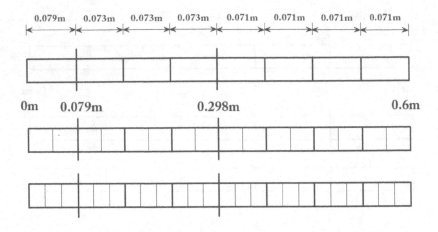

**Figure 9.9**
Fundamental mesh for the third mode h-version FEM, and the first two refinements

### 9.1.4.3 Third mode

The nodal positions for the third mode of a uniform rotating beam are determined to exist at $\eta = 0.503$ and $\eta = 0.868$. For the h-version finite element analysis, we place a node at each of the singularity points, $x = 0.079m$ and $x = 0.298m$. The fundamental mesh for this mode is formed by dividing the region between the singularities into elements of nearly equal size. The fundamental mesh and the refined mesh for $m = 2$ and $m = 3$ are given in Fig. 9.9. The total number of elements in the domain after $m$ refinements is given by $n = 8m$. The code converges to the exact value of non-dimensional frequency up to four decimal places, ($\mu = 61.6972$) for $n = 96$. The convergence rate and the residue of the nodal displacements is shown in Fig. 9.10.

### 9.1.4.4 Fourth mode

The nodal positions for the fourth mode are identified at $\eta = 0.358, 0.644$ and $0.906$, respectively. The singularities occur at $x = 0.057m, 0.213m$ and $0.385m$. By placing a node at each of these points, we form the fundamental mesh for mode 4, as given in Fig. 9.11. For this case, $n = 11m$. Convergence of non-dimensional frequency, $\mu$ to four decimal places of accuracy ($\mu = 120.9019$) happens for $n = 121$. The convergence of frequency and residue of nodal displacements is given in Fig. 9.12.

It is evident that by placing the element nodes at the singularity locations, any h-version FEM code can be verified using the flexural stiffness $EI_1(x)$ derived in this section. For all cases, monotonic convergence is obtained to the frequencies and mode-shapes of a non-rotating beam.

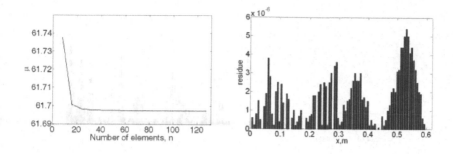

## Figure 9.10

Convergence characteristics and residue of nodal displacements for third mode with h-version finite element code

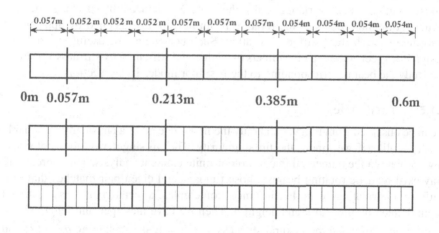

## Figure 9.11

Fundamental mesh for the fourth mode h-version FEM, and the first two refinements

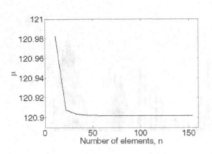
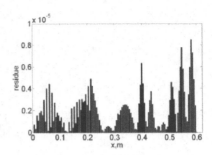

**Figure 9.12**
Convergence characteristics and residue of nodal displacements for fourth mode with h-version finite element code

### 9.1.5   REALISTIC BEAMS

It is of interest to manufacture rotating beams with the spectra $j$th mode eigen pair of non-rotating beams. In the previous sections, we proved that there exists a rotating beam, with the same $j^{th}$ frequency and mode-shape of a uniform non-rotating beam whose length and mass variation is also the same. Mathematically speaking, the centrifugal stiffening effects of the rotating beam have been cancelled by appropriate tailoring of the flexural stiffness variation. Such beams may be useful in structural design if we seek to counter the effects of centrifugal stiffening by stiffness variation. We study the beams corresponding to the first four modes in this section.

#### 9.1.5.1   First mode

It can be observed from Fig. 9.2 that only the first mode gives a physically reasonable $EI_1(x)$ profile, if we ignore the tip singularity. The tip singularity does not cause any problem in the numerical integration or finite element analysis. Therefore, there may exist realistic rotating beams whose fundamental eigen-pair matches that of a uniform non-rotating beam. For instance, contemplate a rectangular cross-sectional beam defined by breadth $b$ and height $h$. Then we have mass per unit length of the beam as $m = \rho bh$ and the flexural stiffness as $EI_1 = E\frac{bh^3}{12}$. Assuming $b = b(x)$ and $h = h(x)$, $m = 1.8816 kg/m^3$, $E = 2 \times 10^{11} Pa$, $\rho = 7840 kg/m^3$, we determine the required beam with following cross section dimensions.

$$h(x) = \sqrt{\frac{12\rho EI_1(x)}{Em}} \qquad (9.29)$$

$$b(x) = \sqrt{\frac{Em^3}{12\rho^3 EI_1(x)}} \qquad (9.30)$$

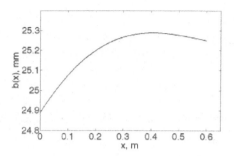 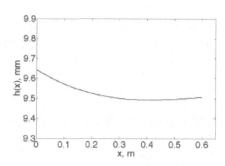

**Figure 9.13**
Variation of height and breadth of a rotating beam equivalent to a uniform non-rotating beam for the first mode

Substituting for $EI_1(x)$, the flexural stiffness function for the first mode, from Eqs. (9.11)–(9.13), we derive a variation for $b(x)$ and $h(x)$ as shown in Fig. 9.13. The top and the front views of a symmetric beam with these dimensions are given in Fig. 9.14. Note that the height and the breadth of this beam do not differ much from the values of the corresponding non-rotating beam. For the corresponding non-rotating beam, $b = 24mm$ and $h = 10mm$. Therefore, the rotating beam with same first mode eigen-pair is physically possible.

### 9.1.5.2 Second mode

The expressions $EI_1(x)$ obtained from Eqs. (9.11)–(9.19), for modes higher than the first, are very complicated and have a very high degree of non-linearity in their variation. Rotating beams with the $EI_1(x)$ variation corresponding to the higher modes will be challenging to manufacture. Moreover, as is evident from Fig. 9.2, the value of $EI_1(x)$ flares up near the singularities. The value of $EI_1(x)$ tends to $\pm\infty$ on either side of these singularities. Thus, it is physically impossible to make such beams, though the flexural stiffness can be used as test functions by avoiding the singularities via numerical integration or nodal placement. However, approximations for $EI_1(x)$ can be determined by selecting a finite number of points in the domain of the beam and satisfying the value of $EI_1(x)$ at these points. Then we can create a function for $EI_1(x)$ passing through these points, by constructing interpolation functions between these points. Cubic Hermite interpolating polynomials are used because they preserve the shape and the monotonicity of the curve. Fig. 9.15 shows three different approximations to the function $EI_1(x)$ for the second mode. Beams A and B are continuous beams based on cubic interpolation polynomials. Beam A has a sharp variation near the point of singularity and beam B has a mild variation. Beam

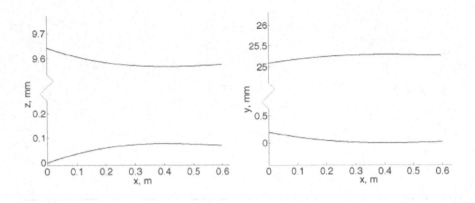

**Figure 9.14**
Top and side views of a rotating beam equivalent to a uniform rotating beam for the
first mode

C is modeled as a step beam, where the value of $EI_1$ changes abruptly at the point
of singularity. For this beam, the function $EI_1(x)$ is replaced by its mean values on
either sides of the singularity. The second-mode frequencies of these beams are computed using the p-version FEM. The computed frequencies are shown in Table 9.5.
The percentage deviation of these frequencies from the exact value is less than 0.1%.

The corresponding mode-shapes, $\phi$, of these beams are normalized, with $\phi(L) = 1$
and compared with the exact mode-shapes of Fig. 9.16. The difference in the value of
$\phi$ and the exact mode-shape is computed as a vector of residues, $-R''$, at 600 equidistant points in the domain. The norm of the vector, $-R''$, calculated by $\sqrt{\sum_i R(i)^2}$ is
very small and is shown in Table 9.5. Beam B shows the best match for frequency and
Beam A provides the best match for the mode-shape. However, the stepped beam C
also displays reasonable results and shows the possibility of countering the centrifugal force effects in a simple manner. Therefore, we observe that by approximating
the flexural stiffness functions derived in this section of the chapter, realistic rotating
beams whose mode-shapes and frequencies match with that of second mode of the
uniform non-rotating beam can be determined.

Next, we study the dimensions of these beams if they have rectangular cross-sections.
The height and breadth variations, $h(x)$ and $b(x)$ computed from Eqs. (9.29)–(9.30)
are shown in Fig. 9.17. We can observe that the variation in the dimensions is
reasonable and Beam C, and to some extent Beam B, are physically feasible. Higher
thickness at the root and higher breadth in the outboard region is needed for Beam C.
Note that the non-rotating beam has dimensions $b = 24mm$ and $h = 10mm$. By pro-
viding a small level of non-uniformity, the centrifugal effects for the second mode
are countered.

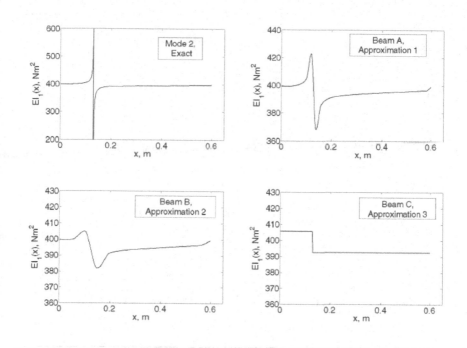

**Figure 9.15**
Realistic Beams: Approximations to the exact function for the second mode

**Table 9.5**

**Frequencies and Mode-shapes of the Approximate Beams for Mode 2**

| Approximation | Estimated Frequency (rad/sec) | %deviation from Exact | Norm of Residues of Mode-shape |
|---|---|---|---|
| Beam A | 892.4341 | 0.0021 | $8.45 \times 10^{-4}$ |
| Beam B | 892.4176 | 0.0003 | $1.73 \times 10^{-3}$ |
| Beam C | 893.2886 | 0.0979 | $3.89 \times 10^{-1}$ |

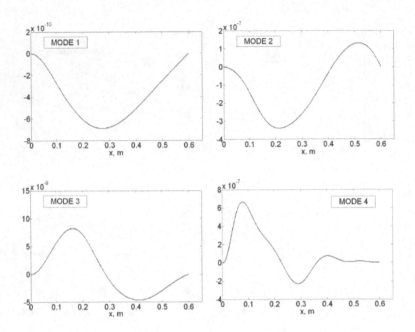

**Figure 9.16**
Residue of Mode-shapes obtained from p-version FEM

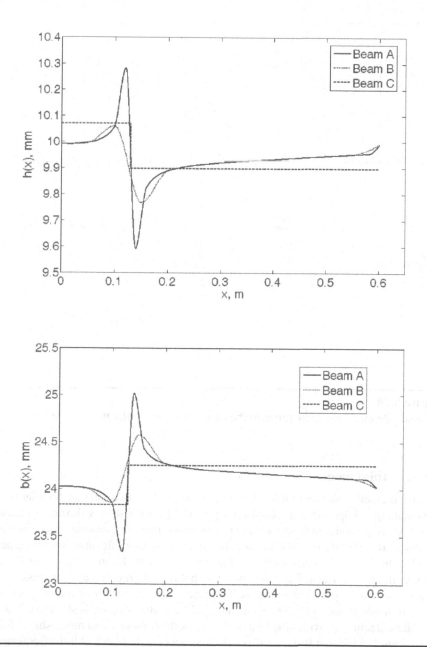

**Figure 9.17**

Variation of height and breadth of a rotating beam equivalent to a uniform non-rotating beam for the second mode

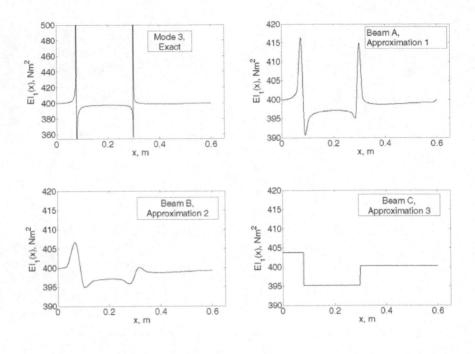

**Figure 9.18**
Realistic Beams: Approximations to the exact function for the third mode

### 9.1.5.3 Third mode

The flexural stiffness function for the third mode displays two internal singularities, thus making it impossible to construct a physical beam with this flexural stiffness variation. Approximations to the function as above may be construed, resulting in beams A, B and C (Fig. 9.18). Beam A is closer to the exact flexural stiffness function but displays a steeper variation at the singular points. Beam B varies mildly at the singular points. Beam C is a step beam with three different flexural stiffness constants along its length. The estimated frequencies and the norm of the residues of the third mode-shape, with 600 equidistant points in the domain, is shown in Table 9.6. Here, Beam A provides the best match for both frequency and mode-shape, followed closely by Beam B. Beam C provides an acceptable match. All the frequency deviations are less than or equal to 0.1%.

The height and breadth variation of these beams with square cross sections is shown in Fig. 9.19. Again, we observe that by adding a small level of non-uniformity in the dimensions of the rotating beam, the effect of centrifugal stiffening for the third mode can be repulsed.

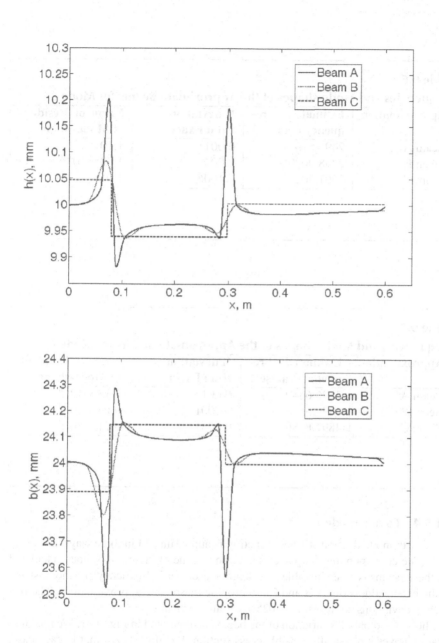

**Figure 9.19**
Variation of height and breadth of a rotating beam equivalent to a uniform non-rotating beam for the third mode

**Table 9.6**

**Frequencies and Mode-shapes of the Approximate Beams for Mode 3**

| Approximation | Estimated Frequency (rad/sec) | %deviation from Exact | Norm of Residues of Mode-shape |
|---|---|---|---|
| Beam A | 2498.8163 | 0.0011 | $1.10 \times 10^{-3}$ |
| Beam B | 2498.9089 | 0.0048 | $1.48 \times 10^{-3}$ |
| Beam C | 2501.3019 | 0.1006 | $1.06 \times 10^{-2}$ |

**Table 9.7**

**Frequencies and Mode-shapes of the Approximate Beams for Mode 4**

| Approximation | Estimated Frequency (rad/sec) | %deviation from Exact | Norm of Residues of Mode-shape |
|---|---|---|---|
| Beam A | 4896.6340 | 0.0015 | $6.13 \times 10^{-4}$ |
| Beam B | 4896.5547 | 0.0001 | $9.63 \times 10^{-4}$ |
| Beam C | 4897.8586 | 0.0251 | $9.21 \times 10^{-3}$ |

#### 9.1.5.4    Fourth mode

The fourth mode flexural stiffness function is approximated in three ways, as in Fig. 9.20. The corresponding frequencies and the residue of the mode-shapes calculated for these beams is given in Table 9.7. Beam A is closer in frequency and mode-shape to the exact values. Beam B and C too provide reasonable approximations. The frequency deviations are less than 0.025% for all the beams.

The height and breadth variation of the beams is illustrated in Fig. 9.21. We note that Beam C gives a physically feasible cross section. We also oberve that in each case, there are $N - 1$ discontinuities corresponding to the singularities for mode N. Furthermore, the dimensions of the stepped beam are close to the non-rotating uniform beam.

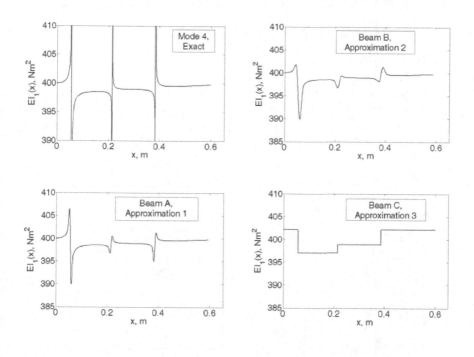

**Figure 9.20**
Realistic Beams: Approximations to the exact function for the fourth mode

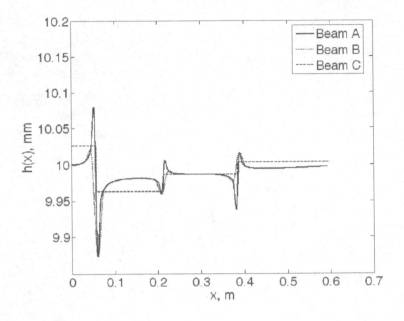

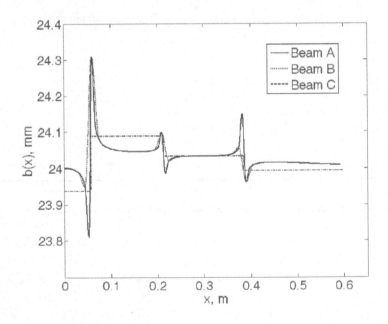

**Figure 9.21**
Variation of height and breadth of a rotating beam equivalent to a uniform non-rotating beam for the fourth mode

### 9.1.5.5 Destiffening effects

We have explicated that physically realizable rotating beams with same eigen pair of non-rotating beams exist for the first four modes. Similar beams can be discovered for higher modes. By using the cross sections obtained in this work, any mode can be destiffened to abate the effect of centrifugal stiffening. This can be useful in design where it is sometimes necessary to move one mode away from a particular value of frequency being a multiple of the rotation speed. To investigate this situation, we find the stiffness reduction required for a uniform rotating beam to match the frequencies of the non-uniform beam for the first four modes. We reduce the uniform beam stiffness from a value of $EI_2 = 400Nm^2$ to $EI_1^{(r)}$, so that for a given mode, the frequency reduction is identical to that obtained using the derived flexural stiffness of the stepped beam, $EI_1^{(C)}$ for modes 2, 3 and 4 and the exact beam, $EI_1^{(exact)}$ for mode 1.

Table 9.8 compares the first four frequencies of the two beams. Note that the uniform rotating beam with stiffness constant of $400Nm^2$ has first four natural frequencies as 148.2320, 897.5588, 2503.8612 and 4901.8567 radians per second, respectively. We observe from Table 9.8 that the frequencies of higher modes such as mode 4 can be reduced with less effect on the lower mode frequencies $(1, 2$ and $3)$, compared to the uniform beam. For the lower modes such as mode 1, the exact beam yields a larger reduction in the frequencies for the higher modes $(2, 3$ and $4)$. In general, if we target mode $j$ for matching with the non-rotating frequency, the modes below $j$ will show less reduction than the uniform beam and for modes greater than $j$, the reduction will be greater. Thus the beam profiles can be applied for tailoring beams to achieve targeted frequencies.

The stiffness constant of the uniform beam, denoted as $EI_1^{(r)}$, is compared with the stiffness variation of the non-uniform beams in Fig. 9.22. It is seen from Fig. 9.22 that the derived flexural stiffness functions suggest an approach to destiffen the rotating beams without compromising for the stiffness of the beam at the root. In fact, for modes 2, 3 and 4, the root stiffness is higher than that of the baseline rotating beam $(400Nm^2)$. Since the root stiffness is critical for handling stresses in cantilever beams, being able to decrease the frequencies while not having to reduce root stiffness is a desirable feature.

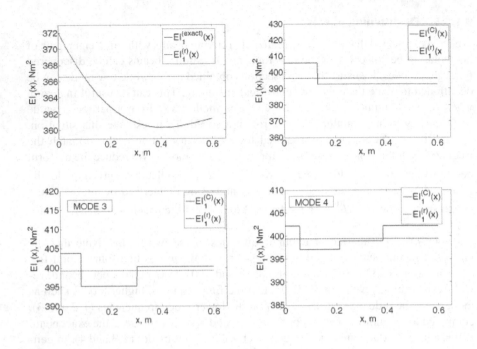

**Figure 9.22**

Destiffening effects of flexural stiffness variations for various modes compared with uniform rotating beam ($EI_1^{(r)}$)

**Table 9.8**

**Frequencies and Mode-shapes of the Approximate Beams for Mode 4**

| Mode | $EI_1(x)$ | $\omega_1(rad/sec)$ | $\omega_2(rad/sec)$ | $\omega_3(rad/sec)$ | $\omega_4(rad/sec)$ |
|------|-----------|---------------------|---------------------|---------------------|---------------------|
| Mode 1 | $EI_1^{(exact)}(x)$ | **142.4015** | 856.1476 | 2386.1690 | 4670.2321 |
|        | $EI_1^{(r)}(x)$ | **142.4015** | 859.6891 | 2397.4121 | 4693.0495 |
| Mode 2 | $EI_1^{(C)}(x)$ | 148.3470 | **893.2886** | 2488.1736 | 4873.0676 |
|        | $EI_1^{(r)}(x)$ | 147.5737 | **893.2886** | 2491.8591 | 4878.3147 |
| Mode 3 | $EI_1^{(C)}(x)$ | 148.0982 | 897.1842 | **2501.3019** | 4894.2987 |
|        | $EI_1^{(r)}(x)$ | 148.0813 | 896.6487 | **2501.3019** | 4896.8352 |
| Mode 4 | $EI_1^{(C)}(x)$ | 148.0853 | 897.2456 | 2502.4406 | **4897.8586** |
|        | $EI_1^{(r)}(x)$ | 148.0827 | 896.8270 | 2501.8171 | **4897.8586** |

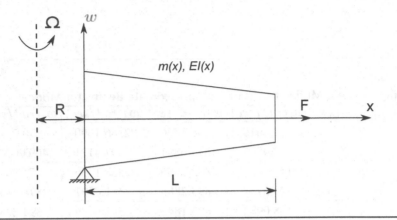

**Figure 9.23**
Schematic of a rotating beam with pinned-free boundary condition

## 9.2   ROTATING BEAMS AND NON-ROTATING BEAMS WITH SHARED EIGENPAIR: PINNED-FREE BOUNDARY CONDITION

The idea of finding common ground between rotating and non-rotating beam was investigated in the previous section. Equivalent rotating beams, with cantilever (fixed-free) boundary conditions, which have the same mode shape and frequency as a uniform non-rotating beam for a particular mode were found. Therefore, the beams are isospectral for a given mode. Consequently, the derived flexural stiffness functions were used to validate both the p-version and h-version of FEM by comparing the results with the exact solutions of non-rotating beams and also with prior literature. In this section, we address the important problem of pinned-free beam. Such beams are important for practical structures such as articulated helicopter rotor blades and robotic arms.

In this section, we discover equivalent rotating beams with pinned-free boundary conditions which have the same frequency and mode shape of a uniform non-rotating beam with the same boundary conditions. The derived flexural stiffness functions can serve as test-functions for checking approximate methods, and also can be used for the design of realistic beams which will help to counter the centrifugal stiffening effects in these hinged rotating beams. Furthermore, the vibration testing of rotating beams are an important engineering problem which involves a challenging experimental setup and sophisticated instrumentations. So in lieu of setting up expensive test rigs for rotating beams, we can use the equivalent realistic non-rotating beams for a given mode, thus making the process of measurement of the vibration frequencies more amenable.

### 9.2.1   MATHEMATICAL FORMULATION

Consider a long slender beam which is rotating about a vertical axis, with uniform angular velocity $\Omega$ and with dimensions as shown in Fig. 9.23. The governing equation for free vibration of this beam is given by [104]

$$(EI_1(x)w'')'' + m(x)\ddot{w} - (T(x)w')' = 0 \tag{9.31}$$

where $T(x) = \int_x^L m(x)\Omega^2(R+x)dx + F$, $EI_1(x)$ is the flexural stiffness of the beam, $m(x)$ is the mass per unit length of the beam, $w(x,t)$ is the bending displacement, $R$ is the hub radius, $L$ is the length of the beam, and $F$ is the axial force applied at the free end of the beam. For our study, we consider $R = F = 0$.

Now we consider a non-rotating beam of the same length $L$, same uniform mass distribution $m(x)$ and same boundary conditions as that of the rotating beam given by the Eq. (9.31). This beam is modeled by the 4th order differential equation [139]

$$(EI_2(x)w'')'' + m(x)\ddot{w} = 0 \tag{9.32}$$

where $EI_2(x)$ is the flexural stiffness of the non-rotating beam.

A rotating beam will be again defined as equivalent, in the $i^{th}$ mode, to a corresponding non-rotating beam of same length and mass distribution if and only if the $i^{th}$ mode-shapes and the $i^{th}$ frequencies of the two beams are exactly the same. In such rotating beams, the effect of rotation in increasing the frequency of the $i^{th}$ mode is countered by a characteristic variation of flexural stiffness along their length to give natural frequencies corresponding to a non-rotating beam. By this definition, let the $i^{th}$ mode-shape and frequency of these two beams be the same and let them be equal to $\phi_i$ and $\omega_i$, respectively. The bending displacement $w_i$ for the $i^{th}$ mode is given by

$$w_i = \phi_i e^{i\omega_i t} \tag{9.33}$$

Clearly, the mode-shape $\phi_i$ satisfies both the governing differential equations given by Eqs. (9.31) and (9.32). Substituting Eq. (9.33) in Eqs. (9.31) and (9.32) we get

$$(EI_1(x)\phi_i'')'' - m(x)\omega_i^2 \phi_i - (T(x)\phi_i')' = 0 \tag{9.34}$$

$$(EI_2(x)\phi_i'')'' - m(x)\omega_i^2 \phi_i = 0 \tag{9.35}$$

Subtracting Eq. (9.35) from Eq. (9.34) results in

$$(EI_1(x)\phi_i'')'' - (EI_2(x)\phi_i'')'' - (T(x)\phi_i')' = 0 \tag{9.36}$$

Equation (9.36) links the flexural stiffness of a rotating beam and a non-rotating beam with the same $i^{th}$ eigenpair. Note that the expression for $\phi_i$ is different for different modes, therefore different expressions for $EI_1(x)$ are derived for different modes. Specifically, $EI_1(x)$ in terms of $EI_2(x)$ is given by

$$EI_1(x) = \frac{\int(\int((EI_2(x)\phi_i'')'' + (T(x)\phi_i')')dx)dx}{\phi_i''} \tag{9.37}$$

Contemplating a non-rotating beam with uniform mass distribution and uniform flexural stiffness variation, we have $m(x) = m$ and $EI_2(x) = EI_2$. Thus Eq. (9.37) becomes

$$EI_1(x) = \frac{\int(\int(EI_2\phi_i^{(4)} + (T(x)\phi_i')')dx)dx}{\phi_i''} \tag{9.38}$$

The mode shapes of an uniform pinned-free beam can be solved exactly and is represented by the beam function [30]

$$\phi_i(x) = C_i \left( \sin(\beta_i x) + \left( \frac{\sin(\beta_i L)}{\sinh(\beta_i L)} \right) \sinh(\beta_i x) \right) \quad (9.39)$$

where $\beta_i$'s represent the solutions to the frequency equation $\tan(\beta_i L) = \tanh(\beta_i L)$ and $C_i$'s are arbitrary constants. The frequencies $\omega_i$'s are obtained using the expression given by [139]

$$\omega_i = (\beta_i L)^2 \sqrt{\frac{EI_2}{mL^4}} \quad (9.40)$$

where $(\beta_1 L)^2 = 15.4182$, $(\beta_2 L)^2 = 49.9649$, $(\beta_3 L)^2 = 104.2477$, and $(\beta_4 L)^2 = 178.2697$ are the exact values of the first four frequencies, up to four decimal places, corresponding to the first four elastic modes of the pinned-free beam. The rigid body mode frequency is zero, which is not considered in our problem, since this mode has no relation with the elastic deflection of the beam.

## 9.2.2 FLEXURAL STIFFNESS FUNCTIONS (FSF'S)

Here $EI_1(x)$ is the flexural stiffness function (FSF) of a rotating beam equivalent in the $i^{th}$ mode to its corresponding non-rotating beam. Note that since $\phi_i$ is different for different modes, the FSF's are also different for different modes. Substituting Eq. (9.39) in Eq. (9.38), we can easily derive the flexural stiffness variations $EI_1(x)$ for each mode such that the frequency and mode-shape for the corresponding mode of the beam are the same as that of a uniform non-rotating beam. The properties of the uniform non-rotating beam considered is given in Table 9.1 [157]. The first four natural frequencies of this beam, up to four places of decimal, are 624.4500 rad/s, 2023.6183 rad/s, 4222.1174 rad/s and 7220.0707 rad/s, respectively. The corresponding non-dimensional frequencies ($\mu = \sqrt{(m\omega^2 L^4/EI_2)}$) are 15.4182, 49.9649, 104.2477 and 178.2697, respectively. The functions $EI_1(x)$ obtained from Eq. (9.38) are given in Eq. (9.41) for the first four mode shapes. Here $(EI_1)_i$ corresponds to the $i^{th}$ mode.

$$(EI_1)_i = \frac{N_i}{D_i} \quad (9.41)$$

where

$$N_1 = \sinh(z_1)(L^2 m\Omega^2 (0.0139374x^2 - 0.0121295L^2)$$
$$-0.429781 EI_2) - 15.4182 EI_2 \sin(z_1)$$
$$+0.564858L^4 m\Omega^2 \sin(z_1) - 0.254673L^3 mx\Omega^2 \cos(z_1)$$
$$-0.00709899L^3 mx\Omega^2 \cosh(z_1) - 0.5L^2 mx^2\Omega^2 \sin(z_1) + c_1 x + c_2$$

$$(9.42)$$

$$D_1 = -15.4182 \sin(z_1) - 0.429781 \sinh(z_1)$$

$$(9.43)$$

$$N_2 = \sinh(z_2)(0.0601635EI_2 + L^2m\Omega^2(0.0005779L^2$$
$$-0.000602058x^2)) - 49.9649EI_2\sin(z_2)$$
$$+0.520014L^4m\Omega^2\sin(z_2) - 0.141471L^3mx\Omega^2\cos(z_2)$$
$$+0.0001703L^3mx\Omega^2\cosh(z_2) - 0.5L^2mx^2\Omega^2\sin(z_2) + c_1x + c_2$$

$$(9.44)$$

$$D_2 = 0.0601635\sinh(z_2) - 49.9649\sin(z_2)$$

$$(9.45)$$

$$N_3 = \sinh(z_3)(L^2m\Omega^2(0.000026017x^2$$
$$-0.0000255L^2) - 0.0054244EI_2) - 104.248EI_2\sin(z_3)$$
$$+0.509593L^4m\Omega^2\sin(z_3) - 0.0979415L^3mx\Omega^2\cos(z_3)$$
$$-5.0963 \times 10^{-6}L^3mx\Omega^2\cosh(z_3)$$
$$-0.5L^2mx^2\Omega^2\sin(z_3) + c_1x + c_2$$

$$(9.46)$$

$$D_3 = -104.248\sin(z_3) - 0.00542448\sinh(z_3)$$

$$(9.47)$$

$$N_4 = \sinh(z_4)(0.0004EI_2 + L^2m\Omega^2(1.11169 \times 10^{-6}L^2$$
$$-1.12431 \times 10^{-6}x^2)) - 178.27EI_2\sin(z_4)$$
$$+0.505609L^4m\Omega^2\sin(z_4) - 0.0748964L^3mx\Omega^2\cos(z_4)$$
$$+1.68413 \times 10^{-7}L^3mx\Omega^2\cosh(z_4)$$
$$-0.5L^2mx^2\Omega^2\sin(z_4) + c_1x + c_2$$

$$(9.48)$$

$$D_4 = 0.00040086\sinh(z_4) - 178.27\sin(z_4)$$

$$(9.49)$$

Here $z_1 = 3.9266x/L$, $z_2 = 7.0686x/L$, $z_3 = 10.2102x/L$, $z_4 = 13.3518x/L$ and $c_1$ and $c_2$ are arbitrary constants resulting from the integration. The resulting FSF's are quite complex containing trigonometric, hyperbolic and polynomial functions. We can also observe that as $\Omega \to 0$, $EI_1 \to EI_2$. The derived $EI_1(x)$ variations for the pinned free boundary conditions are very much different from that derived for the cantilever boundary condition in the earlier section on the cantilever beam in this chapter.

Let us consider a non-rotating beam with properties shown in Table 9.1. The constants $c_1$ & $c_2$ are taken as zero. We want to determine the flexural stiffness of the

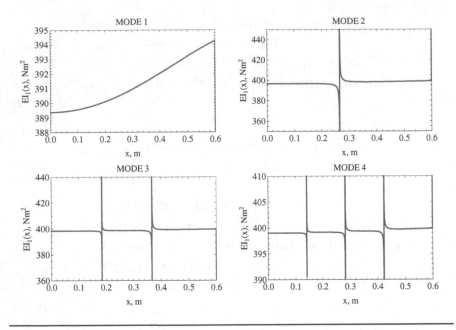

**Figure 9.24**
$EI_1(x)$ variation for first four modes for a beam with properties shown in Table 9.1

rotating beams with a shared eigenpair for $\Omega = 360$ RPM. The variations of $EI_1(x)$ versus $x$ are shown in Fig. 9.24 for the first four modes. The uniform non-rotating beam has a corresponding flexural stiffness of $EI_2 = 400$ Nm$^2$. Note that all the FSF's have singularities on both ends due to the moment boundary condition $M = EIw'' = 0$ at both the pinned end $(x = 0)$ and the free end $(x = L)$ of the beam. But we will see later in the section that these tip singularities do interfere in the finite element analysis of these beams. Even if the tip singularities are excluded, the FSF corresponding to the $i^{th}$ mode has $(i-1)$ internal singularities. This occurs because of the existence of the nodal points in the beam shape functions. If the nodal points are at $x_1, x_2, \ldots, x_i$, then the singularities will manifest at $L - x_1, L - x_2, \ldots, L - x_i$. However, the first mode has a flexural stiffness distribution $EI_1(x)$ which is physically feasible and inseting this function into a rotating beam simulation should yield the non-rotating beam frequency for the fundamental mode. Since the effect of rotation is to stiffen the beam, the rotating beam has a lower stiffness, especially near the root of the beam. Note that $EI_2 = 400$ Nm$^2$ and $EI_1(x)$ ranges between 389 Nm$^2$ and 395 Nm$^2$.

## 9.2.3   NUMERICAL SIMULATION USING P-VERSION FEM

The formulation in the previous section proves that the derived FSF's should give the same frequencies and mode-shapes as that of the corresponding uniform non-rotating beam. For a non-rotating uniform beam, the frequencies and mode-shapes can be solved exactly to any desired level of accuracy. So if we interject the functions $EI_1(x)$

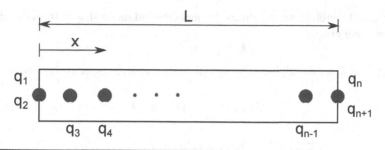

**Figure 9.25**
Beam element for p-version finite element

obtained in Eqs. (9.41–9.49) into a rotating beam finite element code, then we will get the same frequencies and mode-shapes as that of a corresponding uniform non-rotating beam [169]. This process can be used to verify the veracity of the solutions obtained from the FEM code and the rate of convergence of the finite element mesh.

For our case, we will apply the p-version of the finite element method [104]. Only one element is used for the whole beam. This p-version is used here because it can contend with non-uniform variations of flexural stiffness, and also due to reduced order of the eigenvalue problem. The displacement $w$ is expressed in terms of a $n$-degree polynomial function as

$$w = \sum_{i=0}^{n} p_i x^i \qquad (9.50)$$

The vector of generalized co-ordinates is given defined by $P = \{p_i\}$. A set of $n-3$ nodes $\{\alpha_1, \alpha_2, \ldots, \alpha_{n-3}\}$ are defined inside the interior of the beam at equidistant points as shown in Fig. 9.25. We now define a vector $Q = \{q_i\}$ such that $q_1 = w(0), q_2 = w'(0), q_3 = w(\alpha_1), q_4 = w(\alpha_4), \ldots, q_{n-1} = w(\alpha_{n-3}), q_n = w(L)$, and $q_{n+1} = w'(L)$.

Using these relations and Eq. (9.50), we obtain a relation between P and Q. We then determine the interpolating functions $\psi_i$ for the $(n-1)$ noded element by satisfying the condition $Q_j = \delta_{ij}$, where $\delta$ is known as the Kronecker delta function, and thus obtain the corresponding constants $p_{ij}$. The interpolation functions is thus expressed by

$$\psi_i = \sum_{j=0}^{n} p_{ij} x^j \qquad (9.51)$$

The expressions for the KE and PE for a rotating beam are expressed by [122]

$$T = \frac{1}{2} \int_0^L m(x) \dot{w}^2 dx \qquad (9.52)$$

$$U = \frac{1}{2} \int_0^L EI_1(x)(w'')^2 dx + \frac{1}{2} \int_0^L T(x)(w')^2 dx \qquad (9.53)$$

Substituting the expression of $w$, i.e. Eq. (9.50), into Eqs. (9.52) and (9.53) and applying Lagrange's equations of motion, we obtain the global stiffness and mass matrices, and therefore we obtain the frequencies and mode shapes from the eigen-value problem given by

$$[K]\{x\} = \omega^2 [M]\{x\} \tag{9.54}$$

where $M$ and $K$ are the global mass and stiffness matrices respectively, written as

$$(M_{ij})_n = \int_0^L m(x)\psi_i\psi_j dx \tag{9.55}$$

$$(K_{ij})_n = \int_0^L EI_1(x)\psi_i''\psi_j'' dx + \int_0^L T(x)\psi_i'\psi_j' dx \tag{9.56}$$

We check the correctness of this FEM code implementation for two different non-dimensional rotating speeds ($\lambda = \sqrt{m\Omega^2 L^4/EI_2} = 0$ & 12) by comparing the results predicted with the flexural stiffness generated with exact values, which are calculated by solving the transcendental equation of a pinned-free beam. The first four frequencies computed for $\lambda = 0$ & 12 from the p-FEM code are $\omega_1 = 15.4182$, $\omega_2 = 49.9649$, $\omega_3 = 104.2477$ & $\omega_4 = 178.2697$, which is an exact match up to the fourth place of decimal. The beam used has the properties shown in Table 9.1 and a rotating speed of $\Omega = 360$ RPM. Since this method is prone to ill-conditioning, so we develop the global mass and stiffness matrices symbolically using the computer software MATHEMATICA. Only in the last step the numerical values are given as inputs and the eigenvalues are obtained.

The FSF's were given by Eqs. (9.41–9.49). By injecting these derived $EI_1(x)$ in Eqs. (9.55) & (9.56) we can verify the veracity of the p-FEM code. Analytical integration of Eqs. (9.55) & (9.56) is very onerous for the FSF's as they are complicated functions with singularities. Therefore, numerical integration is applied to evaluate the mass and stiffness matrices. We use Gaussian-Quadrature technique, which is an approximation of the definite integral of a function, usually formulated as a weighted sum of function values at specified points within the domain of integration. For our analysis we select 35 points, and as illustrated in Fig. 9.26, we gaurantee that none of the singularities lie within $\pm 0.005$ m of any of the Gaussian nodes. This approach of circumventing singularities is typically used to address singularities using numerical methods [24, 25, 112, 142]. Using 35 Gauss points we can obtain an accurate evaluation of the integrals as will be revealed by the numerical results. The same number of Gauss points can be used for any beam length.

The p-FEM code converges with alacrity. The convergence of the code for the first four non-dimensional natural frequencies, up to four places of decimal accuracy, was achieved by using polynomial orders of $n = 10, 12, 14$ and $15$. The beneficial part of these derived stiffness functions is that, for a rotating beam with given FSF's, they give the exact value of non-dimensional frequencies of a uniform non-rotating beam ($\mu = \sqrt{m\omega^2 L^4/EI_2}$) for the first four modes. The time required for the computation ranges from 2-15 minutes. The time required to compute the results of the p-FEM at convergence for the first four modes with $\lambda = 0$ are 2 mins 14 sec, 3 mins 32 sec, 5 mins 51sec & 7 mins 19 sec, respectively, with a PC configuration of Pentium (R) 4 CPU, 1.7 GHz and 2 GB RAM. The same for $\lambda = 12$ are 4 mins 8 sec, 7 mins 15 sec, 11 mins 50 sec & 13 mins 55 sec, respectively.

'+' – Singularities , '*' – Gauss Points

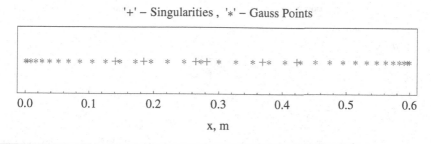

**Figure 9.26**
Gauss-quadrature points of order 35 for the domain $(0, 0.6)$ and the singularity points for the flexural stiffness functions

The rate of convergence is plotted in Fig. 9.27. We obtain the displacement of the nodes from the eigenvectors $\{x\}$ from Eq. (9.54). The mode shapes are then found by using interpolating polynomial technique, to interpolate the displacements between the nodal values. The resulting mode shapes obtained at convergence are in good accord with the actual mode shapes, as we can see from Fig. 9.28.

The p-FEM code generates a rigid body mode and a corresponding rigid body frequency corresponding to the first mode, but since they do not play any role in the elastic deflection of the beam, we have considered from the second mode onwards. Hence the actual second mode becomes the first elastic mode. So the first four mode shapes refer to the mode shapes corresponding to the first four elastic modes.

Thus we have shown that the derived flexural stiffness functions give rotating beams with the same eigenpair as that of a corresponding uniform non-rotating beam. In the first mode, there are no internal singularities present, so a physically feasible equivalent rotating beam exists. The higher modes have internal singularities, but even then it is was possible to integrate them using Gauss quadrature numerical integration techniques. We can see that the FSF's derived in this section have theoretical significance and can be applied as test functions to check the implementation of a rotating beam finite element codes. They could also be used for targeted destiffening of rotating free-pinned beams, as shown next.

## 9.2.4   REALISTIC DESTIFFENED BEAMS

For application purposes, we want to manufacture actual beams with the same $i^{th}$ mode eigenpair as that of a corresponding non-rotating beam. We have demonstrated mathematically that there exists a uniform rotating beam whose $i^{th}$ mode frequencies and mode shapes are the same as that of a uniform non-rotating beam with the same mass distribution. Therefore, the centrifugal stiffening effects have been decimated by determining an appropriate flexural stiffness variation for the rotating beam. This is important from a design point of view, if we seek to ameliorate the centrifugal effects by varying the flexural stiffness. In the following four subsections, realistic beams corresponding to the first four modes, which are generally the dominant

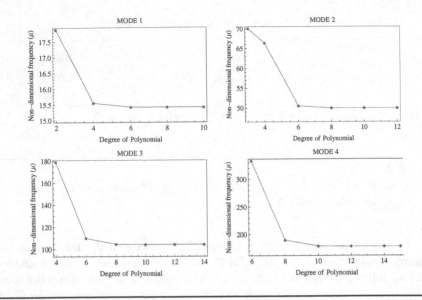

**Figure 9.27**
Convergence of the p-version finite element code for the first four modes using derived flexural stiffness functions

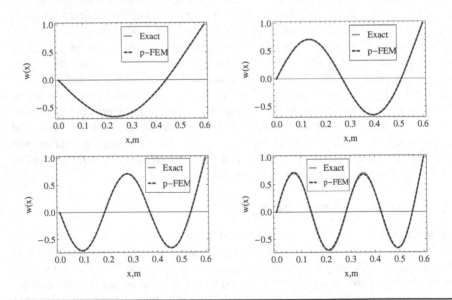

**Figure 9.28**
Comparison of the normalized mode-shapes obtained from the p-FEM code ($EI_1(x)$, rotating beam) and the exact ($EI_2$, uniform beam) mode-shapes

modes from a structural dynamics point of view, are determined.

### 9.2.4.1 First mode

From Fig. 9.24, we note that there exists a physically reasonable variation of $EI_1(x)$ for the first mode, if we disregard the end singularities, which do not play any role in both numerical integration techniques and also in the finite element formulation. So there might exist realistic rotating pinned-free beams whose fundamental eigenpair matches that of a uniform non-rotating beam. Let us for example take a beam with rectangular cross-section given by breadth $b$ and height $h$. The mass per unit length is given by $m = \rho bh$ and flexural stiffness is given by, $EI_1 = E(bh^3/12)$. Assuming that the breadth and height varies as $b = b(x)$ and $h = h(x)$, respectively, and that $m = 1.8816$ kg/m, $E = 2 \times 10^{11}$ Pa and $\rho = 7840$ kg/m$^3$, we get the variations of height and breadth as follows

$$h(x) = \sqrt{\frac{12\rho EI_1(x)}{Em}} \tag{9.57}$$

$$b(x) = \sqrt{\frac{Em^3}{12\rho^3 EI_1(x)}} \tag{9.58}$$

By substituting the value of $EI_1(x)$ that we had got from Eqs. (9.41)–(9.43), we can get the desired variations of $h(x)$ and $b(x)$, which is shown in Fig. 9.29(a). For the corresponding uniform non-rotating beam, the values are $b = 24$ $mm$ and $h = 10$ $mm$. Now we can observe that the height and breadth of this rotating beam does not differ significantly from the values of the corresponding non-rotating beam. Thus a physically possible rotating beam exists for the first mode eigenpair. Figure 9.30 displays how the height and breadth of the realistic beam will vary if the rotation speed is varied.

### 9.2.4.2 Second mode

As discussed earlier, the variation of $EI_1(x)$ for the higher modes are very complex and highly non-linear. Furthermore, the functions have internal singularities where the function blows up and approaches $\pm\infty$ on either side of the singularities. Hence it is physically impossible to manufacture such a rotating beam, though they can be applied mathematically as test functions. However, we can propose some approximate functions by satisfying $EI_1(x)$ at a finite number of points within the domain, and then passing an interpolating polynomial through these points. The polynomial chosen is the Hermite cubic as it preserves both the shape and also the monotonicity of the original function. Another way to approximate is to form a step beam, where the average of the function $EI_1(x)$ is taken on either side of the singularities. Here there is an abrupt change of the stiffness at the singularities. As examples, we have created three beams A,B and C as shown in Fig. 9.31. Beam A has a sharp variation at the singularities, Beam B has a mild variation and Beam C is the step beam approximation. These approximate functions are then put into the p-FEM code to calculate the frequencies which are shown in Table 9.9. The percentage deviation is less than or equal to 0.01%, which is quite good. The mode shapes are normalized with $\phi(L) = 1$

**Breadth**                                          **Height**

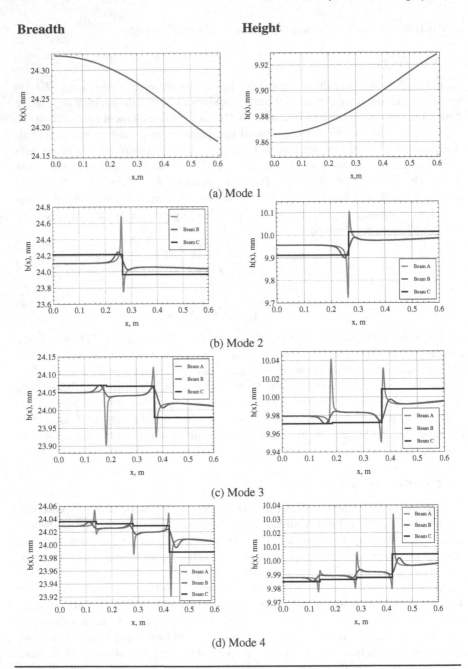

(a) Mode 1

(b) Mode 2

(c) Mode 3

(d) Mode 4

**Figure 9.29**
Variation in height and breadth of a rotating pinned-free beam equivalent to a uniform non-rotating beam for the first four modes

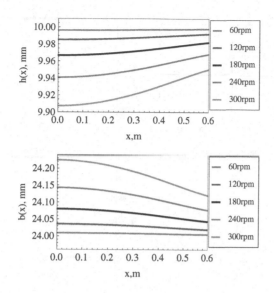

**Figure 9.30**
Height and breadth variation for different rotating speeds, for the first mode

and are compared with the exact mode-shapes. The difference between these two are calculated as a vector residue, $\{R\}$, at large number of equidistant points in the domain. The norm is then calculated as $\sqrt{\sum_i R(i)^2}$, which is found to be very small. The calculated values are shown in Table 9.9. We can see that both Beam A and Beam B give very good matches. Beam C also gives reasonable results and hence they can be a simple means of countering the centrifugal effects of the rotating beam.

The height and breadth variations, which are calculated by Eqs. (9.57) & (9.58), are also shown in Fig. 9.29(b). We can infer that dimensions of the three different beams are reasonable. If we now recall that the dimensions of the uniform non-rotating beams are as $h = 10mm$ and $b = 24mm$, then we can see that by slight variation of the height and breadth, we can counter the centrifugal effects of the second mode.

### 9.2.4.3 Third mode

The flexural stiffness function for the third mode has two internal singularities, hence constructing a physical beam exactly from this function is onerous. But using similar approximation techniques as applied in the previous subsection, we can create the beams A (steeper variation), beam B (mild variation) & beam C (stepped beam) as shown in Fig. 9.32. The frequencies and norms of the mode shapes for these three beams are shown in Table 9.10. The frequency deviation are less than 0.01%.

The height and breadth variation for rectangular cross-section is plotted in Fig. 9.29(c). Through small variations in the dimensions, we can counter the centrifugal effects of the third mode.

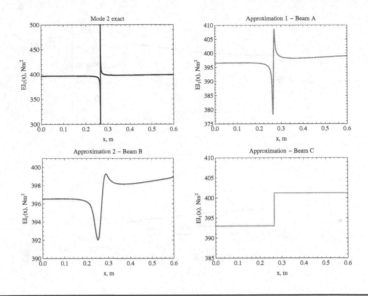

**Figure 9.31**
Realistic beams: approximation to the exact flexural stiffness function for the second
mode

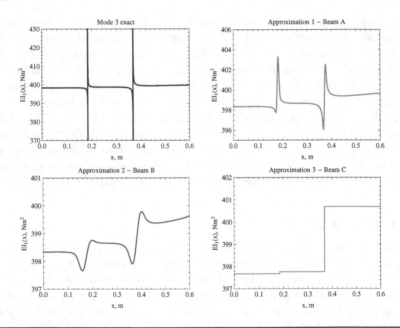

**Figure 9.32**
Realistic beams: approximation to the exact flexural stiffness function for the third
mode

#### 9.2.4.4 Fourth mode

The flexural stiffness function for the fourth mode has three internal singularities, proving again that construction of a physically reasonable beam exactly from this function is very challenging. Therefore, it is again approximated as Beam A, Beam B and Beam C as illustrated in Fig. 9.33. The frequencies and norms calculated for these three beams are given in Table 9.11. The deviation in frequency is less that 0.01%. The height and breadth variation is given in Fig. 9.29(d). Here also we see that the dimensions are proximate to the corresponding non-rotating beam.

### 9.2.5   EFFECT OF ROTATION SPEED

The effect of rotation speed on the height and breadth variation for the first mode of the physically realistic beam is shown in Fig. 9.30. The figure informs us that not only the variations are very subtle for low speeds, but we can also observe that the dimensions are closer to the dimensions of the uniform non-rotating beam. As the rotation speed approaches zero, the dimensions of the physically realistic rotating beam for the first mode approaches the dimensions of the uniform non-rotating beam. For higher rotation speed, the variation in height and breadth also increases. But even then we can notice that this dimensional variations are in the order of $10^{-2}$ mm. Thus a slight variation in the height and breadth of a rotating pinned-free beam can be used to mitigate the centrifugal effects of rotation.

### 9.2.6   REDUCED STIFFNESS FOR TARGETED DESTIFFENING

In the previous section, we have proved that there exists, for the first four modes, physically feasible rotating pinned-free beams with the same eigenpair of a corresponding uniform non-rotating pinned-free beam. During design, it might be sometimes necessary to move one mode away from a particular frequency being a multiple of the rotating speed. By using the cross sections obtained during this work, we can destiffen any of the modes to alleviate the effect of centrifugal stiffening. To study this effect, we determine the stiffness reduction required for a uniform rotating beam

---

**Table 9.9**

**Frequencies and Mode-shapes of the Approximate Beams for Mode 2**

| Approximation | Estimated Frequency (rad/s) | % Deviation from Exact Beam | Norm of Residues of Mode Shape |
|---|---|---|---|
| Beam A | 2023.596 | 0.0012 | $2.9 \times 10^{-4}$ |
| Beam B | 2023.475 | 0.0071 | $1.7 \times 10^{-3}$ |
| Beam C | 2023.793 | 0.0086 | $3.9 \times 10^{-2}$ |

to match the frequencies of a non-rotating beam for the first four modes. Let us suppose that we reduce the uniform beam stiffness from $EI_2 = 400\,Nm^2$ to $EI_1^{(r)}$ so that for a given particular mode the reduced frequency becomes identical to that obtained using the derived flexural stiffness of the stepped beam, $EI_1^{(C)}$ for modes 2-4, and the exact beam, $EI_1^{(exact)}$, for mode 1.

A comparison of the generated results are given in Table 9.12. Note that the flexural stiffness of the uniform non-rotating beam is $400\,Nm^2$, and the corresponding first four natural frequencies are $624.45\,rad/s$, $2023.618\,rad/s$, $4222.1174\,rad/s$, & $7220.0707\,rad/s$ respectively. It can be observed from the table that frequencies of the higher modes can be reduced with less effects on the lower mode frequencies compared with the uniform beam. As for the lower modes, the frequency reduction has a greater effect on the higher mode frequencies. So in closure, we can state that if we target to destiffen a particular mode, say $i$, then the frequencies of modes lower than $i$ will show lesser reduction in frequency that the modes higher than $i$. By using these beam profiles, we can manufacture beams for achieving targeted frequencies. Figure 9.34 shows a comparison between the reduced stiffness, $EI^{(r)}$; and the stiffness variation in the non-uniform beams.

## 9.3   SUMMARY

In the present chapter, we have shown that there exist rotating beams with the same eigen pair as corresponding uniform non-rotating beams, for a given mode.

The first section addresses cantilever beams. The flexural stiffness, $EI_1(x)$ in this rotating beam varies differently for different mode-shapes and the beams have the same length and mass per unit length as the uniform non-rotating beam. The flexural stiffness function, $EI_1(x)$ of the equivalent rotating beams can be applied to test numerical codes written for rotating beams. For the fundamental mode, a physically feasible equivalent rotating beam exists, but for higher modes, the flexural stiffness

---

**Table 9.10**

**Frequencies and Mode-shapes of the Approximate Beams for Mode 3**

| Approximation | Estimated frequency (rad/s) | % Deviation from Exact Beam | Norm of Residues of Mode Shape |
|---|---|---|---|
| Beam A | 4222.106 | 0.0003 | $2.2 \times 10^{-4}$ |
| Beam B | 4222.011 | 0.0025 | $6.2 \times 10^{-4}$ |
| Beam C | 4222.213 | 0.0023 | $1.8 \times 10^{-2}$ |

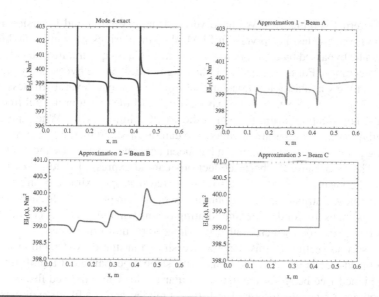

**Figure 9.33**

Realistic beams: approximation to the exact flexural stiffness function for the fourth mode

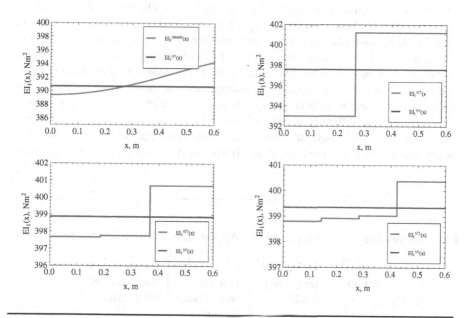

**Figure 9.34**

Comparison between the reduced stiffness $(EI_1^{(r)})$ and the stiffness variation in the non-uniform beams displays internal singularities. We have verified the p-version and h-version FEM for the first four modes. For p-version FEM, the singularities present in the higher modes can be bypassed by appropriate placement of Gauss quadrature points for numerical integration. For h-version FEM, the singularities can be bypassed by placing the nodes at the singularity locations and taking the element flexural stiffness at its midpoint. The codes converge to the expected value of the first four natural frequencies of the non-rotating beam, for which the exact solution exists. When the exact non-rotating beam solutions are obtained, we can rest assured that the governing differential equation of the rotating beam has been accurately solved by the numerical method. This is because the beam deflections can be expressed in terms of mode-shapes as the basis functions. It is also discovered that approximate profiles of the flexural stiffness distributions can be obtained which yield physically feasible beams. Such beams are useful for destiffening rotating beams without requiring considerable stiffness loss at the root, which is otherwise needed by uniform beams.

In the second section of this chapter, we show that there exist rotating beams, having pinned-free boundary conditions, with same eigenpair as of a uniform non-rotating pinned-free beam for a given particular mode. These derived flexural stiffness functions, $EI_1(x)$ are different for different modes, and is a function of mass per unit length ($m$), length ($L$) and rotating speed ($\Omega$) of the rotating beam, and also on the uniform flexural stiffness ($EI_2$) of the uniform non-rotating beam. These derived flexural stiffness functions can be applied for testing the implementation and convergence of numerical codes written for rotating beams. The first mode, generates a physically feasible beam, but all the higher modes contain internal singularities. In the p-FEM, we have addressed these singularities by the use of Gaussian Quadrature points, and carefully selecting them to avoid any of the singularities. The expected convergence of the code is presented for the first four natural frequencies and mode shapes. It is also seen that by approximating these stiffness profile for the higher modes, actual beams with rectangular cross sections, which can be used for mitigating the centrifugal stiffening effects and also for targeted destiffening of the beams, can be manufatured. To determine these equivalent beams, the Euler-Bernoulli beam theory was applied. Although the Euler-Bernoulli theory is a valid approximation

**Table 9.11**

**Frequencies and Mode-shapes of the Approximate Beams for Mode 4**

| Approximation | Estimated frequency (rad/s) | % Deviation from Exact Beam | Norm of Residues of Mode Shape |
|---|---|---|---|
| Beam A | 7220.068 | 0.00004 | $2.9 \times 10^{-4}$ |
| Beam B | 7219.454 | 0.0085 | $3.4 \times 10^{-3}$ |
| Beam C | 7220.129 | 0.0008 | $9.2 \times 10^{-3}$ |

### Table 9.12
### Destiffening Effects of Flexural Stiffness Variations for Various Modes Compared with the with Uniform Rotating Beam $(EI_1^{(r)})$

| Mode | $EI_1(x)$ | $\omega_1$ (rad/s) | $\omega_2$ (rad/s) | $\omega_3$ (rad/s) | $\omega_4$ (rad/s) |
|------|-----------|--------------------|--------------------|--------------------|--------------------|
| Mode 1 | $EI_1^{(exact)}(x)$ | 624.4501 | 2007.2693 | 4181.3782 | 7146.2809 |
|        | $EI_1^{(r)}(x)$ | 624.4501 | 2006.2421 | 4178.7389 | 7141.4551 |
| Mode 2 | $EI_1^{(C)}(x)$ | 629.2181 | 2023.7934 | 4216.0641 | 7202.1130 |
|        | $EI_1^{(r)}(x)$ | 629.8206 | 2023.7934 | 4215.4251 | 7204.2291 |
| Mode 3 | $EI_1^{(C)}(x)$ | 630.2672 | 2027.3147 | 4222.2129 | 7215.2041 |
|        | $EI_1^{(r)}(x)$ | 630.8143 | 2027.0438 | 4222.2129 | 7215.8443 |
| Mode 4 | $EI_1^{(C)}(x)$ | 630.9368 | 2028.0203 | 4225.0120 | 7220.1294 |
|        | $EI_1^{(r)}(x)$ | 631.1809 | 2028.2424 | 4224.7182 | 7220.1294 |

for long and slender beams, but in reality the Timoshenko beam theory has higher fidelity, especially for very high frequency of vibrations. So tergiversating the Euler-Bernoulli theory and applying the Timoshenko beam theory for finding equivalent rotating beams is a possible area for future scientific research.

## 9.4  APPENDIX A

The equation for the $j^{th}$ mode-shape of a uniform non-rotating cantilever beam, from Eq. (9) can be written as

$$\phi_j(x) = C_j \left[\cos(\alpha_j x) - \cosh(\alpha_j x) + K(\sin(\alpha_j x) - \sinh(\alpha_j x))\right] \tag{9.59}$$

where

$$K = \frac{\sin(\alpha_j L) - \sinh(\alpha_j L)}{\cos(\alpha_j L) + \cosh(\alpha_j L)} \tag{9.60}$$

Differentiating Eq. (9.59) twice w.r.t $x$, gives

$$\phi_j''(x) = -C_j \alpha_j^2 \left[\cos(\alpha_j x) + \cosh(\alpha_j x) + K(\sin(\alpha_j x) + \sinh(\alpha_j x))\right] \tag{9.61}$$

Substituting for $x$, $L - y$, in Eq. (9.60) and expanding the resultant expression, we obtain

$$\phi_j''(x) = -C_j \alpha_j^2 \left[a_1 \cos(\alpha_j y) + a_2 K \sin(\alpha_j y) + a_3 \cosh(\alpha_j y) + a_4 K \sinh(\alpha_j y)\right] \tag{9.62}$$

where

$$a_1 = \cos(\alpha_j L) + K \sin(\alpha_j L) \tag{9.63}$$

$$a_2 = \frac{\sin(\alpha_j L)}{K} - \cos(\alpha_j L) \tag{9.64}$$

$$a_3 = \cosh(\alpha_j L) + K \sinh(\alpha_j L) \tag{9.65}$$

$$a_4 = -\frac{\sinh(\alpha_j L)}{K} - \cosh(\alpha_j L) \qquad (9.66)$$

Squaring the characteristic equation, $\cos(\alpha_j L)\cosh(\alpha_j L) = -1$, and using basic trigonometric identities, we obtain,

$$(1 - \sin^2(\alpha_j L))(\cosh^2(\alpha_j L)) = 1 \qquad (9.67)$$

$$\Rightarrow \sin^2(\alpha_j L)\cosh^2(\alpha_j L) = \sinh^2(\alpha_j L) \qquad (9.68)$$

$$\Rightarrow \frac{\sin^2(\alpha_j L)}{\cos^2(\alpha_j L)} = \sinh^2(\alpha_j L) \qquad (9.69)$$

$$\Rightarrow \sin^2(\alpha_j L) = \cos^2(\alpha_j L)\sinh^2(\alpha_j L) \qquad (9.70)$$

Since, $(\alpha_j L)$ is positive for all modes, we have $\sinh(\alpha_j L) > 0$ and $\cosh(\alpha_j L) > 0$. This with $\cos(\alpha_j L)\cosh(\alpha_j L) = -1$, yields $\cos(\alpha_j L) < 0$. From Eqs. (9.68) and (9.70), we obtain $\sin(\alpha_j L)\cosh(\alpha_j L) = \pm\sinh(\alpha_j L)$ and $\sin(\alpha_j L) = \mp\cos(\alpha_j L)\sinh(\alpha_j L)$. The algebraic sign on the R.H.S of these equations is decided by the sign of $\sin(\alpha_j L)$.

Substituting $K$ from Eq. (9.60) in Eq. (9.63), we obtain

$$
\begin{aligned}
a_1 &= \cos(\alpha_j L) + \frac{\sin(\alpha_j L) - \sinh(\alpha_j L)}{\cos(\alpha_j L) + \cosh(\alpha_j L)} \sin(\alpha_j L) \\
&= \frac{\cos^2(\alpha_j L) - 1 + \sin^2(\alpha_j L) - \sin(\alpha_j L)\sinh(\alpha_j L)}{\cos(\alpha_j L) + \cosh \alpha_j L} \\
&= \frac{-\sin(\alpha_j L)\sinh(\alpha_j L)}{\cos(\alpha_j L) + \cosh(\alpha_j L)} \qquad (9.71) \\
&= \frac{-\sin(\alpha_j L)\sinh(\alpha_j L)}{\cos(\alpha_j L) - \frac{1}{\cos(\alpha_j L)}} \\
&= \frac{\sinh(\alpha_j L)\cos(\alpha_j L)}{\sin(\alpha_j L)} \\
&= \mp 1
\end{aligned}
$$

Expanding other coefficients in a similar manner yields, $a_2 = \mp 1$, $a_3 = \pm 1$ and $a_4 = \pm 1$. Hence Eq. (9.62) simplifies as,

$$\phi_j''(x) = \pm C_j \alpha_j^2 [\cos(\alpha_j y) + K\sin(\alpha_j y) - \cosh(\alpha_j y) - K\sinh(\alpha_j y)] \qquad (9.72)$$

Therefore,

$$\phi_j''(x) = \pm \alpha_j^2 \phi(y) \qquad (9.73)$$

Hence, $\phi(y) = 0 \Leftrightarrow \phi_j''(x) = 0$. The assumption $y = L - x$ completes the proof.

## 9.5 APPENDIX B

The $n$ point Gauss quadrature rule for numerical integration over the interval $(-1, 1)$ is given by,

$$\int_{-1}^{1} f(x)dx \approx \sum_{i=0}^{n} w_i f(t_i) \qquad (9.74)$$

The points $t_i$ are the eigenvalues of the matrix $\mathbf{J}$ [176], where matrix $\mathbf{J}$ is defined by,

$$\mathbf{J} = \begin{pmatrix} 0 & a_1 & 0 & \cdots & 0 & 0 \\ a_1 & 0 & a_2 & \cdots & 0 & 0 \\ 0 & a_2 & 0 & \cdots & 0 & 0 \\ \vdots & \vdots & \vdots & \ddots & \vdots & \vdots \\ 0 & 0 & 0 & \cdots & 0 & a_n \\ 0 & 0 & 0 & \cdots & a_n & 0 \end{pmatrix}_{(n+1)\times(n+1)} \qquad (9.75)$$

and the constants $a_i$ are defined by $a_i = \dfrac{i}{\sqrt{4i^2-1}}$. The weights $w_i$ are then given by,

$$w_i = 2\left(V_{1,i}\right)^2 \qquad (9.76)$$

where $\mathbf{V}$ is the matrix of the normalized eigenvectors of $\mathbf{J}$.

The transformation rules for the Gaussian quadrature points and weights for an integral over $(a, b)$ are given by,

$$w_i\big|_{(a,b)} = w_i\big|_{(-1,1)} \times \frac{b-a}{2} \qquad (9.77)$$

$$t_i\big|_{a,b} = t_i\big|_{(-1,1)} \times \frac{b-a}{2} + \frac{a+b}{2} \qquad (9.78)$$

For the domain $(0, 1)$, these formulas become,

$$w_i\big|_{(0,1)} = w_i\big|_{(-1,1)} \times \frac{1}{2} \qquad (9.79)$$

$$t_i\big|_{0,1} = t_i\big|_{(-1,1)} \times \frac{1}{2} + \frac{1}{2} \qquad (9.80)$$

The Gauss quadrature points, sorted in an increasing order, and the corresponding weights with $n = 30$ for the domain $(0, 1)$ is given in Table 9.13.

**Table 9.13**

**Gauss Quadrature Points and Weights of Order 30 for the Domain $(0, 1)$**

| $i$ | $t_i$ | $w_i$ | $i$ | $t_i$ | $w_i$ | $i$ | $t_i$ | $w_i$ |
|---|---|---|---|---|---|---|---|---|
| 1 | 0.0016 | 0.0040 | 11 | 0.2765 | 0.0461 | 21 | 0.7683 | 0.0434 |
| 2 | 0.0082 | 0.0092 | 12 | 0.3236 | 0.0482 | 22 | 0.8103 | 0.0404 |
| 3 | 0.0200 | 0.0144 | 13 | 0.3727 | 0.0498 | 23 | 0.8489 | 0.0369 |
| 4 | 0.0369 | 0.0194 | 14 | 0.4231 | 0.0509 | 24 | 0.8839 | 0.0330 |
| 5 | 0.0587 | 0.0242 | 15 | 0.4743 | 0.0514 | 25 | 0.9148 | 0.0287 |
| 6 | 0.0852 | 0.0287 | 16 | 0.5257 | 0.0514 | 26 | 0.9413 | 0.0242 |
| 7 | 0.1161 | 0.0330 | 17 | 0.5769 | 0.0509 | 27 | 0.9631 | 0.0194 |
| 8 | 0.1511 | 0.0369 | 18 | 0.6273 | 0.0498 | 28 | 0.9800 | 0.0144 |
| 9 | 0.1897 | 0.0404 | 19 | 0.6764 | 0.0482 | 29 | 0.9918 | 0.0092 |
| 10 | 0.2317 | 0.0434 | 20 | 0.7235 | 0.0461 | 30 | 0.9984 | 0.0040 |

# 10 Isospectral Rayleigh Beams

Non-uniform beams can be used to mathematically model many important structures. To accurately model long and slender beams, the Euler-Bernoulli (EB) beam theory is sufficient. However, for short and thick beams, and for accurate frequency prediction of the higher modes of vibration, the Bresse-Timoshenko (BT) beam theory is often preferred. A relatively "simpler theory" was developed by Lord Rayleigh [168] before the Timoshenko beam theory came into being. Rayleigh theory includes the rotary inertia effect but does not take into account the shear deformation [15, 131, 173, 210]. The governing equation of Rayleigh beam is a single fourth-order differential equation in a single variable. In contrast, Timoshenko theory uses the coupled differential equations in two variables (harmonic vibrations). The Rayleigh beam theory also predicts the natural frequencies and mode shapes more accurately compared to the Euler-Bernoulli beam theory, without falling prey to the mathematical complexities of the Timoshenko beam theory.

Inverse problems represent an important class of problems in vibrating systems which involve finding material and geometric properties from known modal parameters and reconstruction of a beam from its spectral data [83]. We have seen in earlier chapters that multiple beams can possess the same spectra for a given boundary condition. The existence of systems that have the same frequencies for a given boundary condition but have different material and geometric properties is of enormous interest in mechanics.

Uncovering isospectral systems are an important subclass of inverse problems. In this chapter, we extend the analysis developed in Chapters 7 and 8 to Rayleigh beams. A transformation is applied to convert the non-dimensional non-uniform Rayleigh beam to a uniform Rayleigh beam, from the $(x, W)$ frame of reference to a hypothetical $(z, U)$ frame of reference. If the material and geometric properties of the beam are specific selected functions of the two introduced auxiliary variables, then the transformation will be achieved and, if the coefficients of the transformed equation matches with the uniform one, then the equivalence is verified. Four specific cases are considered for solving a pair of coupled ODEs and we arrive at the closed form solutions for the mass per unit length, mass moment of inertia and bending stiffness variations of the non-uniform beam which is isospectral to the given uniform beam. Furthermore, we present the constraints under which the boundary conditions can remain unchanged. Examples of beams having a rectangular cross-section are provided to illustrate the application of the analysis. The content of this chapter is adapted from Ref. [21].

## 10.1 MATHEMATICAL FORMULATION

The equation regulating the free vibrations of a non-uniform, in-homogeneous Rayleigh beam of length $L$ with angular frequency $\omega$ and transverse displacement

$W(x)$ is

$$\frac{d^2}{dX^2}\left(EI\frac{d^2W}{dX^2}\right) + \frac{d}{dX}\left(\rho I\omega^2\frac{dW}{dX}\right) - \rho A\omega^2 W = 0, \qquad 0 \le X \le L \qquad (10.1)$$

where $\rho A$ is the mass per unit length, $EI$ is the bending rigidity and $\rho I$ is the mass moment of inertia per unit length. Here $E$, $G$ and $\rho$ indicate the Young's modulus, the shear modulus and the mass density, respectively; $A$ and $I$ denote the area and the area moment of inertia of the cross section, respectively.

For a given uniform beam, flexural stiffness $EI(X) = E_0I_0$, mass per unit length $\rho A(X) = \rho_0 A_0$, and mass moment of inertia per unit length $\rho I(X) = \rho_0 I_0$. We define non-dimensional variables $f$, $g$, $m$ and $x$ as

$$f(x) = \frac{EI(x)}{E_0I_0} \qquad g(x) = \frac{\rho I(x)}{\rho_0 I_0} \qquad m(x) = \frac{\rho A(x)}{\rho_0 A_0} \qquad x = \frac{X}{L}$$

Then, Eqn. (10.1) can be expressed in non-dimensional form as

$$\frac{d^2}{dx^2}\left(f(x)\frac{d^2W}{dx^2}\right) + \eta^2\left(\frac{r_0}{L}\right)^2\frac{d}{dx}\left(g(x)\frac{dW}{dx}\right) - m(x)\eta^2 W = 0, \qquad 0 \le x \le 1$$
$$(10.2)$$

where $\eta$ (non-dimensional natural frequency) and $r_0$ is expressed by

$$r_0 = \sqrt{\frac{I_0}{A_0}} \qquad \eta^2 = \omega^2\left(\frac{\rho_0 A_0 L^4}{E_0 I_0}\right)$$

For reference, a uniform Rayleigh beam will be introduced with displacement function $V(z)$ that satisfies the equation

$$\left(\frac{d^4V}{dz^4}\right) + \eta^2\left(\frac{r_0}{L}\right)^2\frac{d^2V}{dz^2} - \eta^2 V = 0, \qquad 0 \le z \le 1 \qquad (10.3)$$

The second term on the left-hand side of the above equation is present in the Rayleigh formulation and does not appear in the Euler-Bernoulli formulation discussed in earlier chapters [115]. The non-dimensional governing Equation (i.e. Eqn. (10.2)) which is in the $(x, W)$ frame, can be transformed into the $(z, U)$ frame, using the following transformations [86]

$$W(x(z)) = q(z)U(z) \qquad (10.4)$$

$$x = \int_0^z \frac{1}{p(z)}\,dz \qquad (10.5)$$

where $p$ and $q$ are the auxiliary variables involved in the transformation.

Eqn. (10.5) implies that

$$z = \int_0^x p(z(x))\,dx \Rightarrow \frac{d}{dx} = p(z)\frac{d}{dz} \Rightarrow \frac{1}{p(z)} = \frac{dx}{dz} \qquad (10.6)$$
$$x = 0 \Leftrightarrow z = 0 \quad \text{and} \quad x = 1 \Leftrightarrow z = z_0$$

where $z_0$ is defined such that the above relation is valid.

$f$, $g$ and $m$ are functions of $p$ and $q$ and are selected so as to transform Eqn. (10.2) to the desired equation (Eqn. 10.8), and are written as:

$$f = \frac{1}{p^3 q^2} \qquad g = \frac{1}{pq^2} \qquad m = \frac{p\left(\left(\frac{r_0}{L}\right)^2 qq'' - 2\left(\frac{r_0}{L}\right)^2 q'^2 + q^2\right)}{q^4} \tag{10.7}$$

Substituting Eqn. (10.7) into Eqn. (10.2) and applying the transformation, we get

$$\frac{p}{q}\frac{d^4U}{dz^4} + \frac{1}{pq^3}\frac{d^2U}{dz^2}(2pq(2p\frac{d^2q}{dz^2} - \frac{dp}{dz}\frac{dq}{dz}) + q^2(p\frac{d^2p}{dz^2} - 2(\frac{dp}{dz})^2) - 6p^2(\frac{dq}{dz})^2) + \frac{1}{p^2q^4}\frac{dU}{dz}$$
$$(2p^2q\frac{dq}{dz}(\frac{dp}{dz}\frac{dq}{dz} - 8p\frac{d^2q}{dz^2}) + 2pq^2(-p\frac{dp}{dz}\frac{d^2q}{dz^2} + p(2p\frac{d^3q}{dz^3} - \frac{d^2p}{dz^2}\frac{dq}{dz}) + (\frac{dp}{dz})^2\frac{dq}{dz}) + q^3(p^2\frac{d^3p}{dz^3} -$$
$$5p\frac{d^2p}{dz^2}\frac{dp}{dz} + 4(\frac{dp}{dz})^3) + 12p^3(\frac{dq}{dz})^3) + \frac{1}{p^2q^4}U(6p^2(\frac{dq}{dz})^2(p\frac{d^2q}{dz^2} + \frac{dp}{dz}\frac{dq}{dz}) - 2pq(2p\frac{dp}{dz}\frac{d^2q}{dz^2}\frac{dq}{dz} + \tag{10.8}$$
$$p(2p\frac{d^3q}{dz^3}\frac{dq}{dz} + 2\frac{d^2p}{dz^2}(\frac{dq}{dz})^2 + p(\frac{d^2q}{dz^2})^2) - 3(\frac{dp}{dz})^2(\frac{dq}{dz})^2) + q^2(-2p(\frac{dp}{dz})^2\frac{d^2q}{dz^2} - 5p\frac{d^2p}{dz^2}\frac{dp}{dz}\frac{dq}{dz} +$$
$$p^2(p\frac{d^4q}{dz^4} + \frac{d^3p}{dz^3}\frac{dq}{dz} + \frac{d^2p}{dz^2}\frac{d^2q}{dz^2}) + 4(\frac{dp}{dz})^3\frac{dq}{dz})) + \eta^2(\frac{d^2U}{dz^2}\left(\frac{r_0}{L}\right)^2\frac{p}{q} - \frac{p}{q}U) = 0$$

If $A$ and $B$ are expressed as

$$A = -\frac{2\frac{dp}{dz}\frac{dq}{dz}}{pq} + \frac{\frac{d^2p}{dz^2}}{p} - \frac{2(\frac{dp}{dz})^2}{p^2} + \frac{4\frac{d^2q}{dz^2}}{q} - \frac{6(\frac{dq}{dz})^2}{q^2} \tag{10.9}$$

$$B = \frac{4p'^3q'}{p^3q} + \frac{6p'^2q'^2}{p^2q^2} + \frac{6p'q'^3}{pq^3} - \frac{5p'q'p''}{p^2q} - \frac{4q'^2p''}{pq^2} - \frac{2q''}{q^2} - \frac{2p'^2q''}{p^2q} -$$
$$\frac{4p'q'q''}{pq^2} + \frac{6q'^2q''}{q^3} + \frac{p''q''}{pq} + \frac{q'p'''}{pq} - \frac{4q'q'''}{q^2} + \frac{q''''}{q} \tag{10.10}$$

then, Eqn. (10.8) can be rewritten as

$$\left(\frac{d^4U}{dz^4}\right) + \frac{d}{dz}\left(A\frac{dU}{dz}\right) + BU + \eta^2\left(\frac{r_0}{L}\right)^2\frac{d^2U}{dz^2} - \eta^2U = 0, \qquad 0 \le z \le z_0 \tag{10.11}$$

If $A, B = 0$, then Eqn. (10.3) is identical to Eqn. (10.11), thus transforming the non-uniform beam equation to a uniform beam equation.

Thus, the non-uniform beams are isospectral to a given uniform beam if

$$A = 0 \quad \text{and} \quad B = 0 \tag{10.12}$$

These equations are coupled fourth-order ODEs and are onerous to solve. However, for four special cases: (i) $q = q_0$ a constant, (ii) $p\frac{dq}{dz} = k$, a constant, (iii) $pq^2 = k$, a constant and (iv) $p = p_0$ a constant, we find that the coupled ODEs are amenable to an analytical solution.

## 10.1.1   CASE-1: $Q = C_1$, A CONSTANT

When $q = c_1$, $B = 0$ is automatically satisfied. Setting $A = 0$ yields

$$p\frac{d^2p}{dz^2} - 2\left(\frac{dp}{dz}\right)^2 = 0 \tag{10.13}$$

Solving for $p$, we get

$$p(z) = \frac{1}{\beta + \alpha z} \tag{10.14}$$

$\alpha$ and $\beta$ are arbitrary constants present in the general solution.
Substituting Eqn. (10.14) into Eqn. (10.5) and then solving for $z$ we have

$$z = \left\{ \frac{\sqrt{\beta^2 + 2\alpha x} - \beta}{\alpha}, -\frac{\beta + \sqrt{\beta^2 + 2\alpha x}}{\alpha} \right\} \tag{10.15}$$

For positive $\alpha$ and $\beta$, we mandate $z$ to be positive, thus

$$z = \frac{\sqrt{\beta^2 + 2\alpha x} - \beta}{\alpha} \tag{10.16}$$

Substituting $p(z)$, $q(z)$ and Eqn. (10.16) into Eqn. (10.7), we get

$$\{f(x), g(x), m(x)\} = \left\{ \frac{(\beta^2 + 2\alpha x)^{3/2}}{c_1^2}, \frac{\sqrt{\beta^2 + 2\alpha x}}{c_1^2}, \frac{1}{c_1^2\sqrt{\beta^2 + 2\alpha x}} \right\} \tag{10.17}$$

If $z_0 = 1 \Leftrightarrow x_0 = 1$, then $\alpha$ and $\beta$ are linked by the following expression: $\alpha = 2(1 - \beta)$. For $\beta = \{0.6, 0.75, 0.85, 0.9\}$, $c_1 = 1$, and $\frac{r_0}{L} = 0.09$ we plot the mass, stiffness and mass moment of inertia functions of the non-uniform beams which are isospectral to a uniform beam (Fig. 10.1).

We now apply our analysis to beams having a rectangular cross section. The non-dimensional breadth ($b$) and height ($h$) profiles of the cross sections are related to the mass and mass moment of inertia of the beams by the following equations:

$$m(x) = b(x)h(x); \quad g(x) = b(x)h(x)^3 \tag{10.18}$$

$$\Rightarrow b(x) = \sqrt{\frac{m(x)^3}{g(x)}}; \quad h(x) = \sqrt{\frac{g(x)}{m(x)}} \tag{10.19}$$

Moreover, $\frac{f(x)}{g(x)}$ corresponds to the ratio $\frac{E}{\rho}$. Therefore, using the $m(x)$, $g(x)$ and $f(x)$ functions we can derive the $b$, $h$ and $\frac{E}{\rho}$ profiles of the rectangular beams. These $b$, $h$ and $\frac{E}{\rho}$ profiles of the non-uniform beams, which are isospectral to a uniform beam are plotted in Fig. 10.2. These results show that the height, breadth and the ratio of modulus and density vary along the $x$ axis in a manner which forces the natural frequencies to remain identical to that of the uniform beam.

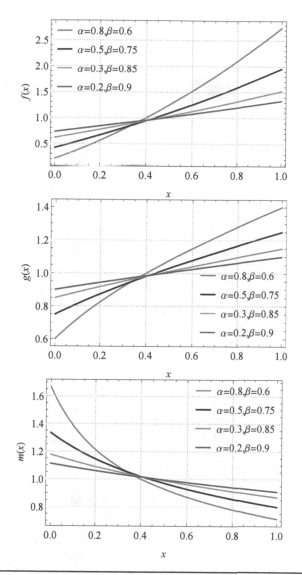

**Figure 10.1**

Mass, bending stiffness and mass moment of inertia functions of a non-uniform beam isospectral to a uniform beam (Case-1: $\beta = \{0.6, 0.75, 0.85, 0.9\}$, $c_1 = 1$, and $\frac{r_0}{L} = 0.09$)

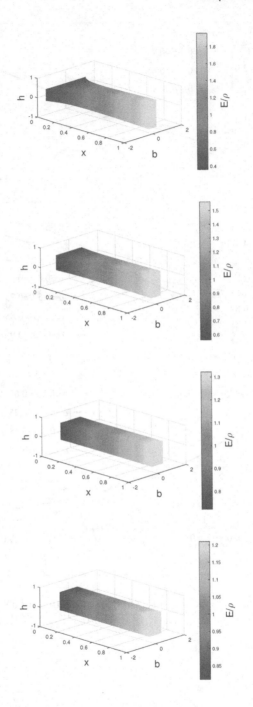

**Figure 10.2**
Breadth, height and ratio of modulus and density distributions of non-uniform beams isospectral to a uniform beam (Case-1: $\beta = \{0.6, 0.75, 0.85, 0.9\}$, $c_1 = 1$, and $\frac{r_0}{L} = 0.09$)

## 10.1.2 CASE-2: $PQ_Z = KPQZ = K$, A CONSTANT

If $p\frac{dq}{dz} = k$, then $B = 0$ is automatically satisfied (Eq. (2.22a) in [86]). Setting $A = 0$ yields

$$\frac{6q''(z)}{q(z)} - \frac{6q'(z)^2}{q(z)^2} - \frac{q^{(3)}(z)}{q'(z)} = 0 \tag{10.20}$$

Solving for $q(z)$, then obtaining $p(z)$ from the constraint, we obtain

$$q(z) = \frac{1}{\alpha + \beta z + \gamma z^2}$$

$$p(z) = -\frac{k\left(\alpha + \gamma z^2 + \beta z\right)^2}{\beta + 2\gamma z} \tag{10.21}$$

$\alpha$, $\beta$ and $\gamma$ are arbitrary constants which emerge in the general solution. Substituting $p(z)$ from Eqn. (10.21) into Eqn. (10.5), we have

$$x = \frac{1}{k(\alpha + z(\beta + \gamma z))} - \frac{1}{\alpha k} \tag{10.22}$$

$$z = \{\frac{-\beta - \sqrt{(\beta + \alpha\beta kx)^2 - 4\alpha^2 kx(\gamma + \alpha\gamma kx)} + \alpha\beta(-k)x}{2(\gamma + \alpha\gamma kx)},$$

$$\frac{-\beta + \sqrt{(\beta + \alpha\beta kx)^2 - 4\alpha^2 kx(\gamma + \alpha\gamma kx)} + \alpha\beta(-k)x}{2(\gamma + \alpha\gamma kx)}\} \tag{10.23}$$

Substituting $p(z)$, $q(z)$ and second expression of Eqn. (10.23) into Eqn. (10.7) with $k = -1$, we obtain

$$f(x) = \frac{(1 - \alpha x)(-(\alpha x - 1)(4\alpha^2\gamma x + \beta^2(1 - \alpha x)))^{3/2}}{\alpha^4}$$

$$g(x) = \frac{\sqrt{-(\alpha x - 1)(4\alpha^2\gamma x + \beta^2(1 - \alpha x))}}{1 - \alpha x} \tag{10.24}$$

$$m(x) = -\frac{\alpha^3(\alpha + 2\gamma r_0^2(\alpha x - 1))}{(\alpha x - 1)^3\sqrt{-(\alpha x - 1)(4\alpha^2\gamma x + \beta^2(1 - \alpha x))}}$$

If $z_0 = 1 \Leftrightarrow x_0 = 1$, then $\alpha$, $\beta$ and $\gamma$ are related by the expression: $\gamma = \frac{-\beta + \alpha^2(-k) - \alpha\beta k}{\alpha k + 1}$. For $\alpha = \{0.3, 0.4, 0.5, 0.55\}$, $\beta = \{0.1, 0.2, 0.3, 0.4\}$, $k = -1$, and $\frac{r_0}{L} = 0.09$ we plot the mass, stiffness and mass moment of inertia functions of the non-uniform beams which are isospectral to a uniform beam as shown in Fig. 10.3.

Applying our analysis to beams with a rectangular cross-section, the $b$, $h$ and $\frac{E}{\rho}$ profiles of the non-uniform beams which are isospectral to the given uniform beam is shown in Fig. 10.4.

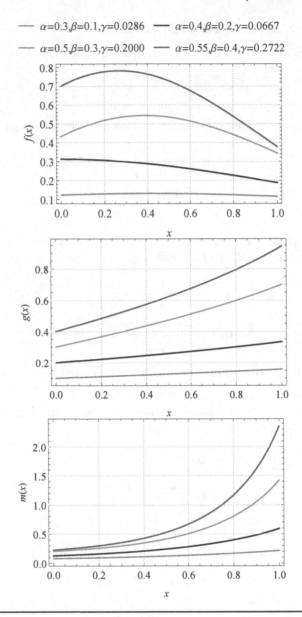

**Figure 10.3**

Mass, bending stiffness and mass moment of inertia functions of a non-uniform beam isospectral to a uniform beam (Case-2: $\alpha = \{0.3, 0.4, 0.5, 0.55\}$, $\beta = \{0.1, 0.2, 0.3, 0.4\}$, $k = -1$, and $\frac{r_0}{L} = 0.09$)

### 10.1.3   CASE-3: $PQ^2 = C$, A CONSTANT

If $pq^2 = c$, then substituting $p$ into $A$, and setting it to zero and simplifying, we get $p\frac{dq}{dz} = a$ constant, which automatically satisfies $B = 0$. From the above two conditions, we obtain

$$\frac{dq}{dz}\frac{1}{q^2} = k \tag{10.25}$$

Solving for $q(z)$ and then deriving $p(z)$ from the constraint, we get

$$q(z) = \frac{1}{-\alpha - kz} \tag{10.26}$$
$$p(z) = c(-\alpha - kz)^2$$

Substituting $p(z)$ from Eqn. (10.26) into Eqn. (10.5) and then solving for $z$, we obtain

$$z = \frac{\alpha^2 cx}{\alpha cx + 1} \tag{10.27}$$

Substituting $p(z)$, $q(z)$ and Eqn. (10.27) into Eqn. (10.7) with $k = -1$, we obtain

$$\{f(x), g(x), m(x)\} = \left\{ \frac{(\alpha cx + 1)^4}{\alpha^4 c^3}, \frac{1}{c}, \frac{\alpha^4 c}{(\alpha cx + 1)^4} \right\} \tag{10.28}$$

If $z_0 = 1 \Leftrightarrow x_0 = 1$, then $\alpha$ and $c$ are related by the expression: $\alpha = \left\{ -\frac{\sqrt{c}\sqrt{c+4}-c}{2c}, \frac{c+\sqrt{c}\sqrt{c+4}}{2c} \right\}$. If $c = \{0.25, 0.5, 0.75, 1\}$ and if $\alpha$ is calculated alternatively from its expressions then the values of $\alpha$ obtained are as follows: $\alpha = \{-1.56155, 2, -0.758306, \frac{1}{2}(\sqrt{5} + 1)\}$. For $k = -1$, $\frac{r_0}{L} = 0.09$, we plot the mass, stiffness and mass moment of inertia functions of the non-uniform beams which are isospectral to a uniform beam as shown in Fig. 10.5. The $b$, $h$ and $\frac{E}{\rho}$ profiles of the non-uniform beams which are isospectral to the given uniform beam is as shown in Fig. 10.6.

### 10.1.4   CASE-4: $P = K$, A CONSTANT

If $p = k$, then $z = kx$ and, if $A = 0$ then

$$\frac{4\frac{d^2q}{dz^2}}{q} - \frac{6\left(\frac{dq}{dz}\right)^2}{(q)^2} = 0 \tag{10.29}$$

Solving for $q(z)$, we have

$$q(z) = \frac{1}{(\beta + \alpha z)^2} \tag{10.30}$$

Substituting $p$ and $q(z)$ into Eqn. (10.7), we get

$$\{f(x), g(x), m(x)\} = \left\{ \frac{(\beta + \alpha kx)^4}{k^3}, \frac{(\beta + \alpha kx)^4}{k}, \right.$$
$$\left. k(\beta + \alpha kx)^8 \left( \frac{1}{(\beta + \alpha kx)^4} - \frac{2\alpha^2 r_0^2}{(\beta + \alpha kx)^6} \right) \right\} \tag{10.31}$$

For $k = \{0.75, 1.25\}$, $\frac{r_0}{L} = 0.09$, $\alpha = 0.75$ and $\beta = 0.75$, we plot the mass, stiffness and mass moment of inertia functions of the non-uniform beams which are isospectral to the uniform beam as shown in Figs. (10.7,10.8). The $b$, $h$ and $\frac{E}{\rho}$ profiles of the non-uniform beams which are isospectral to the given uniform beam is as shown in Fig. 10.9.

## 10.2  BOUNDARY CONDITIONS

While applying the transformation to the differential equation, the boundary conditions also get transformed. The clamped boundary condition remains invariant, but for certain special cases, other boundary conditions can also retain invariance. We present various conditions under which the boundary conditions remain invariant. The aforementioned transformation expresses the derivatives $W_x, W_{xx}, W_{xxx}$ in terms of $U_z, U_{zz}, U_{zzz}$ as follows:

$$W = qU \tag{10.32}$$

$$\frac{dW}{dx} = p\left(q\frac{dU}{dz} + U\frac{dq}{dz}\right) \tag{10.33}$$

$$\frac{d^2W}{dx^2} = (p)^2 q\frac{d^2U}{dz^2} + p\frac{dU}{dz}\left(2p\frac{dq}{dz} + q\frac{dp}{dz}\right) + pU\left(p\frac{d^2q}{dz^2} + \frac{dp}{dz}\frac{dq}{dz}\right) \tag{10.34}$$

$$\frac{d^3W}{dx^3} = (p)^3 q\frac{d^3U}{dz^3} + 3(p)^2\frac{d^2U}{dz^2}\left(p\frac{dq}{dz} + q\frac{dp}{dz}\right) + p\frac{dU}{dz}\left(3p\left(p\frac{d^2q}{dz^2} + 2\frac{dp}{dz}\frac{dq}{dz}\right) + \right.$$
$$\left. q\left(p\frac{d^2p}{dz^2} + \left(\frac{dp}{dz}\right)^2\right)\right) + pU\left(3p\frac{dp}{dz}\frac{d^2q}{dz^2} + p\left(p\frac{d^3q}{dz^3} + \frac{d^2p}{dz^2}\frac{dq}{dz}\right) + \left(\frac{dp}{dz}\right)^2\frac{dq}{dz}\right) \tag{10.35}$$

It can be observed from Eqns. (10.32, 10.33) that

$$\frac{dW}{dx} = 0 \quad \text{and} \quad W = 0 \Leftrightarrow \frac{dU}{dx} = 0 \quad \text{and} \quad U = 0 \tag{10.36}$$

Thus, the clamped end condition for a beam is preserved by the aforementioned transformation for any functions $m, g$ and $f$.

The free and pinned boundary configuration depends on how the shearing force $(V)$ and bending moment $(M)$ transforms. For the harmonic vibration of Rayleigh beams, $M$ and $V$ are given by

$$M = f(x)\frac{d^2W}{dx^2} \qquad V = \frac{d}{dx}\left(f(x)\frac{d^2W}{dx^2}\right) + \eta^2\left(\frac{r_0}{L}\right)^2 g(x)\frac{dW}{dx} \tag{10.37}$$

Applying the transformation to Eqn. (10.37) we obtain

$$f(x)\frac{d^2W}{dx^2} = \frac{d^2U}{dz^2}\frac{1}{(p)(q)} + U\left(p\frac{d^2q}{dz^2} + \frac{dp}{dz}\frac{dq}{dz}\right)\frac{1}{(p)^2(q)^2} + \frac{dU}{dz}\left(2p\frac{dq}{dz} + q\frac{dp}{dz}\right)\frac{1}{(p)^2(q)^2} \tag{10.38}$$

**Table 10.1**

**A Summary of the Boundary Conditions, and the Cases under Which they are Preserved**

| Boundary Condition | | Condition on the Auxiliary Variables |
|---|---|---|
| Clamped | $W = 0 = W_x$ | Cases-(i), (ii), (iii), (iv) |
| Pinned | $W = 0 = W_{xx}$ | Case-(iii) |
| Free | $W_{xx} = 0 = \frac{d}{dx}\left(f(x)\frac{d^2W}{dx^2}\right) + \eta^2\left(\frac{r_0}{L}\right)^2 g(x)\frac{dW}{dx}$ | $p$ and $q = constant$ |
| Antiresonant(i) | $W = 0 = \frac{d}{dx}\left(f(x)\frac{d^2W}{dx^2}\right) + \eta^2\left(\frac{r_0}{L}\right)^2 g(x)\frac{dW}{dx}$ | Case-(iii) |

$$\frac{d}{dx}\left(f(x)\frac{d^2W}{dx^2}\right) + \eta^2\left(\frac{r_0}{L}\right)^2 g(x)\frac{dW}{dx} =$$

$$\frac{d^3U}{dz^3}\frac{1}{q} + \frac{dq}{dz}\frac{d^2U}{dz^2}\frac{1}{(q)^2} + \frac{dU}{dz}(pq(3p\frac{d^2q}{dz^2} - 2\frac{dp}{dz}\frac{dq}{dz}) + (q)^2(p\frac{d^2p}{dz^2} - 2(\frac{dp}{dz})^2) -$$

$$4(p)^2(\frac{dq}{dz})^2)\frac{1}{(p)^2(q)^3} + U(q(p(p\frac{d^3q}{dz^3} + \frac{d^2p}{dz^2}\frac{dq}{dz}) - 2(\frac{dp}{dz})^2\frac{dq}{dz}) - 2p\frac{dq}{dz}(p\frac{d^2q}{dz^2} +$$

$$\frac{dp}{dz}\frac{dq}{dz}))\frac{1}{(p)^2(q)^3} + \eta^2((\frac{r_0}{L})^2 U\frac{dq}{dz}\frac{1}{(q)^2} + (\frac{r_0}{L})^2\frac{dU}{dz}\frac{1}{q})$$

$$(10.39)$$

Free boundary configuration is preserved if the following relations are true

$$f(x)\frac{d^2W}{dx^2} = 0 \Leftrightarrow \frac{d^2U}{dz^2} = 0 \qquad (10.40a)$$

$$\frac{d}{dx}\left(f(x)\frac{d^2W}{dx^2}\right) + \eta^2\left(\frac{r_0}{L}\right)^2 g(x)\frac{dW}{dx} = 0 \Leftrightarrow \frac{d^3U}{dz^3} + \eta^2\left(\frac{r_0}{L}\right)^2\frac{dU}{dz} = 0 \quad (10.40b)$$

For Eqn. (10.40a) to hold, the coefficients of $U$ and $\frac{dU}{dz}$ in Eqn. (10.38) should vanish. The coefficient of $U$ equals zero implies $(p\frac{d^2q}{dz^2} + \frac{dp}{dz}\frac{dq}{dz}) = \frac{d}{dz}(p\frac{dq}{dz})$ is zero. This holds when $p\frac{dq}{dz}$ is a constant. For the coefficient of $\frac{dU}{dz}$ to be zero, $(2pq\frac{dq}{dz} + q^2\frac{dp}{dz}) = \frac{d}{dz}(pq^2)$ should be zero. This holds when $pq^2$ is a constant. Therefore, when $p\frac{dq}{dz}$ and $pq^2$ are constants, Eqn. (10.40a) holds. Eqn. (10.40b) is satisfied only when $p$ and $q$ are constants. This transforms a given uniform beam to a different uniform beam. When $p\frac{dq}{dz}$ and $pq^2$ are constants, Eqn. (10.40b) will have an extra term in the right hand side of the relation, which will add an error in the calculation of natural frequencies.

## 10.3    FINITE ELEMENT VALIDATION

The finite element method is used to compute the natural frequencies of the non-uniform beams. A description of the methodology is available in standard textbooks [36]. The beam is discretized into several finite elements of equal length ($l$) each of which has two nodes, and each node has two degrees of freedom (DOF) - transverse displacement ($w$) and the slope of deflection. Specifically, $w$, along the $i$th element is given by

$$w = H_1 W_i + H_2 \phi_i + H_3 W_{i+1} + H_4 \phi_{i+1} = \lfloor H \rfloor \{d\}_i \qquad (10.41)$$

The Hermite shape functions ($H_1$, $H_2$, $H_3$ and $H_4$) are given by $H_1 = 2\zeta^3 - 3\zeta^2 + 1$, $H_2 = \frac{\zeta^3 - 2\zeta^2 + \zeta}{l}$, $H_3 = -2\zeta^3 + 3\zeta^2$ and $H_4 = \frac{\zeta^3 - \zeta^2}{l}$ , where $\zeta = \frac{x - x_i}{l}$. The expressions for the kinetic energy $T_i$ and potential energy $U_i$ of the $i$th beam element is given by

$$T_i = \frac{1}{2} \int_{x_i}^{x_{i+1}} m(x) \left( \frac{\partial w}{\partial t} \right)^2 dx + \frac{1}{2} \int_{x_i}^{x_{i+1}} g(x) \left( \frac{\partial^2 w}{\partial t \partial x} \right)^2 dx \qquad (10.42a)$$

$$U_i = \frac{1}{2} \int_{x_i}^{x_{i+1}} f(x) \left( \frac{\partial^2 w}{\partial x^2} \right)^2 dx \qquad (10.42b)$$

The elemental mass and stiffness matrices are given by

$$M_{ij} = \int_{x_i}^{x_{i+1}} g(x) H_i' H_j' dx + \int_{x_i}^{x_{i+1}} m(x) H_i H_j dx \qquad (10.43a)$$

$$K_{ij} = \int_{x_i}^{x_{i+1}} f(x) H_i'' H_j'' dx \qquad (10.43b)$$

These elemental mass $M_{ij}$ and $K_{ij}$ stiffness matrices are assembled appropriately to obtain global mass $[M]$ and stiffness $[K]$ matrices. The natural frequencies ($\eta$) and mode shapes ($x$) are then obtained by solving the following eigenvalue problem:

$$[K]\{x\} = \eta^2 [M]\{x\} \qquad (10.44)$$

The non-dimensional frequencies ($\eta$) of the uniform and the non-uniform Rayleigh beams are computed for the following boundary configurations: (i) clamped-pinned, (ii) pinned-pinned and (iii) clamped-clamped. For the non-uniform beam, $f, g$ and $m$ calculated from the case $pq^2 = c$ is selected for the clamped-pinned and pinned-pinned configurations. The frequencies are given in Tables 10.2 and 10.3. For the clamped-clamped configuration, the $f, g$ and $m$ from any of the four cases can be considered. The frequencies for this configuration are provided in Table 10.4.

The mode shape $U(z)$ of the uniform clamped-clamped Rayleigh beam is provided in the Appendix A. The mode shape $W(x)$ of the isospectral non-uniform Rayleigh beam can be determined using the aforementioned transformation from $U(z)$ as follows: First, calculate $W(z)$ using $W(z) = q(z)U(z)$. Then calculate $W(x)$ from $W(z)$ by substituting $z$ in terms of $x$. The first six mode shapes computed using

| Table 10.2 Non-dimensional Frequencies of a Clamped-pinned Beam (500 Elements) | |
| --- | --- |
| Uniform beam [$\frac{r_0}{L} = 0.09$] | Isospectral non-uniform beam [$\frac{r_0}{L} = 0.09$] |
| 14.745 | 14.745 |
| 43.0405 | 43.0405 |
| 78.549 | 78.549 |
| 116.69 | 116.69 |
| 155.383 | 155.383 |
| 193.87 | 193.87 |

| Table 10.3 Non-dimensional Frequencies of a Pinned-pinned Beam (500 Elements) | |
| --- | --- |
| Uniform beam [$\frac{r_0}{L} = 0.09$] | Isospectral Non-uniform Beam [$\frac{r_0}{L} = 0.09$] |
| 9.49728 | 9.49728 |
| 34.3645 | 34.3645 |
| 67.7395 | 67.7395 |
| 104.602 | 104.602 |
| 142.489 | 142.489 |
| 180.426 | 180.426 |

Table 10.4
Non-dimensional Frequencies of a Clamped-clamped Beam (500 Elements).

| Uniform beam [$\frac{r_0}{L} = 0.09$] | Isospectral Non-uniform Beam [$\frac{r_0}{L} = 0.09$] |
| --- | --- |
| 21.322 | 21.322 |
| 52.5959 | 52.5959 |
| 90.0704 | 90.0704 |
| 129.325 | 129.325 |
| 168.719 | 168.719 |
| 207.675 | 207.675 |

the aforementioned transformation and the mode shapes computed using FEM for the non-uniform clamped-clamped beam are given in Fig. 10.10. Similarly, the first six mode shapes determined for the non-uniform clamped-pinned and pinned-pinned beam are provided in Figs. 10.11 and 10.12.

## 10.4 SUMMARY

In this chapter, an analytical procedure for determining non-uniform Rayleigh beams, which are isospectral to a uniform beam, is expounded. A transformation is used to convert the non-uniform beam equation into a uniform beam equation. Analytical expressions for the mass, bending stiffness and mass moment of inertia of such non-uniform beams are determined considering four specific cases. We provide the necessary conditions, on the auxiliary variables, needed to preserve the boundary configurations. Beams having a rectangular cross-section are showcased for all the four cases to illustrate the application of this analysis. The breadth, height and the ratio of modulus and density variations of the rectangular cross-section, with a span, of isospectral non-uniform beams, are determined. The non-dimensional frequencies of the uncovered isospectral non-uniform beams are determined using FEM, and they are found to be identical to that of the given uniform beam for a particular boundary condition.

It is easier to analyze uniform beams. Once the dynamic characteristics of a uniform beam are computed, from the above technique, we can create a non-uniform beam of same spectra to that of a uniform beam and know its dynamic characteristics. Furthermore, recent advancements in machining techniques- such as additive manufacturing and rapid prototyping [106, 121] - facilitate the manufacturing of beams with known breadth, height, and material property variation. Structural identification issues, such as damage and blockage identification, has been conducted with the help of quasi-isospectral operators for rods and symmetric ducts [22, 23]. We can develop similar transformation for various versions of beams which along with the quasi-isospectral operators are an integral part of damage and blockage identification. Finally, the procedure outlined in this chapter can also be extended to find isospectral uniform rotating beams, axially loaded uniform beams and tapered rotating beams.

## APPENDIX A: EXACT SOLUTION AND THE MODE SHAPE:

The exact solution of the uniform Rayleigh beam equation is as follows:

$$
U(z) = c_1 e^{\frac{z\sqrt{-\eta\sqrt{(\frac{r_0}{L})^4\eta^2+4}-(\frac{r_0}{L})^2\eta^2}}{\sqrt{2}}} + c_2 e^{-\frac{z\sqrt{-\eta\sqrt{(\frac{r_0}{L})^4\eta^2+4}-(\frac{r_0}{L})^2\eta^2}}{\sqrt{2}}} +
$$

$$
c_3 e^{\frac{z\sqrt{\eta\sqrt{(\frac{r_0}{L})^4\eta^2+4}-(\frac{r_0}{L})^2\eta^2}}{\sqrt{2}}} + c_4 e^{-\frac{z\sqrt{\eta\sqrt{(\frac{r_0}{L})^4\eta^2+4}-(\frac{r_0}{L})^2\eta^2}}{\sqrt{2}}} \tag{10.45}
$$

Applying the boundary conditions for the clamped-clamped non-uniform beam, we arrive at a system of equations. Upon equating the determinant of the coefficient matrix to zero, we obtain

$$
\left(\frac{r_0}{L}\right)^2 \eta \sin\left(\frac{\sqrt{\eta\left(\sqrt{\left(\frac{r_0}{L}\right)^4 \eta^2 + 4} + \left(\frac{r_0}{L}\right)^2 \eta\right)}}{\sqrt{2}}\right) \sinh\left(\frac{\sqrt{2}}{\sqrt{\frac{\sqrt{\left(\frac{r_0}{L}\right)^4 \eta^2 + 4}}{\eta} + \left(\frac{r_0}{L}\right)^2}}\right) +
$$

$$
2\cos\left(\frac{\sqrt{\eta\left(\sqrt{\left(\frac{r_0}{L}\right)^4 \eta^2 + 4} + \left(\frac{r_0}{L}\right)^2 \eta\right)}}{\sqrt{2}}\right) \cosh\left(\frac{\sqrt{2}}{\sqrt{\frac{\sqrt{\left(\frac{r_0}{L}\right)^4 \eta^2 + 4}}{\eta} + \left(\frac{r_0}{L}\right)^2}}\right) - 2 = 0
$$

$$(10.46)$$

Solving Eqn. (10.46), we obtain the natural frequencies of the uniform clamped-clamped Rayleigh beam. To arrive at the $i$th mode shape we substitute $i$th natural frequency in Eqn. (10.45) and solve for $c_1, c_2, c_3$ and $c_4$. The values of $c_1, c_2, c_3$ and $c_4$ for $\eta = 90.0704$ (third mode) are as follows:

$$c_1 = -0.500178 - 0.349711i, c_2 = -0.500178 + 0.349711i, c_3 = 0.0003567, c_4 = 1$$
$$(10.47)$$

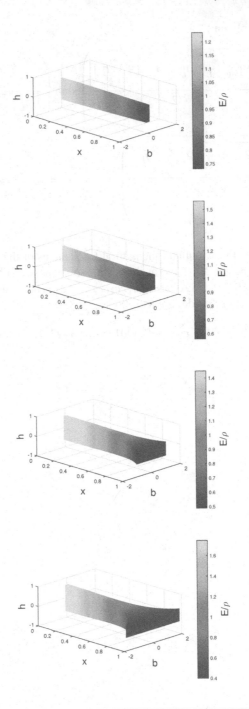

**Figure 10.4**
Breadth, height and ratio of modulus and density distributions of non-uniform beams isospectral to a uniform beam (Case-2: $\alpha = \{0.3, 0.4, 0.5, 0.55\}$, $\beta = \{0.1, 0.2, 0.3, 0.4\}$, $k = -1$, and $\frac{r_0}{L} = 0.09$)

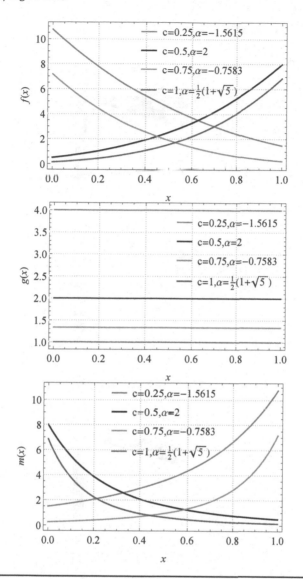

**Figure 10.5**

Mass, bending stiffness and mass moment of inertia functions of a non-uniform beam isospectral to a uniform beam (Case-3: $c = \{0.25, 0.5, 0.75, 1\}$, $\alpha = \left\{-1.56155, 2, -0.758306, \frac{1}{2}\left(\sqrt{5}+1\right)\right\}$, $k = -1$, and $\frac{r_0}{L} = 0.09$)

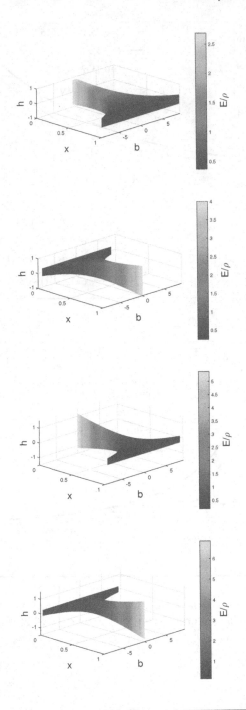

**Figure 10.6**
Breadth, height and ratio of modulus and density distributions of non-uniform beams isospectral to a uniform beam (Case-3: $c = \{0.25, 0.5, 0.75, 1\}$, $\alpha = \left\{-1.56155, 2, -0.758306, \frac{1}{2}\left(\sqrt{5}+1\right)\right\}$, $k = -1$, and $\frac{r_0}{L} = 0.09$)

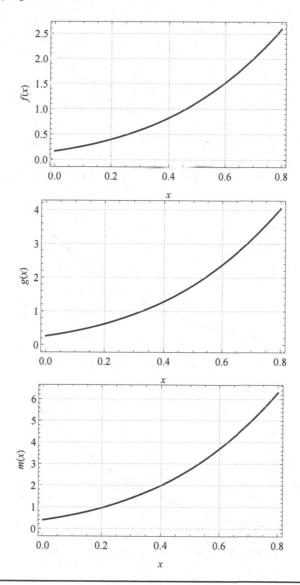

**Figure 10.7**

Mass, bending stiffness and mass moment of inertia functions of a non-uniform beam isospectral to a uniform beam (Case-4: $k = \{0.75\}, \frac{r_0}{L} = 0.09, \alpha = 0.75$ and $\beta = 0.75$)

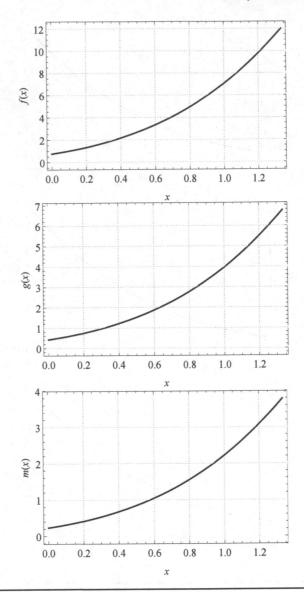

**Figure 10.8**
Mass, bending stiffness and mass moment of inertia functions of a non-uniform beam isospectral to a uniform beam (Case-4: $k = \{1.25\}, \frac{r_0}{L} = 0.09, \alpha = 0.75$ and $\beta = 0.75$)

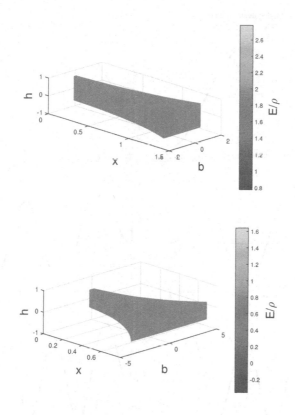

**Figure 10.9**
Breadth, height and ratio of modulus and density distributions of non-uniform beams isospectral to a uniform beam (Case-4: $k = \{1.25, 0.75\}, \frac{r_0}{L} = 0.09, \alpha = 0.75$ and $\beta = 0.75$)

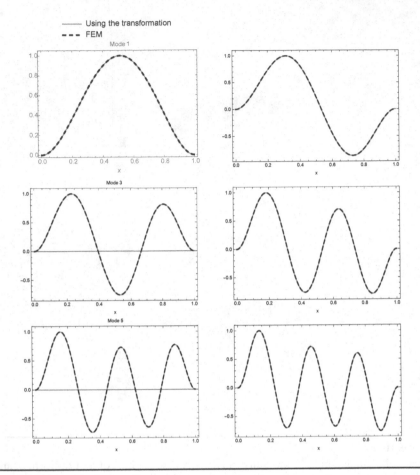

**Figure 10.10**
First six mode shapes calculated using the isospectral transformation and FEM (500 elements) for the clamped-clamped non-uniform beam (Case-2: $\alpha = 0.3$, $\beta = 0.1$)

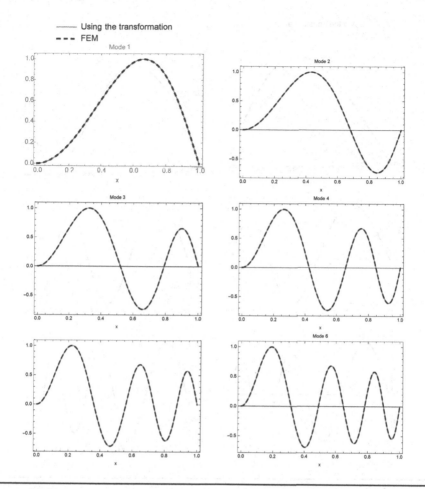

**Figure 10.11**
First six mode shapes calculated using the isospectral transformation and FEM (500 elements) for the clamped-pinned non-uniform beam (Case-3: $c = 0.25$, $\alpha = -1.56155$)

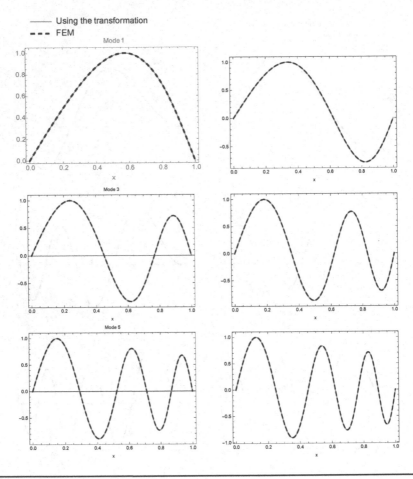

**Figure 10.12**
First six mode shapes calculated using the isospectral transformation and FEM
(500 elements) for the pinned-pinned non-uniform beam (Case-3: $c = 0.25$, $\alpha = -1.56155$)

# Bibliography

1. R. Agarwal, R. Dutta, and R. Ganguli. A numerical method for constructing isospectral discrete systems. *International Journal for Computational Methods in Engineering Science and Mechanics*, 16(6):323–335, 2015.
2. C. L. Amba-Rao. Effect of end conditions on the lateral frequencies of uniform straight columns. *The Journal of the Acoustical Society of America*, 42(4):900–901, 1967.
3. K. E. Atkinson. *An Introduction to Numerical Analysis*. John Wiley & Sons, 2008.
4. R. Attarnejad and A. Shahba. Application of differential transform method in free vibration analysis of rotating non-prismatic beams. *World Applied Sciences Journal*, 5(4):441–448, 2008.
5. R. Attarnejad and A. Shahba. Basic displacement functions for centrifugally stiffened tapered beams. *International Journal for Numerical Methods in Biomedical Engineering*, 27(9):1385–1397, 2011.
6. R. Attarnejad and A. Shahba. Dynamic basic displacement functions in free vibration analysis of centrifugally stiffened tapered beams; a mechanical solution. *Meccanica*, 46(0025-6455):1–15, 2011.
7. N. M. Auciello. On the transverse vibrations of non-uniform beams with axial loads and elastically restrained ends. *International Journal of Mechanical Sciences*, 43(1):193–208, 2001.
8. K. V. Avramov, C. Pierre, and N. Shyriaieva. Flexural-flexural-torsional nonlinear vibrations of pre-twisted rotating beams with asymmetric cross-sections. *Journal of Vibration and Control*, 13(4):329–364, 2007.
9. E. Balagurusamy. *Numerical Methods*. Tata McGraw-Hill Education, 1999.
10. D. V. Bambill, D. H. Felix, and R. E. Rossi. Vibration analysis of rotating Timoshenko beams by means of the differential quadrature method. *Structural Engineering and Mechanics*, 34(2):231–245, 2010.
11. J. R. Banerjee. Free vibration of centrifugally stiffened uniform and tapered beams using the dynamic stiffness method. *Journal of Sound and Vibration*, 233(5):857–875, 2000.
12. J. R. Banerjee and F. W. Williams. Further flexural vibration curves for axially loaded beams with linear or parabolic taper. *Journal of Sound and Vibration*, 102(3):315–327, 1985.
13. J. R. Banerjee. Dynamic stiffness formulation and free vibration analysis of centrifugally stiffened timoshenko beams. *Journal of Sound and Vibration*, 247(1):97–115, 2001.
14. J. R. Banerjee. Free vibration of sandwich beams using the dynamic stiffness method. *Computers and Structures*, 81(18&19):1915–1922, 2003.

15. J. R. Banerjee and D. R. Jackson. Free vibration of a rotating tapered Rayleigh beam: a dynamic stiffness method of solution. *Computers & Structures*, 124:11–20, 2013.

16. J. R. Banerjee, H. Su, and D. R. Jackson. Free vibration of rotating tapered beams using the dynamic stiffness method. *Journal of Sound and Vibration*, 298(4):1034–1054, 2006.

17. V. Barcilon. Inverse problem for the vibrating beam in the free–clamped configuration. *Philosophical Transactions of the Royal Society of London. Series A, Mathematical and Physical Sciences*, 304(1483):211–251, 1982.

18. J. G. P. Barnes. An algorithm for solving non-linear equations based on the secant method. *The Computer Journal*, 8(1):66–72, 1965.

19. T. R. Beal. Dynamic stability of a flexible missile under constant and pulsating thrusts. *AIAA Journal*, 3(3):486–494, 1965.

20. S. Bhadra and R. Ganguli. Aeroelastic optimization of a helicopter rotor using orthogonal array-based metamodels. *AIAA Journal*, 44(9):1941–1951, 2006.

21. K. S. Bhat and R. Ganguli. Isospectrals of non-uniform Rayleigh beams with respect to their uniform counterparts. *Royal Society Open Science*, 5(2):171717, 2018.

22. A. Bilotta, A. Morassi, and E. Turco. Reconstructing blockages in a symmetric duct via quasi-isospectral horn operators. *Journal of Sound and Vibration*, 366:149–172, 2016.

23. A. Bilotta, A. Morassi, and E. Turco. The use of quasi-isospectral operators for damage detection in rods. *Meccanica*, 53, pages 1–27, 2017.

24. B. Biondi and S. Caddemni. Closed form solutions of euler-bernouli beams with singularities. *International Journal of Solids and Structures*, 42(9-10): 3027–3044, 2005.

25. B. Biondi and S. Caddemni. Euler-bernouli beams with multiple singularities in the flexural stiffness. *European Journal of Mechanics A/Solids*, 26(5):789–809, 2007.

26. S. I. Birbil and S. C. Fang. An electromagnetism-like mechanism for global optimization. *Journal of Global Optimization*, 25:263–282, 2003.

27. S. I. Birbil and S. C. Fang. An electromagnetism-like mechanism for global optimization. *Journal of Global Optimization*, 25:263–282, 2003.

28. S. I. Birbil and S. C. Fang. On the convergence of a population-based global optimization algorithm. *Journal of Global Optimization*, 30:301–318, 2004.

29. A. Bokaian. Natural frequencies of beams under compressive axial loads. *Journal of Sound and Vibration*, 126(1):49–65, 1988.

30. A. Bokaian. Natural frequencies of beams under tensile axial loads. *Journal of Sound and Vibration*, 142(3):481–498, 1990.

31. D. Boley and G. H. Golub. A survey of matrix inverse eigenvalue problems. *Inverse Problems*, 3(4):595–622, 1987.

32. G. Borg. Eine umkehrung der sturm-liouvilleschen eigenwertaufgabe. *Acta Mathematica*, 78(1):1–96, 1946.

33. X. Cai, P. V. Jeberg, and H. P. Langtangen. A numerical method for computing the profile of weld pool surfaces. *International Journal for Computational Methods in Engineering Science and Mechanics*, 6(2):115–125, 2005.

34. R. Chandra and I. Chopra. Experimental-theoretical investigation of the vibration characteristics of rotating composite box beams. *Journal of Aircraft*, 29(4):657–664, 1992.

35. T. R. Chandrupatla and A. D. Belegundu. *Introduction to Finite Elements in Engineering*. Prentice-Hall, New Jersey, 2001.

36. T. R. Chandrupatla, A. D. Belegundu, T. Ramesh, and C. Ray. *Introduction to Finite Elements in Engineering*. Prentice-Hall, Englewood Cliffs, NJ, 1991.

37. R. S. Chen. Letter to the editor: Evaluation of natural vibration frequency of a compression bar with varying cross-section by using the shooting method. *Journal of Sound and Vibration*, 201(4):520–527, 1997.

38. X. Chen and M. T. Chu. On the least squares solution of inverse eigenvalue problems. *SIAM J. Numer. Anal.*, 33:2417–2430, 1996.

39. M. Chi, B. G. Dennis, and J. Vossoughi. Transverse and torsional vibrations of an axially-loaded beam with elastically constrained ends. *Journal of Sound and Vibration*, 96(2):235–241, 1984.

40. S-B. Choi and M. S. Han. Vibration control of a rotating cantilevered beam using piezoactuators: experimental work. *Journal of Sound and Vibration*, 277(1):436–442, 2004.

41. J. H. Chou and J. Ghaboussi. Genetic algorithm in structural damage detection. *Computers and Structures*, 79:1335–1353, 2001.

42. M. T. Chu. Constructing a hermitian matrix from its diagonal entries and eigenvalues. *SIAM J. Matrix Anal. Applic.*, 16:207–217, 1995.

43. M. T. Chu. Inverse eigenvalue problems. *SIAM Review*, 40(1):1–39, 1998.

44. J. Cifuentes and R. K. Kapania. A comparison of fem and semi-analytical method in the buckling and vibration of non-prismatic columns under tip force and self-weight. In *56th AIAA/ASCE/AHS/ASC Structures, Structural Dynamics, and Materials Conference*, page 1437, 2015.

45. R. D. Cook, D. S. Malkus, M. E. Plesha, and R. J. Witt. *Concepts and Applications of Finite Element Analysis*. John Wiley & Sons, 2007.

46. E. Crossen, M. S. Gockenbach, B. Jadamba, A. A. Khan, and B. Winkler. An equation error approach for the elasticity imaging inverse problem for predicting tumor location. *Computers & Mathematics with Applications*, 67(1):122–135, 2014.

47. D. Das, P. Sahoo, and K. Saha. Dynamic analysis of non-uniform taper bars in post-elastic regime under body force loading. *Applied Mathematical Modelling*, 33(11):4163–4183, 2009.

48. S. Das, S. Maity, B. Y. Qu, and P. N. Suganthan. Real-parameter evolutionary multimodal optimization – a survey of the state-of-the-art. *Swarm and Evolutionary Computation*, 1:71–88, 2011.

49. C. De Boor and G. H. Golub. The numerically stable reconstruction of a Jacobi matrix from spectral data. *Linear Algebra and its Applications*, 21(3):245–260, 1978.

50. M. A. De Rosa and M. Lippiello. Natural vibration frequencies of tapered beams. *Engineering Transactions*, 57(1):45–66, 2009.

51. B. P. Deepak, R. Ganguli, and S. Gopalakrishnan. Dynamics of rotating composite beams: A comparative study between CNT reinforced polymer composite beams and laminated composite beams using spectral finite elements. *International Journal of Mechanical Sciences*, 64(1):110–126, 2012.

52. R. Diaz and E. Pariguan. On hypergeometric functions and pochhammer k-symbol. *Divulgaciones Matematicas*, 15(2):179–192, 2007.

53. H. Diken. Vibration control of a rotating Euler–Bernoulli beam. *Journal of Sound and Vibration*, 232(3):541–551, 2000.

54. M. Dilena and A. Morassi. Damage detection in discrete vibrating systems. *Journal of Sound and Vibration*, 289(4):830–850, 2006.

55. M. V. Drexel and J. H. Ginsberg. Modal overlap and dissipation effects of a cantilever beam with multiple attached oscillators. *Journal of Vibration and Acoustics*, 123(2):181–187, 2001.

56. W. Du. Dynamic modeling and ascent flight control of Ares-I crew launch vehicle. *Graduate theses and dissertations, Digital Repository*, Iowa State University, 11540:1–167, 2010.

57. R. Dutta, R. Ganguli, and V. Mani. Exploring isospectral spring–mass systems with firefly algorithm. In *Proceedings of the Royal Society of London A: Mathematical, Physical and Engineering Sciences*, page rspa20110119. The Royal Society, 2011.

58. R. Dutta, R. Ganguli, and V. Mani. Exploring isospectral spring-mass systems with firefly algorithm. *Proceedings of the Royal Society A*, 467:3222–3240, 2011.

59. R. Dutta, R. Ganguli, and V. Mani. Exploring isospectral spring-mass systems with firefly algorithm. *Proceedings of the Royal Society A: Mathematical, Physical and Engineering Science*, 467(2135):3222–3240, 2011.

60. R. Dutta, R. Ganguli, and V. Mani. Exploring isospectral cantilever beams using electromagnetism inspired optimization technique. *Swarm and Evolutionary Computation*, 9:37–46, 2013.

61. A. Eddanguir and R. Benamar. A discrete model for transverse vibration of a cantilever beam carrying multi lumped masses: Analogy with the continuous model. In *Design and Modeling of Mechanical Systems*, pages 89–96. Springer, 2013.

62. J. C. Egaña and R. L. Soto. On the numerical reconstruction of a spring-mass system from its natural frequencies. *Proyecciones (Antofagasta)*, 19(1):27–41, 2000.

63. C. R. Farrar, S. W. Doebling, and D. A. Nix. Vibration-based structural damage identification. *Philosophical Transactions of the Royal Society of London. Series A: Mathematical, Physical and Engineering Sciences*, 359(1778):131–149, 2001.

64. E. H. Feng and R. E. Jones. Carbon nanotube cantilevers for next-generation sensors. *Physical Review B*, 83:195412–17, 2011.

65. H. Fletcher. Normal vibration frequencies of a stiff piano string. *The Journal of the Acoustical Society of America*, 36(1):203–209, 1964.
66. G. E. Forsythe. Generation and use of orthogonal polynomials for data-fitting with a digital computer. *Journal of the Society for Industrial & Applied Mathematics*, 5(2):74–88, 1957.
67. D. Fouskakis and D. Draper. Stochastic optimization: a review. *International Statistical Review*, 70:315–349, 2002.
68. Draper, D, and D. Fouskakis. "A case study of stochastic optimization in health policy: problem formulation and preliminary results." *Journal of Global Optimization* 18, no. 4 (2000): 399-416.
69. A. E. Galef. Bending frequencies of compressed beams. *Journal of the Acoustical Society of America*, 44(2):643–643, 1968.
70. R. Ganguli. *Structural Health Monitoring*. Springer, 2020.
71. S. S. Ganji, A. Barari, and D. D. Ganji. Approximate analysis of two-mass-spring systems and buckling of a column. *Computers & Mathematics with Applications*, 61(4):1088–1095, 2011.
72. K. Ghanbari. On the isospectral beams. In *Electronic Journal of Differential Equations Conference*, volume 12, pages 57–64, 2005.
73. G. M. L. Gladwell. The inverse problem for the Euler-Bernoulli beam. *Proceedings of the Royal Society of London A: Mathematical, Physical and Engineering Sciences*, 407(1832):199–218, 1986.
74. G. M. L. Gladwell. On isospectral spring-mass systems. *Inverse Problems*, 11(3):591–602, 1995.
75. G. M. L. Gladwell. Inverse finite element vibration problems. *Journal of Sound and Vibration*, 221(2):309–324, 1999.
76. G. M. L. Gladwell. *Inverse Problems in Vibration*. Martinus Nijhoff Publishers, Dordrecht, 2002.
77. G. M. L. Gladwell. Isospectral vibrating beams. *Proceedings of the Royal Society of London. Series A: Mathematical, Physical and Engineering Sciences*, 458(2027):2691–2703, 2002.
78. G. M. L. Gladwell. *Inverse problems in vibration*, volume 119. Springer Netherlands, second edition, 2005.
79. G. M. L. Gladwell and A. Morassi. On isospectral rods, horns and strings. *Inverse Problems*, 11(3):533, 1995.
80. G. M. L. Gladwell and A. Morassi. A family of isospectral Euler-Bernoulli beams. *Inverse Problems*, 26(3):035006, 2010.
81. G. M. L. Gladwell and N. B. Willms. The reconstruction of a tridiagonal system from its frequency response at an interior point. *Inverse Problems*, 4(4):1013, 1988.
82. G. M. L. Gladwell and B. R. Zhu. Inverse problems for multidimensional vibrating systems. *Proceedings of the Royal Society of London. Series A: Mathematical and Physical Sciences*, 439(1907):511–530, 1992.
83. G. M. L. Gladwell, A. H. England, and D. Wang. Examples of reconstruction of an Euler-Bernoulli beam from spectral data. *Journal of Sound and Vibration*, 119(1):81–94, 1987.

84. G. M. L. Gladwell and A. Morassi. On isospectral rods, horns and strings. *Inverse Problems*, 11:533–554, 1995.

85. M. Gol Alikhani, N. Javadian, and R. Tavakkoli-Moghaddam. A novel hybrid approach combining electromagnetism-like method with solis and wets local search for continuous optimization problems. *Journal of Global Optimization*, 44:227–234, 2009.

86. H. P. W. Gottlieb. Isospectral Euler-Bernoulli beams with continuous density and rigidity functions. *Proceedings of the Royal Society of London. A. Mathematical and Physical Sciences*, 413(1844):235–250, 1987.

87. H. P. W. Gottlieb. Iso-spectral operators: Some model examples with discontinuous coefficients. *Journal of Mathematical Analysis and Applications*, 132(1):123–137, 1988.

88. H. P. W. Gottlieb. Harmonic frequency spectra of vibrating stepped strings. *Journal of Sound and Vibration*, 108(1):63–72, 1986.

89. H. P. W. Gottlieb. Multi-segment strings with exactly harmonic spectra. *Journal of Sound and Vibration*, 118(2):283–290, 1987.

90. H. P. W. Gottlieb. Density distribution for isospectral circular membranes. *SIAM Journal on Applied Mathematics*, 48, pages 948–951, 1988.

91. H. P. W. Gottlieb. Isospectral Euler-Bernoulli beams with continuous density and rigidity functions. *Proceedings of The Royal Society A*, 413:235–250, 2002.

92. H. P. W. Gottlieb. Isospectral circular membranes. *Inverse Problems*, 20:155–161, 2004.

93. W. B. Gragg and G. W. Stewart. A stable variant of the secant method for solving nonlinear equations. *SIAM Journal on Numerical Analysis*, 13(6):889–903, 1976.

94. R. O. Grossi and M. V. Quintana. The transition conditions in the dynamics of elastically restrained beams. *Journal of Sound and Vibration*, 316(1):274–297, 2008.

95. P. K. Gudla and R. Ganguli. Error estimates for inconsistent load lumping approach in finite element solution of differential equations. *Applied Mathematics and Computation*, 194(1):21–37, 2007.

96. P. K. Gudla and R. Ganguli. An automated hybrid genetic-conjugate gradient algorithm for multimodal optimization problems. *Applied Mathematics and Computation*, 167:1457–1474, 2005.

97. J. B. Gunda and R. Ganguli. New rational interpolation functions for finite element analysis of rotating beams. *International Journal of Mechanical Sciences*, 50(3):578–588, 2008.

98. J. B. Gunda and R. Ganguli. Stiff-string basis functions for vibration analysis of high speed rotating beams. *Journal of Applied Mechanics*, 75(2):024502, 2008.

99. J. B. Gunda, A. P. Singh, P. P. S. Chabbra, and R. Ganguli. Free vibration analysis of rotating tapered blades using Fourier-p superelement. *Structural Engineering and Mechanics*, 27(2):243–257, 2007.

100. J. B. Gunda and R. Ganguli. New rational interpolation functions for finite element analysis of rotating beams. *International Journal of Mechanical Sciences*, 50(3):578–588, 2008.

101. E. Haber, U. M. Ascher, and D. Oldenburg. On optimization techniques for solving nonlinear inverse problems. *Inverse Problems*, 16(5):1263, 2000.

102. S. M. Hashemi and M. J. Richard. Natural frequencies of rotating uniform beams with coriolis effects. *Journal of Vibrations and Acoustics-Transactions of the ASME*, 123(4):444–455, 2001.

103. G. M. Hensley and R. H. Plaut. Three-dimensional analysis of the seismic response of guyed masts. *Engineering Structures*, 29(9):2254–2261, 2007.

104. D. Hodges and M. Rutkowski. Free-vibration analysis of rotating beams by a variable-order finite-element method. *AIAA Journal*, 19(11):1459–1466, 1981.

105. R. A. Horn and C. R. Johnson. *Matrix Analysis, 1985*. Cambridge University Press, 1985.

106. D. Hu and R. Kovacevic. Sensing, modeling and control for laser-based additive manufacturing. *International Journal of Machine Tools and Manufacture*, 43(1):51–60, 2003.

107. D. Hua and P. Lancaster. Linear matrix equations from an inverse problem of vibration theory. *Linear Algebra and its Applications*, 246:31–47, 1996.

108. G. C. Hughes and C. W. Bert. Effect of gravity on nonlinear oscillations of a horizontal, immovable-end beam. *Nonlinear Dynamics*, 3(5):365–373, 1992.

109. M. Ji and J. Klinowski. Taboo evolutionary programming: a new method of global optimization. *Proc. R. Soc. Lond. A*, 462:3613–3627, 2006.

110. R. Jiménez, L. Santos, N. Kuhl, and J. Egana. The reconstruction of a specially structured jacobi matrix with an application to damage detection in rods. *Computers & Mathematics with Applications*, 49(11):1815–1823, 2005.

111. R. D. Jiménez, L. C. Santos, N. M. Kuhl, J. C. Egaña, and R. L. Soto. An inverse eigenvalue procedure for damage detection in rods. *Computers & Mathematics with Applications*, 47(4):643–657, 2004.

112. X. Jin, M. Leon, and Q. Wang. A practical method for singular integral equations of the second kind. *Engineering Fracture Mechanics*, 75(5):1005–1014, 2008.

113. S. Kambampati and R. Ganguli. Non-uniform beams and stiff strings isospectral to axially loaded uniform beams and piano strings. *Acta Mechanica*, 226(4):1227–1239, 2015.

114. S. Kambampati and R. Ganguli. Nonrotating beams isospectral to tapered rotating beams. *AIAA Journal*, 54(2):750–757, 2016.

115. S. Kambampati, R. Ganguli, and V. Mani. Determination of isospectral non-uniform rotating beams. *Journal of Applied Mechanics*, 79(6):061016, 2012.

116. S. Kambampati, R. Ganguli, and V. Mani. Non-rotating beams isospectral to a given rotating uniform beam. *International Journal of Mechanical Sciences*, 66:12–21, 2013.

117. S. Kambampati, R. Ganguli, and V. Mani. Rotating beams isospectral to axially loaded nonrotating uniform beams. *AIAA Journal*, 51(5):1189–1202, 2013.

118. I. A. Karnovsky and O. I. Lebed. *Formulas for Structural Dynamics: Tables, Graphs and Solutions*. McGraw-Hill, New York, 2001.

119. R. Kathiravan and R. Ganguli. Strength design of composite beams using gradient and particle swarm optimization. *Composite Structures*, 81:471–79, 2007.

120. E. Kreyszig. *Advanced Engineering Mathematics*. Wiley-India, 2007.

121. J. P. Kruth, M. C. Leu, and T. Nakagawa. Progress in additive manufacturing and rapid prototyping. *CIRP Annals-Manufacturing Technology*, 47(2):525–540, 1998.

122. A. Kumar and R. Ganguli. Rotating beams and nonrotating beams with shared eigenpair. *Journal of Applied Mechanics*, 76:051006, 2009.

123. S. Kumar and R. Kumar. Theoretical and experimental vibration analysis of rotating beams with combined ACLD and stressed layer damping treatment. *Applied Acoustics*, 74(5):675–693, 2013.

124. B. Kundu and R. Ganguli. Analysis of weak solution of Euler–Bernoulli beam with axial force. *Applied Mathematics and Computation*, 298:247–260, 2017.

125. C-F. J. Kuo and S. C. Lin. Modal analysis and control of a rotating Euler-Bernoulli beam part I: control system analysis and controller design. *Mathematical and Computer Modelling*, 27(5):75–92, 1998.

126. D. M. Lauzon and V. R. Murthy. Determination of vibration characteristics of multiple-load-path blades by a modified Galerkin's method. *Computers & Structures*, 46(6):1007–1020, 1993.

127. K. Lee, Y. Cho, and J. Chung. Dynamic contact analysis of a tensioned beam with a moving mass-spring system. *Journal of Sound and Vibration*, 331(11):2520–2531, 2012.

128. S. Lee, B. D. Youn, and B. C. Jung. Robust segment-type energy harvester and its application to a wireless sensor. *Smart Materials and Structures*, 18:095021–33, 2009.

129. S. Y. Lee and Y. H. Kuo. Bending vibrations of a rotating non-uniform beam with an elastically restrained root. *Journal of Sound and Vibration*, 154(3):441–451, 1992.

130. S. Y. Lee and J. J. Sheu. Free vibrations of rotating inclined beam. *Journal of Applied Mechanics-Transactions of the ASME*, 74(3):406–414, 2007.

131. X. F. Li, A.Y. Tang, and L.Y. Xi. Vibration of a Rayleigh cantilever beam with axial force and tip mass. *Journal of Constructional Steel Research*, 80:15–22, 2013.

132. H. P. Lin and S. C. Chang. Free vibrations of two rods connected by multi-spring-mass systems. *Journal of Sound and Vibration*, 330(11):2509–2519, 2011.

133. D. S. Liu and D. Y. Chiou. A coupled IEM/FEM approach for solving elastic problems with multiple cracks. *International Journal of Solids and Structures*, 40(8):1973–1993, 2003.

134. T. Livingston, J. G. Bliveau, and D. R. Huston. Estimation of axial load in prismatic members using flexural vibrations. *Journal of Sound and Vibration*, 179(5):899–908, 1995.

135. S. Lukasik and S. Zak. Firefly algorithm for continuous constrained optimization tasks. *ICCCI. Lec. Notes Art. Intel.*, 5796:97–100, 2009.

136. A. W. Marshall and I. Olkin. Theory of majorization and its applications. *Academic, New York, USA*, 4:4–93, 1979.

137. H. McCallion. *Vibration of Linear Mechanical Systems*. Longman, London, 1973.

138. J. R. McLaughlin. Analytical methods for recovering coefficients in differential equations from spectral data. *SIAM review*, 28(1):53–72, 1986.

139. L. Meirovitch. *Elements of Vibration Analysis*, volume 2. McGraw-Hill, New York, 1986.

140. L. Meirovitch. *Fundamentals of Vibrations*. McGraw-Hill, New York, 2001.

141. Y. V. Mikhlin, A. F. Vakakis, and G. Salenger. Direct and inverse problems encountered in vibro-impact oscillations of a discrete system. *Journal of Sound and Vibration*, 216(2):227–250, 1998.

142. R. K. Miller. On ignoring the singularity in numerical quadrature. *Mathematics of Computation*, 25(115):521–532, 1971.

143. Feldstein, A. and Miller, R. K., 1971. Error bounds for compound quadrature of weakly singular integrals. *Mathematics of Computation*, 25(115), pp.505-520.

144. A. Morassi. Constructing rods with given natural frequencies. *Mechanical Systems and Signal Processing*, 40:288–300, 2013.

145. S. M. Murugan, N. Suresh, R. Ganguli, and V. Mani. Target vector optimization of composite box-beam using real coded genetic algorithm: a decomposition approach. *Structural and Multidisciplinary Optimization*, 33:131–146, 2007.

146. S. Naguleswaran. Lateral vibration of a centrifugally tensioned uniform Euler-Bernoulli beam. *Journal of Sound and Vibration*, 176(5):613–624, 1994.

147. S. Naguleswaran. Vibration and stability of an Euler–Bernoulli beam with up to three-step changes in cross-section and in axial force. *International Journal of Mechanical Sciences*, 45(9):1563–1579, 2003.

148. S. Naguleswaran. Transverse vibration of an uniform Euler–Bernoulli beam under linearly varying axial force. *Journal of Sound and Vibration*, 275(1):47–57, 2004.

149. P. Nath. Confluent hypergeometric function. *Sankhyā: The Indian Journal of Statistics*, 11(2):153–166, 1951.

150. B. Noble. *Applied Linear Algebra, 1969*. Prentice Hall, NJ, 1988.

151. P. Nylen and F. Uhlig. Inverse eigenvalue problems associated with spring-mass systems. *Linear Algebra and its Applications*, 254(1):409–425, 1997.

152. O. Ozdemir and M. O. Kaya. Flapwise bending vibration analysis of a rotating tapered cantilever Bernoulli-Euler beam by differential transform method. *Journal of Sound and Vibration*, 289(1–2):413–420, 2006.

153. Ozgumus, O. O. and Kaya, M. O., 2010. Vibration analysis of a rotating tapered Timoshenko beam using DTM. Meccanica, 45(1), pp.33-42.

154. O. O. Ozgumus and M. O. Kaya. Flapwise bending vibration analysis of double tapered rotating Euler-Bernoulli beam by using the differential transform method. *Meccanica*, 41(6):661–670, 2006.

155. S. K. Panigrahi, S. Chakraverty, and B. K. Mishra. Vibration based damage detection in a uniform strength beam using genetic algorithm. *Meccanica*, 44(6):697–710, 2009.

156. P. M. Pawar and R. Ganguli. *Structural Health Monitoring using Genetic Fuzzy Systems*. Springer, 2011.

157. P. M. Pawar and R. Ganguli. Genetic fuzzy system for damage detection in beams and helicopter rotor blades. *Computer Methods in Applied Mechanics and Engineering*, 192:2031–2057, 2003.

158. G. Piana. *Vibrations and Stability of Axially and Transversely Loaded Structures*. PhD thesis, Politecnico di Torino, 2013.

159. H. U. Rahman, K. Y. Chan, and R. Ramer. Cantilever beam designs for RF MEMS switches. *Journal of Micromechanics and Microengineering*, 20:075042–53, 2010.

160. A. Raich and T. Liszkai. Benefits of an implicit redundant genetic algorithm method for structural damage detection in noisy environments. *Proceedings of Genetic and Evolutionary Computation-GECCO, Lecture Notes in Computer Science*, 2724:2418–2419, 2003.

161. K. K. Raju and G. V. Rao. Free vibration behavior of prestressed beams. *Journal of Structural Engineering*, 112(2):433–437, 1986.

162. K. K. Raju and G. V. Rao. Free vibration behavior of tapered beam columns. *Journal of Engineering Mechanics*, 114(5):889–892, 1988.

163. Y. M. Ram. Inverse mode problems for the discrete model of a vibrating beam. *Journal of Sound and Vibration*, 169(2):239–252, 1994.

164. Y. M. Ram and S. Elhay. Dualities in vibrating rods and beams: Continuous and discrete models. *Journal of Sound and Vibration*, 184(5):759–766, 1995.

165. N. Ramachandran and R. Ganguli. Family of columns isospectral to gravity-loaded columns with tip force: A discrete approach. *Journal of Sound and Vibration*, 423:421–441, 2018.

166. N. Ramachandran and R. Ganguli. Family of non-rotating beams isospectral to rotating beams: A discrete approach. *AIAA Scitech*, page 215, 2019.

167. S. S. Rao and P. Sawyer. Fuzzy finite element approach for analysis of imprecisely defined systems. *AIAA Journal*, 33:2364–2370, 1995.

168. Lord Rayleigh and J.W Strutt. *The Theory of Sound*, volume 2. Macmillan, 1896.

169. J. N. Reddy. *An Introduction to the Finite Element Method*. McGraw-Hill, New York, 1993.

170. C. Richter and H. Lipson. Untethered hovering flapping flight of a 3d-printed mechanical insect. *Artificial Life*, 17(2):73–86, 2011.

171. S. Roundy. Improving power output for vibration-based energy scavengers. *Pervasive Computing*, 4:28–36, 2005.

172. K. Sarkar and R. Ganguli. Rotating beams and non-rotating beams with shared eigenpair for pinned-free boundary condition. *Meccanica*, 48(7):1661–1676, 2013.

173. K. Sarkar, R. Ganguli, D. Ghosh, and I. Elishakoff. Closed-form solutions and uncertainty quantification for gravity-loaded beams. *Meccanica*, 51(6):1465–1479, 2016.

174. K. Sato. Transverse vibrations of linearly tapered beams with ends restrained elastically against rotation subjected to axial force. *International Journal of Mechanical Sciences*, 22(2):109–115, 1980.

175. P. Schneider and R. Kienzler. Teaching nonlinear mechanics: An extensive discussion of a standard example feasible for undergraduate courses. *International Journal for Computational Methods in Engineering Science and Mechanics*, 15(2):172–181, 2014.

176. H. R. Schwarz. *Numerical Analysis: A Comprehensive Introduction*. John Wiley and Sons, New York, 1989.

177. A. Senatore. Measuring the natural frequencies of centrifugally tensioned beam with laser doppler vibrometer. *Measurement*, 39(7):628–633, 2006.

178. J. Senthilnath, S. N. Omkar, and V. Mani. Clustering using firefly algorithm: Performance study. *Swarm and Evolutionary Computation*, 1:164–171, 2011.

179. F. J. Shaker. Effect of axial load on mode shapes and frequencies of beams. *NASA Technical Note TN D-8109*, pages 1–30, 1975.

180. S. K. Sinha. Combined torsional-bending-axial dynamics of a twisted rotating cantilver Timoshenko beam with contact-impact loads at the free end. *Journal of Applied Mechanics*, 74(3):505–522, 2007.

181. L. J. Slater. *Confluent Hypergeometric Functions*. Cambridge University Press, Cambridge, 1960.

182. F. J. Solis and J. B. Wets. Minimization by random search techniques. *Mathematics of Operations Research*, 6:19–30, 1981.

183. R. L. Soto. A numerical reconstruction of a Jacobi matrix from spectral data. *Tamkang Journal of Mathematics*, 20(1):57–63, 1989.

184. V. Srinivas, K. Ramanjaneyulu, and C. Jeyasehar. Multi-stage approach for structural damage identification using modal strain energy and evolutionary optimization techniques. *Structural Health Monitoring*, 10:219–230, 2011.

185. G. W. Stewart. Gershgorin theory for the generalized eigenvalue problem $ax = \lambda bx$. *Mathematics of Computation*, 29(130):600–606, 1975.

186. D. Storti and Y. Aboelnaga. Bending vibrations of a class of rotating beams with hypergeometric solutions. *Journal of Applied Mechanics*, 54(2):311–314, 1987.

187. G. Subramanian and A. Raman. Isospectral systems for tapered beams. *Journal of Sound and Vibration*, 198:257–266, 1996.

188. Subramanian, G. and Balasubramanian, T.S., 1987. Beneficial effects of steps on the free vibration characteristics of beams. *Journal of sound and vibration*, 118(3), pp.555-560.

189. G. Sudheer, P. S. Harikrishna, and Y. V. Rao. Free vibration analysis of tapered columns under self-weight using pseudospectral method. *Journal of Vibroengineering*, 18(7):4583–4591, 2016.

190. C. Sun and K. Liu. Vibration of multi-walled carbon nanotubes with initial axial loading. *Solid State Communications*, 143(4):202–207, 2007.

191. N. Touat, M. Pyrz, and S. Rechak. Updating fe models in dynamics using the centrifugal force algorithm. *International Journal for Computational Methods in Engineering Science and Mechanics*, 13(1):28–36, 2012.

192. K. M. Udupa and T. K. Varadan. Hierarchical finite element method for rotating beams. *Journal of Sound and Vibration*, 138(3):447–456, 1990.

193. G. Vairo. A simple analytical approach to the aeroelastic stability problem of long-span cable-stayed bridges. *International Journal for Computational Methods in Engineering Science and Mechanics*, 11(1):1–19, 2010.

194. K. G. Vinod, S. Gopalakrishnan, and R. Ganguli. Free vibration and wave propagation analysis of uniform and tapered rotating beams using spectrally formulated finite elements. *International Journal of Solids and Structures*, 44(18):5875–5893, 2007.

195. L. N. Virgin. *Vibration of Axially-loaded Structures*. Cambridge University Press, New York, 2007.

196. B. K. Wada, C. P. Kui, and R. J. Glaser. Extension of ground-based testing for large space structures. *Journal of Spacecraft and Rockets*, 23:184–188, 1986.

197. A-P. Wang and Y-H. Lin. Vibration control of a tall building subjected to earthquake excitation. *Journal of Sound and Vibration*, 299(4):757–773, 2007.

198. C. C. Wang and H. T. Yau. Application of the differential transformation method to bifurcation and chaotic analysis of an AFM probe tip. *Computers & Mathematics with Applications*, 61(8):1957–1962, 2011.

199. G. Wang and N. M. Wereley. Free vibration analysis of rotating blades with uniform tapers. *AIAA Journal*, 42(12):2429–2437, 2004.

200. J. H. Wang and Yin Z. Y. A ranking selection-based particle swarm optimizer for engineering design optimization problems. *Structural and Multidisciplinary Optimization*, 37:131–147, 2008.

201. L. Wang and L. P. Li. An effective differential evolution with level comparison for constrained engineering design. *Structural and Multidisciplinary Optimization*, 41:947–963, 2010.

202. X. J. Wang, L. Gao, and C. Y. Zhang. Electromagnetism-like mechanism based algorithm for neural network training. *Lecture Notes in Computer Science*, 5227:40–45, 2008.

203. Y. Wang, Z. Cai, Y. Zhou, and Z. Fan. Constrained optimization based on hybrid evolutionary algorithm and adaptive constraint-handling technique. *Structural and Multidisciplinary Optimization*, 37:395–413, 2009.

204. D. J. Wei, S. X. Yan, Z. P. Zhang, and X-F. Li. Critical load for buckling of non-prismatic columns under self-weight and tip force. *Mechanics Research Communications*, 37(6):554–558, 2010.

205. M. Whorton, C. Hall, and S. Cook. Ascent flight control and structural interaction for the Ares-I crew launch vehicle. In *48th AIAA/ASME/ASCE/AHS/ASC Structures, Structural Dynamics, and Materials Conference*, page 1780, 2007.

206. F. W. Williams and J. R. Banerjee. Flexural vibration of axially loaded beams with linear or parabolic taper. *Journal of Sound and Vibration*, 99(1):121–138, 1985.

207. A. D. Wright, C. E. Smith, R. W. Thresher, and J. L. C. Wang. Vibration modes of centrifugally stiffened beams. *Journal of Applied Mechanics*, 49:197, 1982.

208. L. Wu, C. Xie, and C. Yang. Aeroelastic stability of a slender missile with constant thrust. *Procedia Engineering*, 31:128–135, 2012.

209. P. Wu, K-J. Yang, and B-Y. Huang. A revised EM-like mechanism for solving the vehicle routing problems. *Proceedings of the Second International Conference on Innovative Computing, Information and Control*, 2001.

210. L-Y. Xi, X-F. Li, and G-J. Tang. Free vibration of standing and hanging gravity-loaded Rayleigh cantilevers. *International Journal of Mechanical Sciences*, 66:233–238, 2013.

211. J. B. Yang, L. J. Jiang, and D. Ch. Chen. Dynamic modelling and control of a rotating Euler–Bernoulli beam. *Journal of Sound and Vibration*, 274(3):863–875, 2004.

212. L. Yang, Z. C. Deng, J. N. Yu, and G. W. Luo. Optimization method for the inverse problem of reconstructing the source term in a parabolic equation. *Mathematics and Computers in Simulation*, 80(2):314–326, 2009.

213. X. S. Yang. *Nature-Inspired Metaheuristic Algorithms*. Luniver Press, 2008.

214. X. S. Yang. Firefly algorithms for multimodal optimization. in: Stochastic algorithms: Foundations and applications. *SAGA. Lec. Notes Comp. Sc.*, 5792:169–178, 2009.

215. X. S. Yang. Firefly algorithm, stochastic test functions and design optimisation. *Int. J. Bio-Inspired Comp.*, 2:78–84, 2010.

216. A. Yavari and S. Sarkani. On applications of generalized functions to the analysis of Euler-Bernoulli beam-columns with jump discontinuities. *International Journal of Mechanical Sciences*, 43:1543–1562, 2001.

217. A. Yavari, S. Sarkani, and J. N. Reddy. On nonuniform Euler-Bernoulli and Timoshenko beams with jump discontinuities: application of distribution theory. *International Journal of Solids and Structures*, 38:8389–8406, 2001.

218. F. H. Yeh and W. H. Liu. Free vibrations of nonuniform beams with rotational and translational restraints and subjected to an axial force. *Journal of the Acoustical Society of America*, 85(3):1368–1371, 1989.

219. H. H. Yoo and S. H. Shin. Vibration analysis of rotating cantilever beams. *Journal of Sound and Vibration*, 212(5):807–828, 1998.

220. Yoo, H.H., Park, J.H. and Park, J., 2001. Vibration analysis of rotating pre-twisted blades. *Computers & Structures*, 79(19), pp.1811–1819.

221. H. Youssef, S. M. Sait, and H. Adiche. Evolutionary algorithms, simulated annealing and tabu search: a comparative study. *Engineering Applications of Artificial Intelligence*, 14:167–181, 2001.

222. Y. Zhang, G. Liu, and X. Han. Transverse vibrations of double-walled carbon nanotubes under compressive axial load. *Physics Letters A*, 340(1):258–266, 2005.

223. A. Zhou, B. Y. Qu, H. Li, S. Z. Zhao, P. N. Suganthan, and Q. Zhang. Multi-objective evolutionary algorithms: A survey of the state of the art. *Swarm and Evolutionary Computation*, 1:32–49, 2011.

224. W. D. Zhu and C. D. Mote. Dynamic modeling and optimal control of rotating Euler-Bernoulli beams. In *Proceedings of the 1997 American Control Conference (Cat. No.97CH36041)*, volume 5, pages 3110–3114. IEEE, 1997.

# Index

Printed in the United States
by Baker & Taylor Publisher Services